# Topics in Fluorescence Spectroscopy

Volume 6
Protein Fluorescence

## Topics in Fluorescence Spectroscopy

Edited by JOSEPH R. LAKOWICZ

---

Volume 1: Techniques
Volume 2: Principles
Volume 3: Biochemical Applications
Volume 4: Probe Design and Chemical Sensing
Volume 5: Nonlinear and Two-Photon-Induced Fluorescence
Volume 6: Protein Fluorescence

# Topics in Fluorescence Spectroscopy

## Volume 6
## Protein Fluorescence

Edited by

JOSEPH R. LAKOWICZ
*Center for Fluorescence Spectroscopy and*
*Department of Biochemistry and Molecular Biology*
*University of Maryland School of Medicine*
*Baltimore, Maryland*

Springer Science+Business Media, LLC

The Library of Congress cataloged the first volume of this title as follows:

Topics in fluorescence spectroscopy/edited by Joseph R. Lakowicz. p. cm.
Includes bibliographical references and index.
Contents: v. 1. Techniques
1. Fluorescence spectroscopy. I. Lakowicz, Joseph R.

QD96.F56T66 1991 91-32671
543'.0858—dc20 CIP

ISBN 978-1-4757-8194-6 ISBN 978-0-306-47102-5 (eBook)
DOI 10.1007/978-0-306-47102-5


Originally published by Kluwer Academic/Plenum Publishers, New York in 2000
Softcover reprint of the hardcover 1st edition 2000

http://www.wkap.nl

10 9 8 7 6 5 4 3 2 1

A C.I.P. record for this book is available from the Library of Congress

# Dedication

*Dedicated to Professor Ludwig Brand for his Pioneering Contributions to Protein Fluorescence*

This volume of Topics in Fluorescence is dedicated to Prof. Ludwig Brand for his numerous contributions to our understanding of protein fluorescence. Prof. Brand was born in Vienna, Austria. He moved to England as a child before the second world war and came to Boston shortly after the war. Dr. Brand received his Ph.D. from the University of Indiana in 1959 in the field of Biochemistry. Dr. Brand studied with Professor H. R. Mahler at Brandeis University and then with Professor Ephraim Katchalski at the Weizmann Institute. After these studies he joined the Johns Hopkins University where he has remained to this day.

Dr. Brand has made extensive contributions to the development of time-resolved fluorescence. He pioneered the use of time-correlated single photon counting for measurements of time dependent spectral relaxation and excited state reactions. He also accomplished the first and most definitive resolution of the emission from two tryptophan residues in a two tryptophan protein. Dr. Brand's enthusiasm for the field of fluorescence and his consistent good humor have provided effective training for many individuals now using fluorescence to study biochemical and cellular phenomena. Please join us in wishing Dr. Brand continued health and productivity.

J. R. Lakowicz, *Baltimore, Maryland*
J. B. A. Ross, *New York, New York*

# Contributors

*Herbert C. Cheung* • Department of Biochemistry and Molecular Genetics, University of Alabama at Birmingham, Birmingham, Alabama 35294-2041.

*Sabato D'Auria* • Institute of Protein Biochemistry and Enzymology, C.N.R., Naples 80125, Italy.

*Wen-Ji Dong* • Department of Biochemistry and Molecular Genetics, University of Alabama at Birmingham, Birmingham, Alabama 35294-2041.

*Maurice R. Eftink* • Department of Chemistry, The University of Mississippi, Oxford, Mississippi 38677.

*Yves Engelborghs* • Laboratory of Biomolecular Dynamics, University of Leuven, Heverlee B-3001, Belgium.

*Alan Fersht* • Cambridge Center for Protein Engineering, Cambridge University, Cambridge CB2 1EW, United Kingdom.

*Alessandro Finazzi Agrò* • Department of Experimental Medicine and Biochemical Science, University of Rome, Rome 00133, Italy.

*Ari Gafni* • Department of Biological Chemistry, Biophysics Research Division, and Institute of Gerontology, The University of Michigan, Ann Arbor, Michigan 48109.

*Jacques Gallay* • Applied Electromagnetic Radiation Laboratory, University of Paris-Sud, Orsay 91898, France.

*Rudi Glockshuber* • Institute for Molecular Biology and Biophysics, Honggerberg Technical University, Zurich CH-8093, Switzerland.

*Ignacy Gryczynski* • Center for Fluorescence Spectroscopy, University of Maryland at Baltimore, Baltimore, Maryland 21201.

*Jacques Haiech* • Department of Pharmacology and Physicochemistry of Molecular and Cellular Interactions, Louis Pasteur University, Illkirch 67401, France.

*Jens Hennecke* • Institute for Molecular Biology and Biophysics, Honggerberg Technical University, Zurich CH-8093, Switzerland.

*Rhoda Elison Hirsch* • Department of Medicine (Hematology) and Department of Anatomy & Structural Biology, Albert Einstein College of Medicine of Yeshiva University, Bronx, New York 10461.

*Marie-Claude Kilhoffer* • Department of Pharmacology and Physicochemistry of Molecular and Cellular Interactions, Louis Pasteur University, Illkirch 67401, France.

*Joseph R. Lakowicz* • Center for Fluorescence Spectroscopy, University of Maryland at Baltimore, Baltimore, Maryland 21201.

*Linda A. Luck* • Department of Chemistry, Clarkson University, Potsdam, New York 13699-5605.

*Giampiero Mei* • Department of Experimental Medicine and Biochemical Science, University of Rome, Rome 00133, Italy.

*Nicola Rosato* • Department of Experimental Medicine and Biochemical Science, University of Rome, Rome 00133, Italy.

*J. B. Alexander Ross* • Department of Biochemistry and Molecular Biology, Mount Sinai School of Medicine, New York, New York 10029-6574.

*Mosè Rossi* • Institute of Protein Biochemistry and Enzymology, C.N.R., Naples 80125, Italy.

*Kenneth W. Rousslang* • Department of Chemistry, University of Puget Sound, Tacoma, Washington 98416-0062.

*Elena Rusinova* • Department of Biochemistry and Molecular Biology, Mount Sinai School of Medicine, New York, New York 10029-6574.

*Alain Sillen* • Laboratory of Biomolecular Dynamics, University of Leuven, Leuven B-3001, Belgium.

*Jana Sopková* • Applied Electromagnetic Radiation Laboratory, University of Paris-Sud, Orsay 91898, France.

*Duncan G. Steel* • Departments of Physics and Electrical Engineering and Computer Science, Biophysics Research Division, and Institute of Gerontology, The University of Michigan, Ann Arbor, Michigan 48109.

*Vinod Subramaniam* • Department of Molecular Biology, Max Planck Institute for Biophysical Chemistry, Gottingen D-37077, Germany.

*Michel Vincent* • Applied Electromagnetic Radiation Laboratory, University of Paris-Sud, Orsay 91898, France.

# Preface

The intrinsic or natural fluorescence of proteins is perhaps the most complex area of biochemical fluorescence. Fortunately the fluorescent amino acids, phenylalanine, tyrosine and tryptophan are relatively rare in proteins. Tryptophan is the dominant intrinsic fluorophore and is present at about one mole % in protein. As a result most proteins contain several tryptophan residues and even more tyrosine residues. The emission of each residue is affected by several excited state processes including spectral relaxation, proton loss for tyrosine, rotational motions and the presence of nearby quenching groups on the protein. Additionally, the tyrosine and tryptophan residues can interact with each other by resonance energy transfer (RET) decreasing the tyrosine emission. In this sense a protein is similar to a three-particle or multi-particle problem in quantum mechanics where the interaction between particles precludes an exact description of the system. In comparison, it has been easier to interpret the fluorescence data from labeled proteins because the fluorophore density and locations could be controlled so the probes did not interact with each other.

From the origins of biochemical fluorescence in the 1950s with Professor G. Weber until the mid-1980s, intrinsic protein fluorescence was more qualitative than quantitative. An early report in 1976 by A. Grindvald and I. Z. Steinberg described protein intensity decays to be multi-exponential. Attempts to resolve these decays into the contributions of individual tryptophan residues were mostly unsuccessful due to the difficulties in resolving closely spaced lifetimes. Also, interactions between the residues caused the total decay to differ from the sum of the contributions from each residue. In fact, the early resolution of two individual tryptophan residues in a protein by J. B. A. Ross, L. Brand and co-workers in 1981 still represents one of the most definitive results, and one verified in multiple other laboratories. A significant obstacle in resolving intrinsic protein fluorescence was the non-exponential decay of tryptophan itself. It is surprising to recognize that this issue was clarified around 1980.

In the mid 1980's there was a rush to study proteins which contained a single tryptophan residue. This was an attempt to remove the confounding interactions between residues. This effort led to some success. We learned that a tryptophan residue can display single exponential decay in certain proteins, and the local polarity can range from completely buried to completely exposed to water. Additionally, we learned that the indole side chains could be held rigid or could be very free to rotate in different single tryptophan proteins. M. Eftink and others pointed out there is no significant correlation between the emission maxima, quantum yields and lifetimes of single tryptophan proteins. The study of single tryptophan proteins could remove interaction between the residues, but could not remove the specific local interactions in the protein which had dramatic effects on each tryptophan residue.

A detailed understanding of protein fluorescence started to emerge from the advances in structural biology and the capabilities of molecular biology. Many laboratories have published detailed analyses of multi-tryptophan proteins in which all the trp residues are removed, and then replaced one by one in an attempt to determine the spectral properties of each residue. These studies revealed that changes in a single nearby amino acid could dramatically affect the emission spectrum of a nearby residue. We learned that amino acid side chains from residues such as histidine or lysine can quench nearby tryptophan. In some cases the spectral properties of the wild type proteins could be explained by the sum of the emission from the single trp mutants. In other cases the properties of the wild type proteins could not be explained as a simple summation of the mutant protein data. Such studies revealed interactions between the trp residues which could not be found from studies of the wild type proteins. When we now see the complexities of a protein containing just two or three trp residues, it is understandable that intrinsic protein fluorescence was difficult to interpret without studies of mutant proteins.

The present volume of *Topics in Fluorescence Spectroscopy* is intended to begin a new era in protein fluorescence. The individual chapters are devoted to one or just a few proteins for which detailed information on each trp residue has been obtained. I asked the authors to describe how each trp residue is affected by its local environment, and how the data can be correlated with the three dimensional structure. The detailed interactions described in these chapters will eventually evolve to a quantitative understanding of protein fluorescence. With such knowledge the fluorescence spectral properties will become increasingly useful for understanding the structure, function and dynamics of proteins.

In closing I thank all the authors for their cooperation and diligence in summarizing their fluorescence studies which advance our understanding of intrinsic protein fluorescence as a quantitative tool in structural biology.

Joseph R. Lakowicz
*Baltimore, Maryland*

# Contents

**7. The Conformational Flexibility of Domain III of Annexin V is Modulated by Calcium, pH and Binding to Membrane/ Water Interfaces**

Jaques Gallay, Jana Sopkova, and Michael Vincent

# 1

# Intrinsic Fluorescence of Proteins

## Maurice R. Eftink

### 1.1. Introduction

Fluorescence spectroscopy has long been one of the most useful biophysical techniques available to scientists studying the structure and function of biological molecules, particularly proteins. The pioneering work by Weber,[1,2] Teale,[2,3] Konev,[4] Burstein,[5] Brand[6] and their numerous proteges and colleagues[7–12] has demonstrated that proteins are capable of emitting prompt luminescence when excited with ultraviolet light. Further, this body of work has shown that protein fluorescence can reveal a variety of information, such as the extent of rotational motional freedom, the exposure of amino acid side chains to quenchers, and intramolecular distances. Chapters in this volume will go into detail about particular applications. This introductory chapter gives an overview, summarizes some patterns, and highlights what I think are important recent contributions and open questions.

### 1.2. Overview

The applications of fluorescence have grown and the advantages of the method are significant, making it one of the most widely used methods in a biochemist's or molecular biologist's arsenal. As a technique, fluorescence requires very limited quantities of material. In a typical fluorescence measurement, only nanomoles of the analyte is required, with the lower limit being single molecules in certain experimental designs. For proteins, tyrosine

---

*Maurice R. Eftink* • Department of Chemistry, The University of Mississippi, Oxford, MS 38677.

*Topics in Fluorescence Spectroscopy, Volume 6: Protein Fluorescence*, edited by Joseph R. Lakowicz. Kluwer Academic / Plenum Publishers, New York, 2000

and tryptophan residues provide intrinsic fluorescence probes. The fluorescnece of tryptophan almost always dominates, in proteins having both types of aromatic residues, and tryptophan is much more sensitive to its microenvironment than is tyrosine. Consequently, the vast majority of studies of intrinsic protein fluorescence focus on the tryptophan residues. Since there are usually few tryptophan residues per protein, this means that the method senses only these few points in the structure of a protein. Recent advances in molecular biology are making it almost routine to be able to add or delete tryptophan residues from specific positions in a protein. Alternatively, extrinsic fluorescence probes can be covalently or non-covalently attached to a protein, thus enabling a variety of fluorescence properties to be introduced;[13] also, other intrinsic fluorophores exist in some proteins.[14]

As mentioned above, an important property of fluorescence is that this signal is very environmentally sensitive, thus making this method useful for gaining information about protein structures. For example, the emission spectrum of the indole side chain of tryptophan is very sensitive to the polarity of its environment, providing a convenient probe to distinguish native and unfolded states of proteins. This environmental sensitivity is a consequence of the fact that the fluorescence emission of a fluorophore competes with other molecular processes that occur on the time scale of the emission process. That is, photon emission can occur on the same nanosecond time scale as the rotational and translational motion of small molecules and protein side chains. Consequently, the dipolar relaxation of polar groups and water around an excited state of a fluorophore can cause red shifts in the fluorescence, the collision with quenching groups or molecules can deactivate the excited state, and rotational motion of the fluorophore on the emission time scale can lead to measurable depolarization of the emitted light. Resonance energy transfer from a donor (D) fluorophore to an acceptor (A) can also occur on a time scale that is competitive with the emission process, when the D $\rightarrow$ A distance is sufficiently close and orientation of the electronic dipoles is not prohibitive. Such energy transfer measurements can be analyzed to obtain the D $\rightarrow$ A distance, which can be a very useful type of structural information, particularly for large multi-protein complexes, where crystal or nmr structures may not be possible.[15]

This environmental and motional sensitivity of fluorescence is experimentally realized by the fact that the method is multi-dimensional in nature. Fluorescence intensity can be measured as a function of excitation or emission wavelengths to obtain spectra. Intensity can be measured as a function of time to obtain fluorescence decay profiles. Intensity can be measured as a function of quencher (or other added agent, such a protons or co-solvent) to obtain information about dynamic accessibility and other proximal relationships. Intensity can be measured as a function of polarizer angle to obtain information about the rotational motion of the fluorophore. And these dimensional axes can be used in combination, for example, with measurements of intensity versus polarizer angle and time (time resolved anisotropy decays) or intensity versus wavelength and quencher concentration. This multi-dimensional nature of fluorescence is of great utility and partially overcomes the one significant disadvantage of the method, which is that the emission signals of similar fluorophores (e.g., tryptophan residues in a protein) are not resolved along the wavelength axis and are only sometimes resolved along the time, quencher concentration, and polarizer angle "experimental axes". It usually is necessary to combine these axes, and/or to study mutant proteins with different numbers of tryptophan residues, in order to assign the emission spectra and decay times of individual tryptophan residues. And such a resolution of individual spectra for individual tryptophan residues is often not tractable, particularly when the number of emitting sites is three or more.

Another major advantage of fluorescence is that the technique can be adapted to a variety of instrumental configurations. Essentially, what is required is to be able to get light in and light out of a sample. Besides the standard right angle detection geometry with rectangular cuvettes, fluorescence measurements can be made in capillaries, stopped-flow cells, high pressure cells, and microscope slides, to name a few arrangements. The rapidity of the measurements is also important, since this allows relatively high signal-to-noise data to be obtained with convenient measurements times, which can be so short as to be used in transient kinetics experiments.

Whereas fluorescence is intrinsically sensitive to competing nanosecond processes, thus making fluorescence useful for gaining information about protein dynamics and low resolution structural information (e.g., D $\rightarrow$ A distances), perhaps the most frequent application of fluorescence is as a probe for conformational transitions of proteins, including protein unfolding transitions (equilibrium and kinetics of), ligand binding, and protein-protein association processes.[16,17,18] These applications enable thermodynamic and kinetics information to be obtained. The key to these applications is the existence of a difference in some fluorescence signal for the different states of the protein. Provided that such a fluorescence difference exists, regardless of the cause of the fluorescence difference, the thermodynamic or kinetic data can be obtained. The experimental advantages of fluorescence (wide concentration range, rapid measurement time, various instrumental configurations) add to the value of the method for these thermodynamics and kinetics applications.

There has been a great deal of effort aimed at understanding the fundamental basis for the fluoresence properties of proteins, including attempts to correlate fluorescence lifetimes and anistropy decays with molecular dynamics calculations. But perhaps a more useful point of view, especially for the new user of this method, is to consider patterns in the fluorescence properties of a large set of single tryptophan containing proteins. In the following pages I will summarize some of these useful patterns, and in doing so will comment on applications of the method. I will go very lightly on the underlying principles, since these have been covered in other chapters in this volume and elsewhere.[7–12] Finally, I will also discuss some very recent advances and current topics of research in the field.

## 1.3. Patterns in Protein Fluorescence

When fluorescence was beginning to be used as a tool to study proteins, it was immediately clear that the emission maximum of the tryptophan residues would be a useful signature.[2] Though as mentioned above, the fluorescence contribution of individual tryptophan residues is greatly overlapped, it was found that the emission maximum of proteins ranged from less than 330 nm to above 350 nm. This range of emission maxima, which we now know can extend to as low as 308 nm for a tryptophan residues (e.g., in azurin (19)), has been found to be a fairly good and convenient measure of the solvent exposure of tryptophan residues in proteins. Whereas local electrostatic charge may play a role as well (20, 21, see below), the pattern that has emerged is that tryptophan residues buried in apolar core regions of proteins have a blue emission maximum, as low as 308 nm, and that tryptophan residues that are exposed to solvent water have a red emission of approximately 350 nm. Partial exposure of residues gives rise to an intermediate emission maxima. (Emission from tyrosine residues can also be observed in proteins, particularly in cases where there are no tryptophans, and there can be other intrinsic or extrinsic fluorescence probes attached to proteins. However, in this article I will comment only on the fluorescence of tryptophan residues in proteins.)

An early analysis by Burstein and coworkers[5] of the range of fluorescence properties of proteins led to the proposal that tryptophan residues can be grouped into one of four or five types of residues, with respect to their spectroscopic properties. These groups being those residues that are fully solvent exposed ($\lambda_{max} \approx 350$ nm), partially exposed on the surface of a protein ($\lambda_{max} \approx 340$ nm), buried within a protein but interacting with a neighboring polar groups ($\lambda_{max} \approx 315$ to $330$ nm), and completely buried in an apolar core ($\lambda_{max} \approx 308$ nm). An extension of this model has the various residue types being assigned to have certain fluorescence quantum yields and band width. However, there were only a few single-tryptophan containing proteins available at that time and this grouping was based primarily on data for multi-tryptophan containing proteins.

As more an more single tryptophan containing proteins have been discovered or have been created by mutagenesis, the model of having only a few classes of residues breaks down. Shown in Figure 1.1 is a plot of the fluorescence quantum yield versus emission wavelength for over 40 such single-tryptophan proteins. First is can be seen that the emission maximum of tryptophan residues does not cluster into a few groups along the x-axis. Second, there does not appear to be a pattern with respect to fluorescence quantum yield and emission maximum. That is, blue fluorescing tryptophan can have either low or high quantum yields. For red fluorescing tryptophans, the range of quantum yields appears to be a bit narrower. However, the pattern that emerges is that there is no pattern. Each tryptophan residues appears to have different properties.

An obvious question is why does an internal tryptophan residue (if we accept the notion that the emission maximum gives a reasonably good indication of whether a tryptophan residue is internal or solvent exposed, which appears to be a pretty dependable interpretation) have such a range of quantum yields. We generally assume that a very blue fluorescence is attributed to an indole ring being completely surrounded by apolar side chains, even to the extent that the imino NH of indole is not able to hydrogen bond

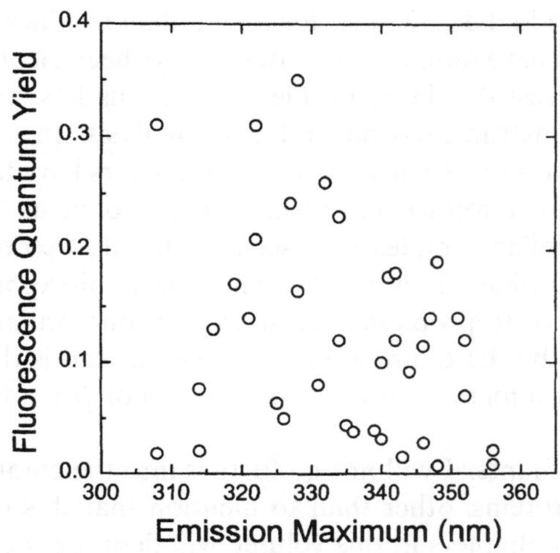


Figure 1.1. Relationship between tryptophan fluorescence quantum yield and emission maximum for several single-tryptophan containing proteins. A list of the proteins used to construct this and other plots can be obtained from www.olemiss.edu/depts/chemistry/Faculty/Eftink/.

with another polar group. Possible explanations will be discussed in a later section, but a simple answer is that internal tryptophan residues are still able to experience quenching reactions that lead to a low quantum yield. These may be energy transfer interactions with metal ions or chromophores that are located sufficiently close for such a quenching mechanism, with the indole ring still being completely surrounded by apolar groups. Disulfide bonds may also be stacked against an internal indole ring and this may lead to a quenching reaction. Also, it has recently been suggested that a phenyl ring, when stacked perpendicularly against an indole ring, can lead to quenching.[22] This will be discussed later, but such a mechanism could account for the low quantum yield of an internal tryptophan residue. If a tryptophan residue is located closer to the surface or is in contact with polar amino acid backbone or side chain groups, we expect its emission maximum to fall into the 320–340 nm range. Local electric field may also play an important role in determining the emission maximum (see a following section, 20, 21). Some of these polar functional groups (e.g., protonated His, peptide groups and amide side chains, Cys side chains) can lead to quenching reactions, whereas others do not. These intramolecular quenching reactions may be inefficient, but the fixed close proximity can result in a significant degree of quenching, even for a very weak quenching functional group.[23]

The fluorescence decay profiles of tryptophan residues in proteins are invariably found to be multi-exponential. There have been numerous studies aimed at accurately determining the number (e.g., three, four, five, etc.) and value of individual decay times for tryptophan residues in proteins. Only in a very few cases have mono-exponential decays been clearly found.[19,24] The desire to characterize the decay profiles of proteins has spurred impressive developments in instrumentation and data analysis. In view of the complexity of these fluorescence decays, some researchers have taken an alternate approach of fitting fluorescence intensity decay data as a distribution of decay times. A similar complexity is seen for the fluorescence decay of the amino acid tryptophan in water,[25,26] which is a bi-exponential. This bi-exponential decay of tryptophan is caused by intramolecular quenching reactions, particularly by the $\alpha$-ammonium side chain, and is thought to involve the existence of rotameric states around the $\alpha$-$\beta$ or $\beta$-$\gamma$ side chain bonds of tryptophan.[25,26]

In this brief chapter I will not go further into the complexity of tryptophan decays in proteins, other than to mention that this complexity exists. Some of the other chapters in this volume will describe the decay profiles of particular proteins. However, it can be interesting to look at overall patterns. Shown in Figure 1.2 is a plot of the mean fluorescence lifetime, $\langle\tau\rangle$ (defined as $\Sigma\alpha_i\tau_i$, where $\alpha_i$ is the amplitude of decay time $\tau_i$), for single tryptophan

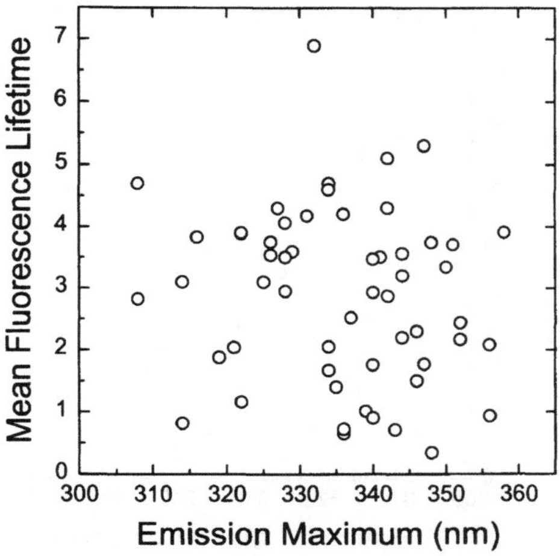


Figure 1.2. Relationship between the mean fluorescence decay time and the emission maximum for several single-tryptophan containing proteins.

containing proteins versus emission maximum. Just as with the quantum yield, there is no pattern for this mean lifetime. A mean lifetime can be as short as ~0.1 ns, in cases where there is a strong intramolecular quenching reaction (e.g., energy transfer to a heme), and individual $\tau_i$ can be as long as 16 ns.[27]

The ratio of the mean fluorescence lifetime divided by the quantum yield is the natural lifetime (actually a mean natural lifetime). Shown in Figure 1.3 are such natural lifetimes for the single tryptophan proteins. In principle, tryptophan should have a natural lifetime in the range of 15–20 ns, a value that might depend on environment. However, the calculated natural lifetimes for proteins ranges over a very wide range of 10 ns to 160 ns. The higher values are related to cases in which the fluorescence quantum yield is much lower than expected from the value of the mean lifetime. This might be explained as being due to a phenomenon called static quenching,[28] which means some process that results in a complete loss of fluorescence without there being a concomitant decrease in the observed fluorescence lifetime. The molecular origin of such static quenching processes is not always known, but the pattern in Figure 1.3 shows that such quenching does exist.

The above three figures each show that individual tryptophan residues in proteins have their own characteristic fluorescence properties and that there are no distinct classes into which residues can be easily grouped.

Another fluorescence property that can be easily measured in the laboratory is the exposure of a tryptophan residue to solute quenchers, such as

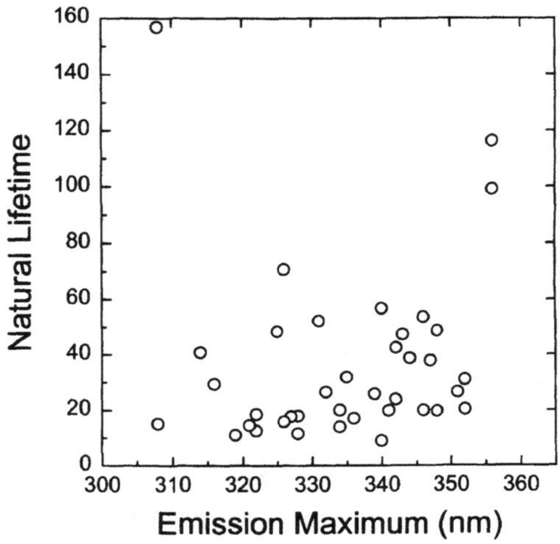


Figure 1.3. Relationship between the natural lifetime and the emission maximum for several single-tryptophan containing proteins.

acrylamide and iodide.[29] Here we do see patterns. Shown in Figures 1.4A and 4B are plots of the quenching rate constant, $k_q$, for acrylamide and iodide, versus emission wavelength for a group of single tryptophan proteins. As would be anticipated, bluer emitting tryptophans are less exposed to these solute quenchers and have smaller $k_q$ values; redder emitting tryptophan residues have larger $k_q$ values. The difference between acrylamide and iodide is that the latter is more selective as a quencher, as indicated by a log-log plot of the $k_q$ for these two quenchers (Figure 1.5). A slope of 1.7 indicates the higher selectivity of iodide for surface tryptophan residues. A similar comparison of acrylamide and oxygen as quenchers shows that oxygen is less selective as a solute quencher.

The rotational correlation time, $\phi$, of a tryptophan residue can be determined from time resolved fluorescence anisotropy measurements.[30] $\phi$ values are very useful due to their relationship to protein structure. As shown in Figure 1.6, the long $\phi$ value for a tryptophan residue in a protein correlates very well with the molecular weight of the protein. This makes the measurement of a $\phi$ value useful for determining such things as whether a protein is in a monomeric or dimeric state. Fluorescence anisotropy decays usually are described by a long rotational correlation time and one or more short rotational correlation times. The latter are typically described in terms of rapid segmental rotation of the tryptophan residue within a cone.[31]

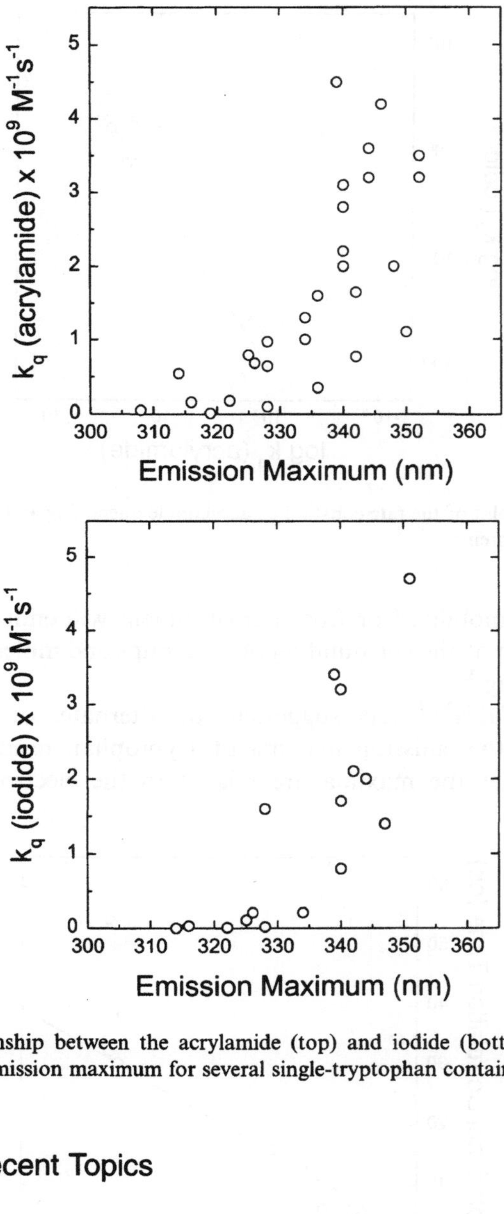


Figure 1.4. Relationship between the acrylamide (top) and iodide (bottom) quenching rate constants and the emission maximum for several single-tryptophan containing proteins.

## 1.4. Some Recent Topics

The classical explanation of the range of emission maxima for tryptophans in proteins is that the maxima are related to the solvent exposure of the residues, with the ability of polar functional groups to reorient during the nanosecond decay time to also be of importance. That is, a tryptophan

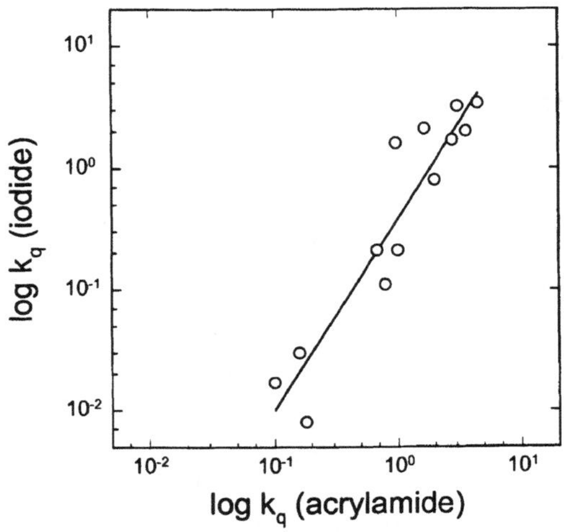


Figure 1.5. Log-log plot of the rate constant of acrylamide quenching and iodide quenching of single-tryptophan proteins.

residue in an immobilized or frozen environment will emit blue due to the limited relaxation of the surrounding polar groups and molecules around the excited indole ring.[32]

Recently, Callis[20,21] has suggested an alternate, or supplementary, explanation for the emission maxima of tryptophan residues in proteins. He suggested that the maxima are related to the electrostatic charge in

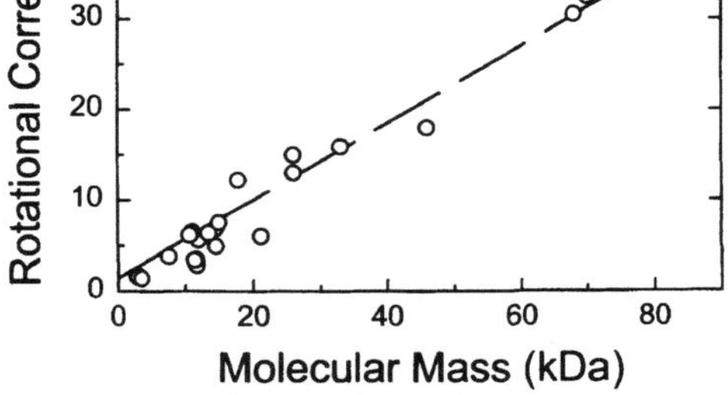


Figure 1.6. Relationship between the long rotational correlation time and the molecular weight for several single-tryptophan containing proteins.

the environment of the tryptophan residue. By using hybrid quantum mechanical–molecular dynamics calculations, starting with the crystal structure coordinates for proteins to calculate the expected electric field around tryptophan residue, Callis found an interesting correlation between the experimental and theoretical emission maxima for a set of proteins. The basis of the correlation is that there is a large change in the electronic dipole moment of the indole ring upon excitation to its excited singlet state, with the pyrrole ring becoming more positive. The local electrostatic field is thus predicted to be able to either stabilize or destabilize the excited state, leading to red or blue shifts. This leads to the prediction that a tryptophan's emission maximum should change in a predictable manner upon addition or removal of a charge group in the immediate vicinity of a tryptophan residue (e.g., protonating a nearby side chain functional group or binding a metal ion).

Another set of recent studies of general and related interest are the characterization of specific intramolecular quenching reactions in proteins by amino acid side chains. We have long known that protonated histidines, cystine, cysteine, and tyrosine residues, and perhaps protonated amino groups can act as intramolecular quenchers. However, Barkley and coworkers[23] have recently provided quantitative data to describe the quenching efficiency of various amino acid side chains, the peptide bond itself, and the different states of protonation of carboxylic acids, alkyl amines, phenol, and imidazole groups. This work clarifies the magnitude and mechanism of possible intramolecular quenching reactions.

Perhaps most unexpected is a series of studies that has implicated aromatic residues, phenylalanine and tyrosine, as having very specific quenching mechanisms for tryptophanyl fluorescence. It had been observed that certain buried tryptophan residues have a very low quantum yield, show short decay times, and show a ten-fold or more increae in their fluorescence intensity upon unfolding of the protein. Among these proteins are immunophilins[33] and homeodomain proteins.[22] The crystal structure of these proteins (or their homologs) shows that the indole rings of these single tryptophan residues participate in NH . . . $\pi$ hydrogen bond with an adjacent aromatic side chain of phenylalanine or tyrosine. This NH . . . $\pi$ hydrogen bond involves the perpendicular positioning of the the indole imino group and the $\pi$ cloud of the second residue. Evidence from these proteins and model studies indicates that this NH . . . $\pi$ interaction can lead to significant quenching and the possibility of this type of quenching can explain why buried and blue tryptophan residues can have a wide range of quantum yields.

The importance of these intramolecular quenching reactions and the local electrostatic field is that they provide explanations for the pattern, or lack thereof, shown in Figures 1.1 and 1.2. The intramolecular quenching reactions are also the ultimate cause of the non-exponential decay that is characteristic of tryptophan residues in proteins. Depending on the environment of a tryptophan residue, it will experience its individual and very asymmetric local electrostatic field and will experience different quenching side chains. If there is flexibility in the motion of side chain groups on the nanosecond time scale, then these quenching groups can undergo intramolecular diffusion, possibly colliding with the excited indole ring and quenching its fluorescence. The intramolecular quenching reactions may not require actual collision; that is, there is reason to believe that there is a distance dependence to quenching reactions that involve electron transfer. Consequently, collisions may not be required, but any motion can still modulate the process, thus becoming a mechanism for heterogeneity in the fluorescence decay. The existence of distinct side chain rotamers, around the tryptophan side chain (or the side chain of a specific quenching residue), is another point of view for the origin of heterogeneity in the emission of a tryptophan residue.[34]

## 1.5. Open Questions

How far can we go with interpreting protein fluorescence in terms of structural and kinetic details? It is hard to imagine ever being able to collect steady-state and time-resolved fluorescence data and then being able to predict, other than in a general way, the microenvironment of a tryptophan residue in a protein. These microenvironments are too aymmetric and varied and fluorescence parameters are not so revealing about actual neighboring residues. It seems that we will always need to take a look at the crystal structures. Making reasonable predictions of fluorescence properties from the structural coordinates is much more likely.

Still, there are some possibilities, particularly in terms of characterizing conformational changes upon ligand binding, protein subunit associations, or changes in solution conditions. We are developing a more complete understanding of how different amino acid side chains can act as intramolecualar quenchers of tryptophan fluorescence. These quenching reactions have signatures, such as their temperature or deuterium isotope dependence. Also, we are beginning to understand that all sides or edges of an indole ring are not equal and that this can lead to differences in the interactions with its asymmetric microenvironment. For example, in the electrostatic interactions described by Callis,[20] the five-membered pyrrole ring of indole becomes more positively charged in the excited state, so that charges near this end of the aromatic ring will lead to certain spectral shifts, whereas charges near the six-membered benzene ring will lead to opposite shifts. Similarly, we know that protonated ammonium groups can produce proton-transfer quenching reactions specifically at position 4 of the indole ring,[35] we know that hydrogen bonding with indole's imino NH group can be important in determining fluorescence properties, and the above mentioned recent studies predict that very specific indole-benzene geometries can lead to quenching. Thus, some characteristic changes in fluorescence characteristics can potentially provide subtle information about changes in the microenvironment of an indole ring, for example, upon ligand binding.

A number of questions remain, of course. How can we determine the dominant intramolecular quenching reaction for a particular tryptophan? How can we routinely indentify when energy transfer occurs between tryptophan residues? Is the emission maximum of a tryptophan residue determined primarily by the local electrostatic field? Or does the more traditional argument regarding polarity and solvent exposure, or some combination of these two models, provide the best explanation of fluorescence maxima? To what extent does $L_b$ emission, or the transition between $L_b$ and $L_a$ electronic states, contribute to emission and time-resolved fluorescence data? What is the best explanation for the non-exponential decay of tryptophan residues in protein? Ground state heterogeneity (rotamers)? Incomplete dipolar relaxations in the excited state? Excited state reactions, including distance dependent intramolecular eletron transfer reactions or proton transfer reactions? Can we gain any further insights about the very strong intramolecular quenching that leads to "static" quenching?

## 1.6. Summary

These are some thoughts to introduce this volume on protein fluorescence. The following articles will describe several specific protein systems and fluorescence techniques. There will be examples that focus on understanding the fluorescence properties of a protein, articles that exploit fluorescence to gain information about protein dynamics, and articles that apply the fluorescence of tryptophan or other fluorophores to gain kinetic or thermodynamic information. The applications of fluorescence are vast.

## References

1. Weber, G. "Polarization of the fluorescence of macromolecules. Theory and experimental method" *Biochem. J.* **51**, 145–155 (1952); Weber, G. "Rotational Brownian motion and polarization of the fluorescence of solutions" *Adv. Pro. Chem.* **8**, 415–459 (1953).

2. Teale, F. W. J. and Weber, G. "Ultraviolet fluorescence of hte aromatic amino acids" *Biochem. J.* **65**, 467–482 (1957).
3. Teale, F. "The ultraviolet fluorescence of proteins in neutral solution" *Biochem. J.* **76**, 381–388 (1960).
4. Konev, S. V. *Fluorescence and Phosphorescence of Proteins and Nucleic Acids*, Plenum Press, New York (1967).
5. Burstein, E. A., Vedenkina, N. S., and Ivkova, M. N. "Fluorescence and the location of tryptophan residues in protein molecules" *Photochem. Photobiol.* **18**, 263–279 (1973).
6. Beechem, J. M. and Brand, L. "Time-resolved fluorescence in proteins" *Ann. Rev. Biochem.* **54**, 43–71 (1985).
7. Longworth, J. W. "Intrinsic Fluorescence of Proteins" in *Excited States of Proteins and Nucleic Acids*, R. F. Steiner and I. Weinryb, eds, Plenum Press, New York, pp. 319–483 (1971).
8. Demchenko, A. P. *Ultraviolet Spectroscopy of Proteins*, Springer-Verlag, New York (1981).
9. Lakowicz, J. R. *Principles of Fluorescence Spectroscopy*, New York, Plenum Press (1983).
10. *Fluorescence Biomolecules*, edited by D. M. Jameson and G. D. Reinhart, Plenum Press, New York (1989).
11. *Time-Resolved Fluorescence Spectroscopy in Biochemistry and Biology*, edited by R. B. Cundall and R. E. Dale, Plenum Press, New York (1983).
12. Eftink, M. R. "Fluorescence techniques for studying protein structure" *Methods in Biochem. Anal.* **35**, 127–205 (1991).
13. Haughland, R. P. "Covalent fluorescent probes" in *Excited States of Biopolymers*, R. F. Steiner, ed., Plenum Press, New York, pp. 29–58 (1983).
14. Tsien, R. Y. "The green fluorescence protein" *Ann. Rev. Biochem.* **67**, 509–544 (1998).
15. Stryer, L. "Fluorescence energy transfer as a spectroscopic ruler" *Ann. Rev. Biochem.* **47**, 819–846 (1978); Fairclough, R. H. and Cantor, C. R. "The use of singlet-singlet energy transfer to study macromolecular assemblies" *Methods Enzymol.* **48**, 347–379 (1977); Selvin, P. R. "Fluorescence energy transfer" *Methods Enzymol.* **246**, 300–334 (1995).
16. Eftink, M. R. "The use of fluorescence methods to monitor unfolding transitions in proteins" *Biophys. J.* **66**, 482–501 (1994).
17. Eftink, M. R. "The use of fluorescence methods to study equilibrium macromolecule-ligand interactions" *Methods Enzymol.* **278**, 221–257 (1997).
18. Eftink, M. R. and Shastry, M. C. R. "Fluorescence methods for studying kinetics of protein folding reactions" *Methods Enzymol.* **278**, 258–286 (1997).
19. Finazzi-Agro, A., Rotilio, G., Avigliano, L., Guerrieri, P., Boffi, V., and Mondovi, B. "Environment of copper in *Pseudomonas fluorescens* azurin: Fluorimetric approach" *Biochemistry* **9**, 2009–2014 (1970); Szabo, A. G., Stepanik, T. M., Wagner, D. M., and Young, N. M. "Conformational heterogeneity of the copper binding site in azurin" *Biophys. J.* **41**, 233–244 (1983).
20. Callis, P. R. "$^1L_a$ and $^1L_b$ transitions of tryptophan: Applications of theory and experimental observations to fluorecence of proteins" *Methods Enzymol.* **278**, 113–150 (1997).
21. Callis, P. R. and Burgess, B. K. "Tryptophan fluorescence shifts in proteins from hybrid simulations: An electrostatic approach" *J. Phys. Chem.* **101**, 9429–9432 (1997).
22. Nanda, V. and Brand, L. "Low quantum yield of tryptophan reveals presence of a conserved NH . . . $\pi$ hydrogen bond in homeodomains" *J. Mol. Biol.* (in press) (1999).
23. Chen, Y. and Barkley, M. D. "Toward understanding tryptophan fluorescence in proteins" *Biochemistry* **37**, 9976–9982 (1998).

24. James, D. R., Demmer, R. P., Steer, R. P., and Verrall, R. E. "Fluorescence lifetime quenching and anisotropy studies with ribonuclease T1" *Biochemistry* **24**, 5517–5526 (1985).
25. Szabo, A. G. and Rayner, D. M. "Fluorescence decay of tryptophan conformers in aqueous solutions" *J. Amer. Chem. Soc.* **102**, 554–563 (1980).
26. Petrich, J. W., Change, M. C., McDonald, D. B., and Fleming, G. R. "On the origin of the nonexponential fluorescence decay in tryptophan and its derivatives" *J. Amer. Chem. Soc.* **105**, 3824–3832 (1983).
27. Schauerte, J. A. and Gafni, A. "Long-lived tryptophan fluorescence in phosphoglycerate mutase" *Biochemistry* **28**, 3948–3954 (1989).
28. Chen, R., Knutson, J. R., Ziffer, H., and Porter, D. "Fluorescence of tryptophan dipeptides: Correlations with the rotamer model" *Biochemistry* **30**, 5184–5195 (1991).
29. Eftink, M. R. "Fluorescence quenching: Theory and applications" in *Topics in Fluorescence Spectroscopy, Vol. 2 Principles*, J. R. Lakowicz, ed. Plenum Press, New York, pp. 53–126 (1991).
30. Steiner, R. "Fluorescence anisotropy: Theory and Applications" in *Topics in Fluorescence Spectroscopy, Vol. 2 Principles*, J. R. Lakowicz, ed. Plenum Press, New York, pp. 1–52 (1991).
31. Lipari, G. and Szabo, A. "Effect of librational motion on fluorescence depolarization and nuclear magnetic resonance relaxation of macromolecules and membranes" *Bioiphys. J.* **30**, 489–506 (1980).
32. Longworth, J. "Excited state interactions in macromolecules" *Photochem. Photobiol* **7**, 587–592 (1968).
33. Silva, N. D. and Prendergast, F. G. "Tryptophan dynamics of FK506 binding protein: Time-resolved fluorescence and simulations" *Biophys. J.* **70**, 1122–1137 (1996).
34. Willis, K. J. and Szabo, A. G. "Conformation of parathyroid hormone: Time-resolved fluorescence studies" *Biochemistry* **31**, 8924–8931 (1992); Dahms, T. E. S., Willis, K. J., and Szabo, A. G. "Conformational heterogeneity of tryptophan in portein crystal" *J. Amer. Chem. Soc.* **117**, 2321–2326 (1995).
35. Saito, I., Sugiyama, H., Yamamoto, A., Muramatsu, S., and Matsuura, T. "Photochemical hydrogen-deuterium exchange reaction of tryptophan. The role in nonradiative decay of the singlet state" *J. Amer. Chem. Soc.* **106**, 4286–4287 (1984).

# 2

# Spectral Enhancement of Proteins by *in vivo* Incorporation of Tryptophan Analogues

J. B. Alexander Ross, Elena Rusinova, Linda A. Luck, and Kenneth W. Rousslang

## 2.1. Introduction

Tryptophan (Trp) residues in proteins and polypeptides have been used extensively as absorption, fluorescence, and phosphorescence probes for studying structure, dynamics, interactions, and local environments. In particular, changes in fluorescence intensity, emission wavelength maximum, lifetimes, and anisotropy, as well as differential accessibility to quenchers and sensitivity to bound ligands, have made Trp a valuable and widely used spectroscopic tool. Valuable information about, for example, enzyme catalysis or interactions with cofactors and metal ions can be obtained from these spectroscopic observables. Trp, however, is a difficult, if not impossible spectroscopic entity to use to study protein-protein interactions. Most proteins contain Trp, and it is difficult to selectively excite the fluorescence of individual proteins when in a complex. Similarly, Trp is a difficult probe to use effectively for protein-DNA or protein-RNA interactions. The absorption spectra of DNA and RNA essentially completely overlap that of Trp. In addition, the number of nucleic acid bases compared with Trp residues is usually very large. Thus, a DNA or RNA molecule often has a much greater extinction coefficient than a binding protein. Depending upon the concentrations

*J. B. Alexander Ross and Elena Rusinova* • Department of Biochemistry and Molecular Biology, Mount Sinai School of Medicine, New York, New York 10029-6574. *Linda A. Luck* • Department of Chemistry, Clarkson University, Potsdam, New York 13699-5605. *Kenneth W. Rousslang* • Department of Chemistry, University of Puget Sound, Tacoma, Washington 98416-0062.

*Topics in Fluorescence Spectroscopy, Volume 6: Protein Fluorescence*, edited by Joseph R. Lakowicz. Kluwer Academic / Plenum Publishers, New York, 2000

required to measure the interaction, the large extinction of DNA or RNA can cause a significant inner filter effect, which can easily result in misinterpretation of fluorescence data.

Because Trp is not the probe of choice for study of macromolecular interactions, extrinsic probes generally have been used that can be excited at wavelengths where neither Trp nor nucleic acids absorb. The introduction of extrinsic probes, however, requires careful consideration of possible effects on structure and function. Chemical modification can generate different conformational states of the protein as well as alter intermolecular interactions or enzymatic activity. In addition, for detailed molecular interpretations there is always the issue of specificity of labeling.

An alternative to introduction of extrinsic probes by chemical modification is replacement of naturally occurring Trp residues with Trp analogues. This can be accomplished by using recombinant protein expression in cells that are auxotrophs for Trp. The objective is to generate proteins or polypeptides that have spectroscopic features appropriately different from those of the unlabeled macromolecule. The incorporated analogue serves as a site-specific, pseudo-intrinsic probe, and in many cases most or all of the native functional properties are retained.

This chapter describes recent advances in applications of Trp analogues as pseudo-intrinsic probes in biology and biophysics. The Trp analogues discussed here are shown in Figure 2.1. After a brief historical retrospective, an overview is presented on the methods for incorporation, followed by a comparison of different analytical tools and approaches that can be used to quantitate analogue incorporation. Next, the special spectroscopic features

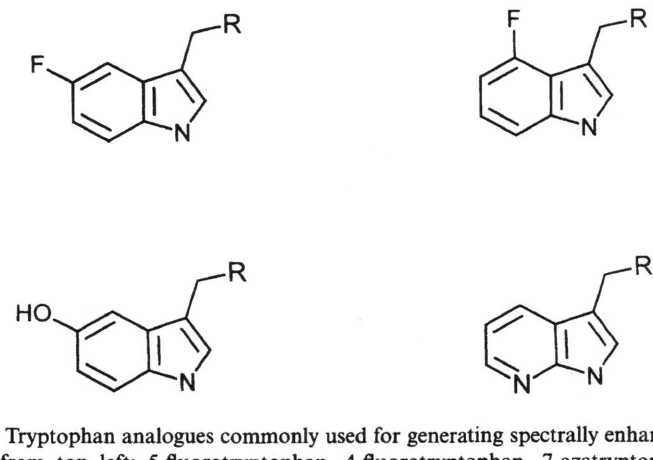


Figure 2.1. Tryptophan analogues commonly used for generating spectrally enhanced proteins. Clockwise from top left: 5-fluorotryptophan, 4-fluorotryptophan, 7-azatryptophan, and 5-hydroxytryptophan.

of these analogues are described as isolated models and after incorporation in a protein. The latter includes several different biophysical applications. While many of the applications to date have focused on protein-DNA interactions, the general principles apply also to protein-RNA and protein-protein interactions.

### 2.1.1. A Brief History

Trp analogues were first used in biological chemistry during the 1950s to elucidate metabolic pathways and the mechanisms involved in protein synthesis.[1–4] It had been noted in several of these reports, however, that many analogues inhibited bacterial growth. Schlesinger[5] reported in 1968 that replacement of Trp by the analogues either 7-azaTrp (7-Atrp) or tryptazan allowed the formation of active alkaline phosphatase in a Trp auxotroph of *Escherichia coli* (*E. coli*), which was in contrast to previous results obtained with histidine analogues.[6] Alkaline phosphatase was synthesized in the auxotroph strain when the cell medium was devoid of inorganic phosphate and either Trp, 7-ATrp or tryptazan was used to supplement the medium.[5] Over the course of the first 30 min, the same rate of protein synthesis was observed in the presence of either Trp or the analogues. The purified enzymes synthesized in the presence of the two analogues exhibited indistinguishable kinetic constants when *p*-nitrophenyl phosphate was used as substrate, although other substrates showed some minor differences in activities. Also, some differences were observed in the protein heat stability. The main differences in physical chemical characteristics, however, were the shapes and intensities of the absorption and fluorescence spectra of the enzymes that had been synthesized in the presence of the analogues. In particular, red-shifted absorption and dramatically altered emission spectra were observed compared to those of the enzymes synthesized in the presence of Trp.

Schlesinger[5] concluded from her results on the effects of the two Trp analogues on alkaline phosphatase, that Trp residues *per se* are not essential for the catalytic activity of this protein. Over a decade later, studies by Foote and coworkers[7] on the effects of 7-ATrp on aspartate transcarbamylase (aspartate carbamoyltransferase; ATCase) showed that a Trp analogue could affect function. Notably, they found that allosteric modulation was enhanced by this analogue. To understand this, they examined an x-ray crystal structure of the enzyme, focusing on Trp-199, which is part of the catalytic chain. To account for the effect upon catalysis, they proposed that when the side chain of Trp-199 residue is replaced with 7-azaindole, the aza ring nitrogen could form a hydrogen bond with the carbamoyl phosphate.

During the 1970s, a major effort was directed towards replacement of certain amino acid residues with their fluorinated analogues for use as $^{19}F$ NMR probes.[8,9] Examples relevant to this review include the fluoro-Trp (FTrp) analogues 4-FTrp, 5-FTrp, and 6-FTrp. Pratt and Ho[10] examined the effects of these analogues incorporated into the *E. coli* enzymes lactose permease, *β*-galactosidase, and D-lactate dehydrogenase. While the analogue 4-FTrp had the least effect on enzyme activity, it was noted that effects on other enzymes were variable.

Significant efforts towards methods for incorporation of FTrp analogues into proteins for $^{19}F$ NMR continued during the 1980s.[9] In retrospect, it seems somewhat surprising that while during the 1970s and 1980s there was considerable interest and many important developments in possible ways to introduce novel fluorescent probes into proteins, for example through selective chemical modification, no further developments appeared in the fluorescence literature along the path opened by Schlesinger.[5] It seems that her results were essentially unnoticed. Nevertheless, investigators in the field of biological fluorescence were clearly considering the general idea of using amino acid analogues to alter the optical properties of proteins. For example, in a 1986 review, Hudson and coworkers[11] suggested that amino acid derivatives with side chains such as azulene or benzo[*b*]thiophene might be useful as substitute fluorophores for Trp. In retrospect, it is clear that a major obstacle was availability of a simple, reliable approach for incorporation of these non-natural amino acids into proteins. An important feature of the earlier successes with alkaline phosphatase[5] and aspartate transcarbamylase[7] was the fact that expression of these particular proteins was under the control of strong, inducible promoters. Thus, it was possible to reduce substantially the toxicity of an analogue by first growing the auxotrophic bacterial cells in the presence of Trp while maintaining expression of these proteins in a repressed state. After accumulating the desired cell density, the analogue could be added and the cells derepressed. In this way, it was possible to achieve relatively high levels of incorporation. Analogue incorporation *in vivo* into proteins lacking inducible promoters does not have this advantage, and the levels of incorporation are typically very low, in some cases undetectable.

The other approaches that have been taken for analogue incorporation are *in vitro*. One well-established methodology guaranteeing 100% incorporation of non-natural amino acids into peptides and small proteins is direct chemical synthesis.[12,13] Another possibly more general solution is *in vitro* transcription-translation using a suppressor RNA amino-acylated with the desired nonnatural amino acid.[14,15] This approach has been used successfully to incorporate 7-ATrp into T4 lysozyme[16] and 5-hydroxyTrp (5-OHTrp) into *β*-galactosidase.[17] The yields from *in vitro* protein synthesis, however, generally fail to achieve those obtained *in vivo*.[18,19]

A general, highly efficient approach for incorporation of Trp analogues *in vivo* into proteins for fluorescence studies of macromolecular interactions was achieved in 1992 in two independent laboratories.[20,21] These two groups took advantage of high-level expression vectors with artificial inducible promoters and used variations of standard methods for protein expression in bacterial auxotrophs to replace protein Trp residues with 5-OHTrp. The proteins were the Y57W mutant of oncomodulin, with expression under control of the OXYPRO promoter,[20] and $\lambda c$I repressor, with expression under control of the *tac* promoter.[21] The basic strategies were similar, and involved essentially three steps. First the bacterial cells were grown in the presence of Trp. Second, prior to induction of expression, the growth medium was replaced with Trp-free medium. Third, after a short period of Trp starvation, the Trp analogue of choice was added to the medium followed by induction under standard conditions. Subsequent protein purification was by standard protocols. In both experiments, mg quantities of analogue-containing protein were obtained, with overall yields essentially equivalent to that obtained when the proteins were expressed with Trp. The efficiency of analogue incorporation differed significantly, however. In particular, expression under control of the *tac* promoter provided much more efficient incorporation. As discussed below, the subsequent experience of many different laboratories with expression of different proteins using different promoters indicates that the efficiency of incorporation is highly promoter dependent.

## 2.2. *In vivo* Analogue Incorporation

Methods for incorporation of non-natural amino acids into proteins and polypeptides by complete chemical synthesis, semi-synthesis, and *in vitro* transcription-translation using analogue-charged suppressor RNAs are covered in recent reviews.[18,19] Another recent review provides a detailed description and discussion of methods for incorporation *in vivo* using recombinant DNA technology.[22]

Incorporation *in vivo* generally follows standard practices for protein expression in bacterial cells using various inducible promoters. A considerable number of recombinant proteins have now been expressed with Trp analogues. The fluorescence and functional characteristics of some of these proteins have been summarized previously.[22] These included, for example, Y57W oncomodulin,[20] Trp *t*RNA synthetase,[23] rat parvalbumin[24] $\sigma^{70}$ subunit of RNA polymerase,[25] a series of mutants of the $\alpha$ subunit of RNA polymerase,[26] and several others that have been reported to the authors of this review by personal communication. Table 2.1 provides an updated

**Table 2.1. Proteins Expressed with Tryptophan Analogs**

| Protein | Trps | Promoter | Analogue | Incorporation | Function |
|---|---|---|---|---|---|
| Y57W oncomodulin (rat)[20] | 1 | OXYPRO | 5-OHTrp | <50% | wild-type |
| $\lambda$ *c*I repressor[21] | 6 | *tac* | 5-OHTrp | >95% | wild-type |
| soluble Tissue Factor[28,35] | 4 | *tac* | 5-OHTrp | <20% | ? |
| soluble Tissue Factor[28,35] | 4 | *tac* | 7-ATrp | <30% | ? |
| soluble Tissue Factor[36] | 4 | *tac* | 5-FTrp | >98% | wild-type |
| Trp tRNA synthetase[23] | 1 | *tac* | 5-OHTrp | >95% | altered |
| Trp tRNA synthetase[23] | 1 | *tac* | 7-ATrp | >95% | altered |
| Trp tRNA synthetase[23] | 1 | *tac* | 4-FTrp | >95% | altered |
| 2 Herpesvirus protein VP16 mutants: F442W, F473W[59] | 1 | *tac* | 5-OHTrp | 50–95% | wild-type |
| 2 Herpesvirus protein VP16 mutants: F442W, F473W[59] | 1 | *tac* | 7-ATrp | 50–95% | wild-type |
| CRP[25,a] | 2 | $\lambda P_L$ | 5-OHTrp | 50–95% | wild-type |
| CRP[a] | 2 | $\lambda P_L$ | 7-ATrp | >98% | wild-type |
| rat parvalbumin F102W[24] | 1 | T7/pLysE | 5-OHTrp | ~50% | wild-type |
| rat parvalbumin F102W[24] | 1 | T7/pLysE | 7-ATrp | ~50% | wild-type |
| rat parvalbumin F102W[24] | 1 | T7/pLysE | X-FTrp X = 4, 5 or 6 | ~50% | wild-type |
| $\alpha$ subunit of RNA polymerase[25] | 1 | T7 | 5-OHTrp | 50–90% | wild-type |
| 11 $\alpha$ subunit mutants: W321F&{W260, . . . ,W270}[b] | 1 | T5 | 5-OHTrp | >95% | wild-type |
| $\sigma^{70}$ subunit RNA polymerase[26] | 4 | T7 | 5-OHTrp | 50–60% | wild-type |
| $\sigma^{70}$ subunit RNA polymerase mutant W314A,W326A[60] | 2 | T5 | 5-OHTrp | 91% | wild-type |
| MyoD[c] | 1 | T5 | 5-OHTrp | >90% | wild-type |
| MyoD[c] | 1 | T5 | 7-ATrp | >90% | wild-type |
| cytidine repressor[a] | 1 | T7 | 5-OHTrp | 30–50% | wild-type |
| phage $\lambda$ lysozyme[61] | 4 | $\lambda P_L$ | 7-ATrp | >98% | wild-type |
| T4 Clamp protein 45[29] | 2 | T7 | 4-FTrp | >95% | wild-type |
| staphylococcal nuclease[57] | 1 | $\lambda P_L$ | 5-OHTrp | 95% | 92% |
| staphylococcal nuclease[57] | 1 | $\lambda P_L$ | 7-ATrp | 98% | 80% |
| staphylococcal nuclease V66W[55] | 2 | $\lambda P_L$ | 5-OHTrp | ? | active |
| staphylococcal nuclease V66W[55] | 2 | $\lambda P_L$ | 7-ATrp | ? | active |
| staphylococcal nuclease V66W[55] | 2 | $\lambda P_L$ | X-FTrp X = 4, 5 or 6 | ? | active |
| TBP[d] | 1 | T7 | 5-OHTrp | 30–50% | ? |
| $NCD_{335-700}$ W370F[e] | 1 | *tac* | 5-OHTrp | ~90% | active |
| $NCD_{335-700}$ W370F[e] | 1 | *tac* | 7-ATrp | >90% | active |
| BirA (biotin repressor)[62] | 7 | *tac* | 5-OHTrp | 85%[f] | active |
| 5 tropomyosin mutants: 90W, 101W, 111W, 122W, 185W[32,33] | 1 | T7 | 5-OHTrp | >90% | all active but 90W |
| tropomyosin mutant 122W[32,33] | 1 | T7 | 7-ATrp | >90% | active |
| annexin V[42] | 1 | T7 | X-FTrp X = 4, 5 or 6 | >95% | variable |

Personal communications from [a]D. F. Senear, [b]T. Heyduk, [c]S. Khotz, [d]M. Brenowitz, [e]D. Stone and R. Mendelson, and [f]D. Beckett.

list of these and other proteins that have been expressed with different Trp analogues.

To minimize co-expression of non-analogue containing protein, it is important to use a non-leaky promoter. However, it should be noted from Table 2.1 that proteins expressed under control of the T7 promoter, which is considered a tight promoter, generally show significantly lower levels of analogue incorporation than that obtained with the *tac* or T5 promoters, for example. This important issue is discussed further in section 2.2.1. Information about the standard molecular biology techniques is available in the laboratory manual by Sambrook *et al.*[27]

### 2.2.1. General Approach for *in vivo* Incorporation of Analogues

Efficient *in vivo* incorporation of Trp analogues can be achieved using a plasmid-based bacterial expression system for the protein of interest provided it can be transferred to and expressed by a Trp auxotrophic cell. Many different laboratories have used the *E. coli* Trp-auxotrophs W3110 TrpA33 (E48M), W3110 TrpA88 (amber mutation), and CY15077 (W3110 *traA2*ΔTrpEA2, a mutation in the *tra* gene and deletion of the Trp operon). These auxotrophic strains are originally from the laboratory of C. Yanofsky at Stanford University. It should be noted, however, that other Trp auxotrophs host cells can be used including eukaryotes. For example, we have experimented with 5-OHTrp and 7-ATrp labeling protocols for proteins expressed in yeast.[28]

Single-step and two-step methods have been described for *in vivo* analogue incorporation using bacteria as host cells.[22] Most laboratories use a two-step method, the elements of which are outlined briefly here. The cells are grown initially in a medium containing essential nutrients and Trp. At a cell density that typically is used for induction of expression of the particular protein, the cells are removed from the growth medium by centrifugation. They then are resuspended in a minimal medium that contains no Trp. Before induction, sufficient time, typically a half hour, is allowed to elapse to exhaust residual Trp pools. The Trp analogue is then added. After about ten minutes the cell culture is induced. Finally, the cells are harvested following the usual time of induction, which is usually 3–6 hours.

As indicated in Table 2.1, proteins have been expressed with Trp analogues under a variety of conditions, using various promoters, including *tac*, T7, T5, as well as temperature-sensitive and oxygen-sensitive promoters. Typically, the purification protocols used have been those developed previously for the non-analogue-containing protein. Tools and approaches for assessing the efficiency of analogue incorporation are discussed in the following section.

The accumulated results from different laboratories show that the efficiency of analogue incorporation is variable and depends upon several factors. In some cases, the efficiency of incorporation is clearly dependent upon the promoter used for regulation of expression, while in others it appears to be a feature of a particular protein-analogue combination. As indicated above, the promoter should not be leaky; it is important to evaluate basal and induced levels of protein expression.[22] "Tight" promoters that generally have provided high levels of incorporation are *tac*, T5, and $\lambda P_L$, which is temperature-sensitive. By contrast, the efficiency of incorporation appears to be low in most cases for proteins expressed with the T7 promoter. A notable exception is the report of greater than 95% incorporation of 4-FTrp into the T4 Clamp protein.[29] The analogue 4-FTrp is non-fluorescent,[30,31] and the estimate of its incorporation was based on residual fluorescence when the sample was excited at 280 nm. Another exception is the report of greater than 90% incorporation of 5-OHTrp and 7-ATrp by several single Trp mutants of tropomyosin expressed under control of the T7 promoter.[32,33] In this case, the estimates of analogue incorporation were based on evaluation of excitation spectra. However, it has been established that the quantum yields of Trp, 5-OHTrp, and 7-ATrp residues can differ considerably depending on the nature of the local environment. This is demonstrated, for example, by the variation in their quantum yields in different solvents (see Table 2.2). Consequently, a fluorescence-based method for estimating incorporation is necessarily qualitative. However, quantitative methods for estimating analogue incorporation have been developed, and these are outlined in section 2.2.2.

The T7 promoter utilizes the highly specific T7 RNA polymerase, which is expressed after induction and prior to expression of the target protein.[34] Thus, T7 RNA polymerase is being synthesized in the presence of analogue. By contrast, other promoters, such as *tac* or T5, can utilize the bacterial RNA polymerase. It may be that analogues such as 5-OHTrp or 7-ATrp, but not fluorinated analogues, compromise either the function or folding of T7 RNA polymerase, thereby lowering the efficiency of expression of the target protein. In our experience, efficient incorporation generally accompanies efficient expression of the target protein.

The differential tolerance of a protein for various Trp analogues is well illustrated by incorporation experiments with recombinant soluble human tissue factor (sTF), a protein that has four Trp residues and has been expressed under control of the *tac* promoter.[28] Two of the Trp residues are buried from solvent, and site-specific mutation on either of these residues to

Table 2.2. Fluorescence Emission Properties of Trp Analogs in Different Solvents

| Solvent | $\varepsilon$ | n | Analog | $\lambda_{max}$, nm | $\phi^a$ | $\tau_{int}$, ns[b] | $\tau_{num}$, ns[c] |
|---|---|---|---|---|---|---|---|
| Dioxane | 2.2 | 1.421 | NATrpA | 332 | 0.30 | 4.53 | 4.36 |
| | | | 5-FTrp | 336 | 0.31 | 3.91 | 3.66 |
| | | | 5-OHTrp | 336 | 0.38 | 5.12 | 5.02 |
| | | | 7-AzaTrp | 363 | 0.39 | 8.36 | 8.02 |
| Ethanol | 25 | 1.361 | NATrpA | 345 | 0.23 | 3.62 | 3.57 |
| | | | 5-FTrp | 340 | 0.13 | 2.21 | 1.86 |
| | | | 5-OHTrp | 337 | 0.30 | 4.29 | 4.15 |
| | | | 7-AzaTrp | 385 | 0.01 | 0.25? | — |
| Acetonitrile | 37 | 1.340 | NATrpA | 340 | 0.33 | 4.79 | 4.76 |
| | | | 5FTrp | 349 | 0.15 | 3.36 | 1.88 |
| | | | 5OHTrp | 333 | 0.27 | 4.24 | 3.9 |
| | | | 7AzaTrp | 373 | 0.31 | 8.67 | 7.64 |
| DMF | 45 | 1.429 | NATrpA | 341 | 0.23 | 4.19 | 4.11 |
| | | | 5FTrp | 355 | 0.16 | 5.02 | 4.32 |
| | | | 5OHTrp | 340 | 0.02 | 0.52? | — |
| | | | 7AzaTrp | 376 | 0.50 | 13.8 | 13.1 |
| pH 7.4 water | ~80 | 1.333 | NATrpA | 356 | 0.14 | 2.96 | 2.91 |
| | | | 5FTrp | 358 | 0.14 | 2.92 | 2.61 |
| | | | 5OHTrp | 341 | 0.22 | 4.12 | 4.03 |
| | | | 7AzaTrp | 415 | 0.01 | 0.65 | 0.6 |

[a]Quantum yields, $\phi$, and emission maxima, $\lambda_{max}$, were calculated from corrected integrated steady-state emission spectra ($\lambda_{ex}$ = 289 nm), assuming a quantum yield of 0.14 for NATrpA in aqueous buffer (pH = 7.4) at 20 °C. The instrument factors for correction of the emission spectra, were generated from corrected emission spectra of tyrosine and tryptophan at pH 7, and of 2-amino pyridine and quinine sulfate in 1 N sulfuric acid, kindly provided by Professor Edward Burstein.
[b]Intensity average lifetime: $\tau_{int} = \Sigma\alpha_i\tau_i^2/\Sigma\alpha_i\tau_i$.
[c]Number average lifetime: $\tau_{num} = \Sigma\alpha_i\tau_i/\Sigma\alpha_i$.

Phe or Tyr reduces protein expression substantially.[35] Expression of the wild-type, four-Trp protein with 5-OHTrp and 7-ATrp gave low yields of protein and poor levels of analogue incorporation.[28] Expression of the wild-type protein with 5-FTrp, by comparison, gave high yields of fully functional protein and there was essentially complete incorporation of this analogue according to spectral and mass analyses, as described below. By contrast, 5-FTrp incorporation by the single-Trp replacement mutants was incomplete, with efficiencies in the range of 60 to 80%, depending upon the levels of protein expression, which were low, a characteristic of these single Trp-site mutants.[36]

### 2.2.2. Analysis of Analogue Incorporation

Several different analytical methods have been used to estimate the degree of Trp analogue incorporation in recombinant proteins, including absorption spectroscopy, mass spectroscopy, and high-performance liquid chromatography (HPLC). Each method utilizes different physical and chemical properties of the various analogues. Thus, each method has different advantages and interferences.

Accurate quantitation of analogue incorporation by absorption spectroscopy depends upon the wavelength and extinction differences in the spectra of the Trp analogues compared with those in the spectra of Trp and Tyr. As described in section 2.3.1, 5-OHTrp, 7-ATrp, and 5-FTrp have significant absorption at wavelengths above 305 nm, where Trp absorption generally becomes negligible. Making the assumption that the spectrum of a protein denatured in high concentrations of guanidinium chloride (typically 6M) is represented by a linear combination of the individual contributions due to Tyr, Trp, and the Trp analogue, the entire absorption spectrum of the protein can be fit by least squares to properly scaled basis-set spectra of these amino acids in the same solvent. This method is referred to as a LINCS analysis.[37] An example is shown in Figure 2.2.

To obtain an accurate estimate of the degree of analogue incorporation by LINCS, proper scaling of the basis set spectra is crucial. To achieve proper scaling of Tyr and Trp, the spectra are measured of a series of model proteins containing different *known ratios* of these aromatic amino acids. The basis-set absorption spectrum of the analogue of interest is then determined

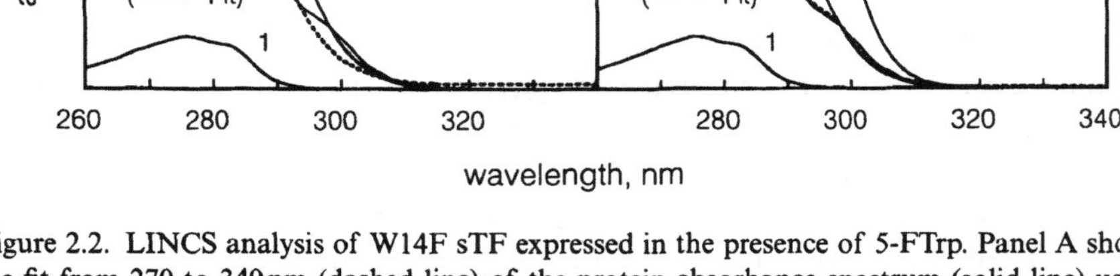


Figure 2.2. LINCS analysis of W14F sTF expressed in the presence of 5-FTrp. Panel A shows the fit from 270 to 340 nm (dashed line) of the protein absorbance spectrum (solid line) using the NATyrA and NATrpA basis sets. Panel **B** shows the corresponding fit (dashed line) when 5-FTrp is included as a third basis set.

from synthetic peptides containing *known ratios* of the analogue, Tyr, and Trp. The absorption of the Trp model compound N-acetyl-tryptophanamide (NATrpA) in guanidinium chloride provides an accurate reference standard for the Trp residue incorporated in a polypeptide chain.[38] Taking the value of 5,500 $cm^{-1} M^{-1}$ for the average extinction coefficient at 280 nm of a Trp residue,[39] Waxman *et al.*[37] calculated the average wavelength-dependent extinctions for the spectrum of each model compound that would recover most accurately the known ratios for the aromatic residues in all peptides or proteins used for the basis set. In the determination of this ratio from synthetic peptides and purified proteins of known Trp and Tyr content, it was noted that the extinction coefficients calculated for Tyr (relative to Trp) in these denatured proteins were generally lower (~20%)[37] than those determined by previous investigators.[40,41] Accurate estimates of incorporation usually can be obtained when the protein is denatured in 6M guanidinium chloride provided there are no interferences from contaminating chromophores or from perturbation of the aromatic residue side chains due to local intramolecular interactions persisting in the denatured state.

Mass spectrometry of the intact protein can provide a sensitive, detailed measure of the degree of incorporation of analogues with additional heavy heteroatoms, such as 4-FTrp, 5-FTrp, or 5-OHTrp, assuming that an analogue containing protein does not differ significantly in its physical and chemical properties from the non-analogue containing protein except in mass. Replacing hydrogen with fluorine increases the mass of the side chain by 18 amu, while addition of oxygen increases the mass by 16 amu.

Electrospray ionization (ESI) has been used to assess incorporation of 5-FTrp into annexin V[42], a single Trp protein, as well as soluble human tissue factor and single-Trp-to-Phe mutants[36] of this protein. While labeling of annexin V and wild type soluble human tissue factor, appeared to be essentially complete, labeling of the single-Trp-to-Phe mutants was less efficient. The mass distribution spectra of soluble human tissue factor and the single-Trp-to-Phe mutants (see Figure 2.3), constructed from mass-to-charge ratio spectra of proteins that were not fully labeled, yielded well-resolved mass peaks corresponding to the expected molecular weights of proteins with four, three, two, one, or none of the Trp residues replaced by 5-FTrp. The overall efficiency of incorporation was assessed by first normalizing the peak heights of each appropriate molecular weight species to the sum of their peak heights. This assumes that peak height is directly proportional to area, which is not necessarily true. However, the possible error introduced by this simplifying assumption is negligible because the spectra are uniformly narrow. Each normalized peak was multiplied by the corresponding fraction of Trp residues replaced (1, 0.75, 0.5, 0.25, or 0), and then the percent incorporation was calculated from their sum. In each case, the total percent

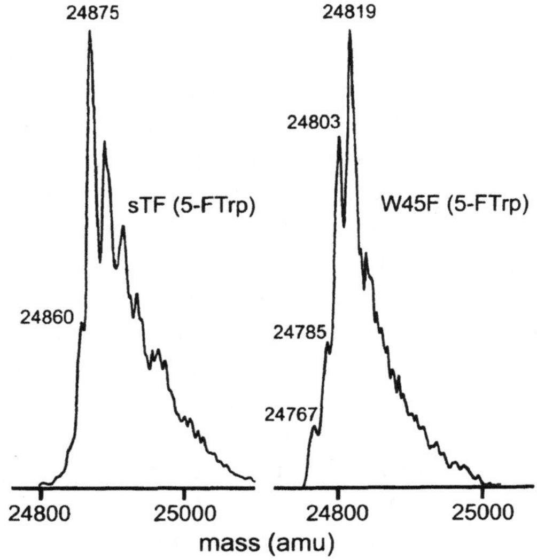


Figure 2.3. Mass spectra of wild-type sTF and the mutant W45F sTF. The expected mass is 24,875 for the wild-type protein when four Trp residues are replaced with 5-FTrp, while the expected mass is 24,818 for the mutant with three Trp residues replaced.

incorporation was within 10% of that obtained from LINCS analysis, described above. Thus, both LINCS and mass spectroscopy can provide comparable information.

The mass spectrum of a protein containing 7-ATrp would be significantly more difficult to analyze in such fashion because the mass difference between the protonated ring carbon and an unprotonated aza nitrogen is only 1 amu. However, HPLC can be a useful alternative for analyzing proteins containing 7-ATrp. Short peptides containing D,L-7-ATrp enantiomers have been separated successfully by reversed-phase HPLC.[43,44] Mendelson and collaborators have demonstrated the potential utility of this approach for resolving analogue and non-analogue proteins containing polypeptide chains of several hundred residues.[45] They incorporated 7-ATrp and 5-OHTrp into a W-370F mutant of the domain comprising residues 335–700 of the nonclaret disjunctional protein (Ncd) motor. This mutant contains a single Trp site. By using standard reversed-phase HPLC conditions (Figure 2.4) it was possible to determine that more than 90% of the Trp was replaced in the motor protein expressed with either of these analogues.

Characterizing the analogue incorporation of a multi-Trp protein by HPLC analysis may be more complex. While the HPLC is carried out under denaturing conditions, solvent composition dependent intramolecular

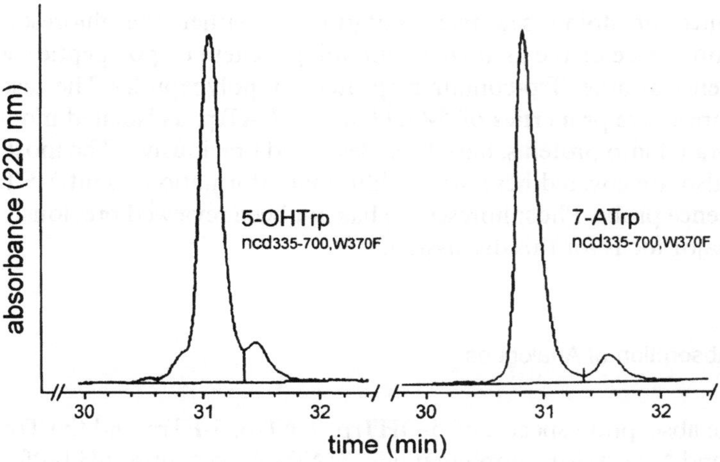


Figure 2.4. Reversed-phase chromatography of the 335–370 domain of ncd expressed with 5-OHTrp (left) and 7-ATrp (right).

interactions might occur at some Trp sites, which in turn could affect the retention times of polypeptide chains containing equivalent numbers of analogue residues but incorporated at different positions. In addition, these partially labeled chains might not be equally represented in the purified sample. This situation could arise, for example, if the presence of an analogue at a particular Trp site affects the efficiency of protein folding during expression. Thus, HPLC analysis, for example of a two-Trp protein sample with molecules containing two, one, and no analogue residues, could yield three or possibly four peaks. Further resolution might be obtained by peptide mapping of labeled and unlabelled proteins samples coupled with quantitative HPLC analysis, as described in a previous review.[22]

## 2.3. Spectral Features of Trp Analogues

The analogues 5-OHTrp, 7-ATrp, 5-FTrp and 4-FTrp have unique absorption and emissive properties, which make them useful in different applications. The analogue 4-FTrp is essentially nonfluorescent at ambient temperatures, making it a "silent" analogue, and it has an absorption spectrum that is blue-shifted compared to that of NATrpA.[30,31] The other analogues share the feature of possessing an absorption spectrum that extends to lower energies than that of Trp. This optical "window" provides the opportunity for observing the absorbance or doing selective excitation of either the fluorescence or phosphorescence of the analogue-containing protein or polypeptide when in the presence of other Trp-containing proteins or polypeptides. The absorption and fluorescence properties of 5-OHTrp and 7-ATrp, as isolated models and incorporated into proteins, have been described previously.[22] The most salient points also are covered here, with additional information about 5-FTrp as a fluorescence probe. Phosphorescence has not been reviewed previously, and it is the major focus for this discussion.

### 2.3.1. Absorption of Analogues

The absorption spectra of 5-OHTrp, 7-ATrp, 5-FTrp and the Trp model compound N-acetyl-tryptophanamide (NATrpA) in neutral pH buffer and in dioxane are compared in Figures 2.5 and 2.6, respectively. These approximate the absorption spectra expected, respectively, for fully exposed or completely buried residue side chains. The spectra in aqueous buffer all show decreased resolution of vibrational structure when compared with the spectra in dioxane. The spectrum of 5-OHTrp has a well-separated, high-intensity band in the region between about 295 and 325 nm. In this wavelength region, which extends beyond Trp absorption, 7-ATrp generally has less extinction than 5-OHTrp. The absorption spectrum of 5-FTrp is the least red-shifted of the three analogues, but it is still possible to carry out selective excitation of this analogue in the presence of Trp.

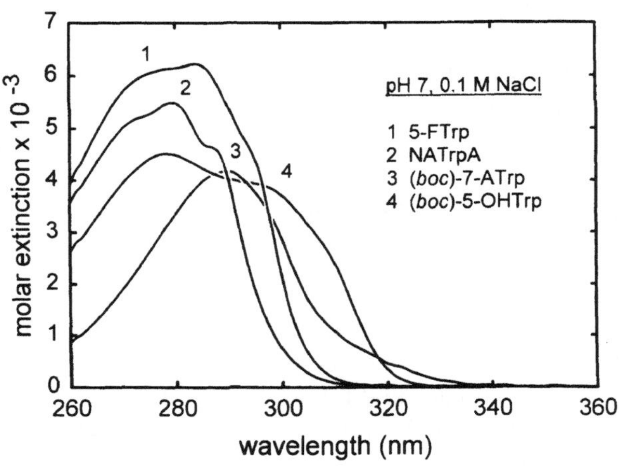


Figure 2.5. Absorption spectra of Trp analogues in neutral pH water.

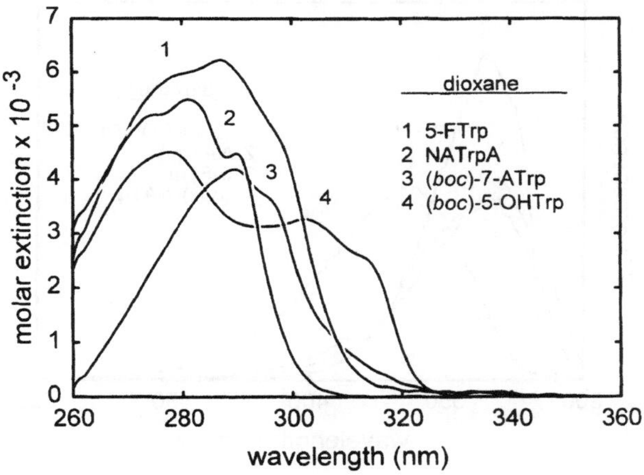


Figure 2.6. Absorption spectra of Trp analogues in dioxane.

### 2.3.2. Fluorescence—Analogue Models

The corrected fluorescence emission spectra of 5-OHTrp, 7-ATrp, 5-FTrp and the Trp model compound N-acetyl-tryptophanamide (NATrpA) in neutral pH buffer and in dioxane are compared in Figures 2.7 and 2.8, respectively; standard emission spectra used to generate the correction factors are shown in Figure 2.9. The emission properties are summarized in Table 2.2. 5-OHTrp fluorescence, when excited at wavelengths longer than 315 nm,

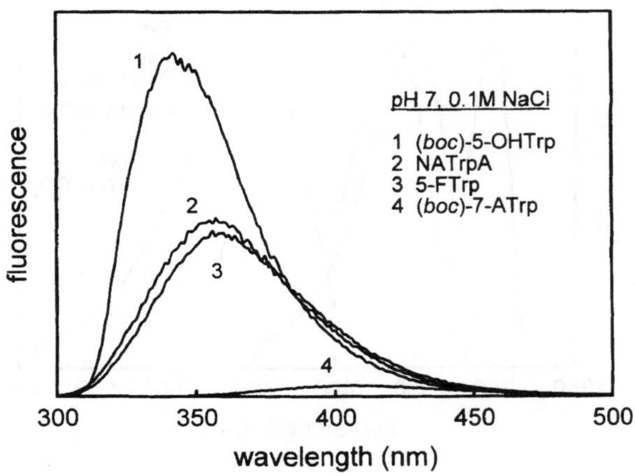


Figure 2.7. Fluorescence spectra of Trp analogues in neutral pH water.

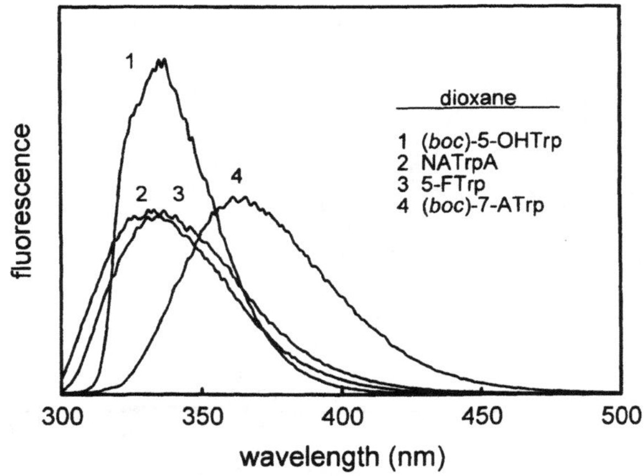


Figure 2.8. Fluorescence spectra of Trp analogues in dioxane.

has a high anisotropy in viscous solutions that is close to the theoretical limit of 0.4, making it an ideal probe for studying molecular dynamics.[22] The wavelength of its emission maximum, which in water is at higher energy than that of Trp, is relatively insensitive to changes in the local environment. By contrast, the fluorescence emission maximum and quantum yield of 7-ATrp is extremely sensitive to the local environment. Its emission in water is at longer wavelengths than that of Trp, and it is strongly quenched. The extinction of 5-FTrp is about 10% greater overall than that of Trp and the

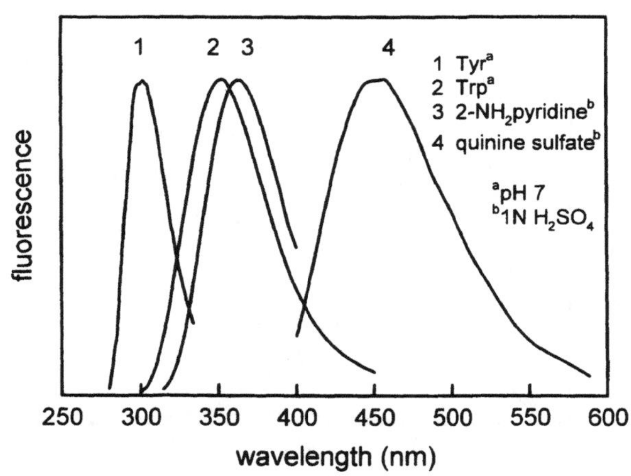


Figure 2.9. Peak normalized standard fluorescence emission spectra from Burstein laboratory used to generate correction factors.

absorbance spectrum has a smaller red-shift than that or either 5-OHTrp or 7-ATrp. The emission of 5-FTrp shows sensitivity towards the local environment similar to that of Trp emission, except shifted to longer wavelengths by a few nm.

### 2.3.3. Fluorescence—Analogue Containing Proteins

As shown in Figure 2.10, using the MyoD homeodomain as an example, the fluorescence emission of 5-OHTrp and 7-ATrp can be excited at wavelengths above 310 nm with minimal contribution from Trp at equivalent concentrations of non-analogue containing protein. This differential absorption, which provides selective excitation of the analogues, has proved particularly useful in investigations of protein-nucleic acid interactions by fluorescence spectroscopy as well as by analytical ultracentrifugation.[46]

It was noted in the foregoing discussion on analogue incorporation that certain Trp analogues are not compatible with certain proteins. Incorporation may be inefficient, protein expression may be low, or there may be perturbation or abolition of function. An example discussed above, is the soluble domain of human tissue factor, which does not express efficiently in the presence of either 5-OHTrp or 7-ATrp. Both x-ray crystal structural data and fluorescence data show that two of the four Trp residues in this domain are deeply buried within the protein matrix in highly constricted environments.[47] The incompatibility may be due to interference with local packing interactions in the case of the 5-hydroxy-indole side chain, or the result of

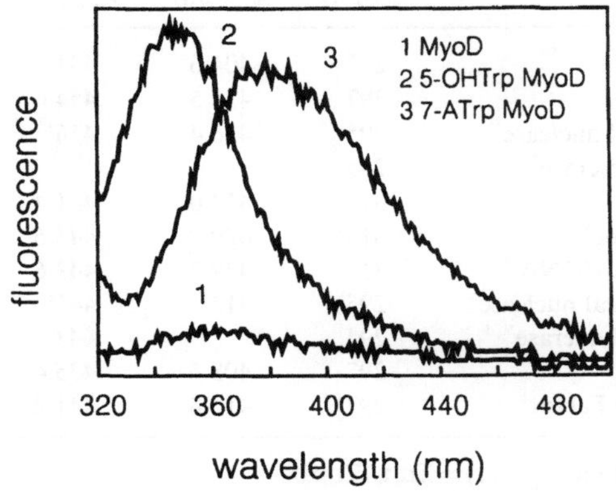


Figure 2.10. Fluorescence emission spectra of MyoD, comparing the single Trp protein expressed in the absence of analogues with proteins expressed with 5-OHTrp or 7-ATrp.

inappropriate hydrogen bond formation in the case of the 7-azaindole side chain. Fluorine, on the other hand, is less bulky than a hydroxyl group and does not participate in hydrogen bonds.[9] Consistent with this, the analogue 5-FTrp is readily incorporated in the soluble domain of human tissue factor, as shown in Figures 2.2 and 2.3. The absorption and fluorescence spectra of the modified protein both are red-shifted. The spectral overlap between the absorption and fluorescence of 5-FTrp is significantly greater than that of Trp. As a result, there is a greater probability of resonance energy transfer among 5-FTrp residues in proteins.[48]

### 2.3.4. Phosphorescence—Analogue Models

The phosphorescence emission spectra of 5-OHTrp, 7-ATrp, 5-FTrp, and NATrpA in neutral pH buffer with 30% (v/v) glycerol at 77 K are compared in Figure 2.11. NATrpA and 5-FTrp show similar structure. 5-OHTrp and 7-ATrp exhibit less well-resolved vibrational bands, particularly noticeable for the 0–0 band, than do either NATrpA[49] or 5-FTrp. The steady-state and time-resolved phosphorescence parameters are summarized in Table 2.3. The phosphorescence quantum yield of 7-ATrp is at least a factor of 10 lower than any of the other model compounds, consistent with the observations of Cioni and coworkers.[50]

Table 2.3. Steady-State and Time-Resolved Phosphorescence for Models and Proteins

| Sample | $\lambda_{ex}$(nm) | $\lambda_{0\text{-}0}$(nm) | $\lambda_{max}$(nm) | $\langle\tau\rangle$ (s)[a] | $\tau$ (s)[b] |
|---|---|---|---|---|---|
| NATrpA[63] | 297 | 404.6 | 431.2 | 6.4 | 6.4 |
| 7-ATrp[c] | 297 | 428.5 | 454.4 | 2.8 | 2.8 |
| 7-ATrp staphylococcal nuclease[56] | 295 | 426.4 | 456[e] | | |
| 7-ATrp $\alpha_2$ RNA polymerase[50,d] | 295 | | | | |
| 5-OHTrp[49] | 315 | 414.0 | 441.4 | 4.9 | 4.9 |
| 5-OHTrp $\lambda$-cI repressor[49] | 315 | 429.2 | 443.6 | 3.6 | 2.4 |
| 5-OHTrp $\lambda$-cI repressor/DNA[51] | 315 | 429.2 | 443.6 | 3.9 | 2.7 |
| 5-OHTrp staphylococcal nuclease[56] | 295 | 413 | 441[e] | | |
| 5-OHTrp $\alpha_2$ RNA polymerase[50,d] | 295 | | 443[e] | | |
| 5-FTrp[c] | 297 | 408.6 | 435.4 | 5.4 | 5.4 |
| 5-FTrp soluble Tissue Factor[c] | 297 | 413.8 | 441.2 | 4.5 | 3.7 |

[a]Intensity average lifetime: $\langle\tau\rangle = \Sigma\alpha_i\tau_i^2/\Sigma\alpha_i\tau_i$.
[b]Number average lifetime: $\tau = \Sigma\alpha_i\tau_i/\Sigma\alpha_i$.
[c]Liu and Rousslang, unpublished data. [d]The estimated temperature was 135 K. [e]These wavelength values are estimates from the published spectra.

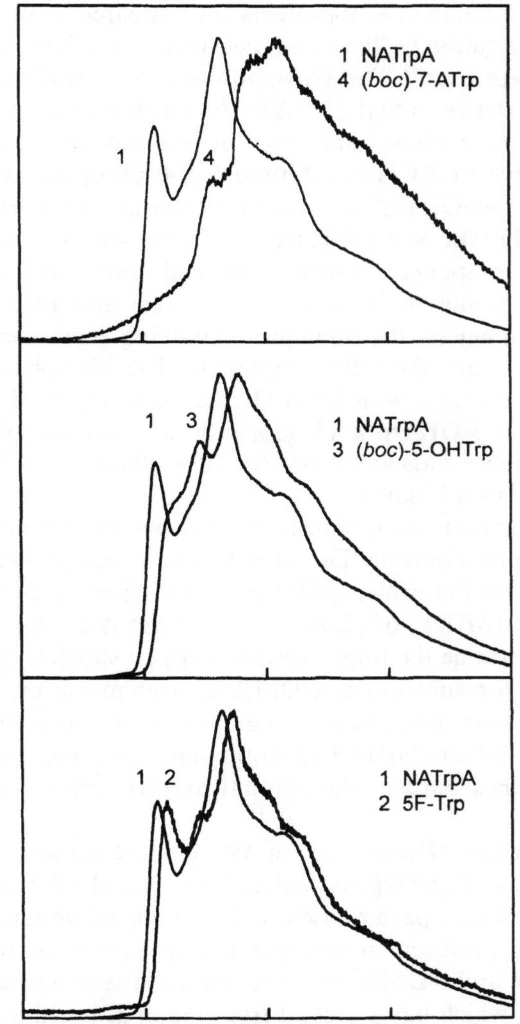


Figure 2.11. Comparison of the phosphorescence emission spectrum of NATrpA with the spectra of 5F-Trp, 5-OHTrp, and 7-ATrp.

### 2.3.5. Phosphorescence—Proteins

Phosphorescence emission parameters and average lifetimes of 7-ATrp, 5-OHTrp, and 5-FTrp in several proteins are compared in Table 2.3. The first triplet state investigation of Trp analogues incorporated into proteins was on 5-OHTrp-$\lambda$ *c*I repressor.[49] The phosphorescence of wild-type $\lambda$ *c*I is red-shifted by 3 nm relative to that of NATrpA, which is characteristic of buried tryptophan, and the phosphorescence of the modified repressor is also red-shifted relative to 5-OHTrp. Although the phosphorescence decays of NATrpA and *tboc*-5-OHTrp are single exponential, the time-resolved emission of both wild-type and 5-OHTrp-$\lambda$ *c*I repressor are multi-exponential, requiring three components whose fractional contributions to the decay are similar. According to both the steady-state and time-resolved phosphorescence parameters, the analogue-containing repressor is structurally indistinguishable from the native repressor. The phosphorescence of the repressor binary complex with DNA also has been reported.[51] The emission characteristics of 5-OHTrp-$\lambda$ *c*I repressor and its complex with DNA are indistinguishable, indicating that the sites of the 5-OHTrp residues are unperturbed by DNA binding.

Aside from conventional optical spectroscopy of Trp and Trp analogue-containing proteins, Optically Detected Magnetic Resonance (ODMR) can be used to measure the triplet splittings,[52] and Microwave-Induced Delayed Phosphorescence (MIDP) of photo-excited triplet states can be employed as a method to determine the three individual triplet sublevel decay times.[53] To provide a basis for subsequent ODMR measurements on 7-ATrp and 5-OHTrp incorporated into staphylococcal nuclease, Ozarowski and coworkers reported the MIDP and ODMR of both analogues, specifying not only the sublevel decay times, but also the spin-lattice relaxation rates connecting the sublevels.[54]

Based upon the ODMR work of Wong and Ozarowski,[55,56] incorporation of 7-ATrp and 5-OHTrp, as well as the 4, 5, and 6-FTrp analogues into the W140 site of wild-type nuclease lead to a modified protein that conserved structure at this position, in agreement with earlier fluorescence work.[57] Phosphorescence and ODMR both showed that the structure of the mutant nuclease, V66W, which has a second tryptophan at position 66, is similarly retained upon incorporation of any of the analogues, with the exception of 7-ATrp. However, structural integrity of both the 140 and 66 sites is lost upon incorporation of 7-ATrp.

The phosphorescence of 7-ATrp and 5-OHTrp was measured as a function of temperature and solvent viscosity to assess their potential for probing the protein environment in $\alpha_2$ RNA Polymerase.[50] The phosphorescence of both analogues was more strongly quenched than that of Trp, when the temperature was raised above the glass transition temperature (~180 K), consistent with the expected shortening of the triplet state lifetime by non-radiative processes. While the phosphorescence of 5-OHTrp could still be measured at 193 K, the phosphorescence of 7-ATrp was undetectable under the same conditions, rendering it of questionable value as a probe of protein structure under ambient conditions. Even though the phosphorescence of 5-OHTrp was severely quenched and the triplet lifetime of 5-OHTrp reduced to 29 $\mu$s in buffer at 274 K, its phosphorescence could still be measured, making it a more promising prospect for investigating protein dynamics near ambient temperatures. However, incorporation of 5-OHTrp into $\alpha_2$ RNA polymerase was incomplete with only 68% replacement, admitting a clear Trp component to the phosphorescence spectrum. Although the protein environment has been known to protect Trp from dynamic quenching, allowing room-temperature phosphorescence to be measured in a variety of proteins, this was not the case in 5-OHTrp $\alpha_2$ RNA polymerase, whose phosphorescence, while still measurable, was unexpectedly low.

Recently, the room-temperature phosphorescence of a series of halogenated Trp analogues was reported by McCaul and Ludescher,[58] in which the 5-FTrp analogue exhibited photo-physical properties similar to those of Trp, making it a promising phosphorescence probe of protein structure and function. Before the analogues prove useful as phosphorescence probes of protein structure in fluid solution, more work needs to be done in order to disclose the mechanism of phosphorescence quenching of the analogues above the glass transition temperature, whether by themselves or when incorporated into proteins.

Preliminary steady-state phosphorescence spectra and decay times have been measured for 5-FTrp and for the 5-FTrp-containing soluble domain of human tissue factor (Liu and Rousslang, unpublished observations) in which more than 95% of the four Trp residues were replaced with 5-FTrp.[36] The 5-FTrp tissue factor steady-state phosphorescence spectrum was red-shifted when compared to that of the model 5-FTrp, indicating that the Trp sites are partly protected from solvent, while the phosphorescence decay was complex as might be expected with multiple Trp residues.

## 2.4. Prospects

The feasibility of incorporating tryptophan analogues into recombinant proteins for investigating protein-protein and protein-nucleic acid interactions by fluorescence spectroscopy was demonstrated in 1992.[20–22] In the intervening few years, this approach has been applied to many different systems, especially the analysis of dynamics and macromolecular assembly in protein-nucleic acid interactions by fluorescence anisotropy and analytical ultracentrifugation.[46] These investigations have demonstrated two significant advantages of using analogues such as 5-OHTrp and 7-ATrp to provide spectroscopic observables for studying these macromolecular interactions. One is that these analogues allow excitation of fluorescence or detection of absorbance at wavelengths near 315 nm, where Trp and nucleic acid bases do not absorb significantly. The other is that in most cases perturbation of function is minimal or not observed.

New areas of investigations involving Trp analogues are emerging. One that seems particularly promising is the complementary use of $^{19}F$ NMR and fluorescence. A unique feature of $^{19}F$ NMR is that spectra can be obtained for individual molecules and assemblies up to 100 kDa.[9] The possibility of fluorine substitutions at different positions on the indole ring provides a unique opportunity to assess solvent accessibility of individual atoms. Combined with fluorescence determination of solvent accessibility, by using collisional quenchers, it is possible to define the spatial relationship of the indole ring with respect to the protein matrix and bulk solvent.[36,48] Another promising area, which has been highlighted in this chapter, is applications of phosphorescence spectroscopy, including optically detected magnetic resonance. Both triplet state spectroscopies have the potential to provide valuable new information about protein local structure. Like fluorescence, the triplet emission of the analogues can be selectively excited in the presence of tryptophan, DNA or RNA. Thus, phosphorescence of spectrally enhanced proteins should also serve as a spectroscopic probe of protein-protein, or protein-DNA interactions. When the mechanisms of phosphorescence quenching at ambient temperatures are better understood, we anticipate that the room-temperature phosphorescence of tryptophan analogues will serve as a useful tool to explore not only protein local microenvironments, but also protein motional dynamics that occur on longer time scales than can be measured by fluorescence.

## Acknowledgments

The authors are indebted particularly to Arthur Szabo, Christopher Hogue, Donald Senear, Thomas Laue, and Robert Mendelson for their contributions towards development of analytical methods and applications involving spectral enhancement of proteins with tryptophan analogues. We also thank Patrik Callis, Ludwig Brand, and Gintaras Diekus for detailed discussions regarding the spectroscopy of tryptophan and tryptophan analogues. In addition, we thank Professor Edward Burstein for providing us with absolute emission spectra of model compounds. J. B. A. Ross gratefully acknowledges support by NIH Grants HL-29019 and CA-63317, L. A. Luck gratefully acknowledges support by U.S. Army grant DAMD17-96-1-6140, and K. W. Rousslang gratefully acknowledges sabbatical support from the University of Puget Sound.

## References

1. Halvorson, H., S. Spiegelman, and R. L. Hinman, *The Effect of Tryptophan Analogs on the Induced Synthesis of Maltase and Protein Synthesis in Yeast.* Arch. Biochim. Biophys., 1955. **55**: pp. 512–525.
2. Pardee, A. B., V. G. Shore, and L. S. Prestidge, *Incorporation of Azatryptophan into Proteins of Bacteria and Bacteriphage.* Biochim.Biophys.Acta, 1956. **21**: pp. 406–407.
3. Davie, E. W., V. V. Konnigsberger, and F. Lipmann, *The Isolation of a Tryptophan-Activating Enzyme.* Arch. Biochim. Biophys., 1956. **65**: pp. 21–38.
4. Sharon, N. and F. Lipmann, *Reactivity of Analogs with Pancreatic Tryptophan Activating Enzyme.* Arch. Biochim. Biophys., 1957. **69**: pp. 219–227.
5. Schlesinger, S., *The Effect of Amino Acid Analogues on Alkaline Phosphatase Formation in Escherichia coli K-12. II. Replacement of Tryptophan by Azatryptophan and by Tryptazan.* J. Biol. Chem., 1968. **243**(14): pp. 3877–3883.
6. Schlesinger, S. and M. J. Schlesinger, *The effect of amino acid analogues on alkaline phosphatase formation in Escherichia coli K-12. I. Substitution of triazolealanine for histidine.* J. Biol. Chem., 1967. **242**(14): pp. 3369–3372.
7. Foote, J., D. M. Ikeda, and E. R. Kantrowitz, *The role of tryptophan in aspartate transcarbamylase.* J. Biol. Chem., 1980. **255**(11): pp. 5154–5158.
8. Sykes, B. D., H. I. Weingarten, and M. J. Schlesinger, *Fluorotyrosine alkaline phosphatase from Escherichia coli: preparation, properties, and fluorine-19 nuclear magnetic resonance spectrum.* Proc. Natl. Acad. Sci. U S A, 1974. **71**(2): pp. 469–473.
9. Danielson, M. A. and J. J. Falke, *Use of 19F NMR to Probe Protein Structure and Conformational Changes*, in *Annual Review of Biophysics and Biomolecular Structure*, R. M. Stroud, et al., Editors. 1996, Annual Reviews, Inc.: Palo Alto. pp. 163–195.
10. Pratt, F. A. and C. Ho, *Incorporation of Fluorotryptophan in Proteins of Escherichia coli.* Biochemistry, 1975. **14**: pp. 3035–3040.
11. Hudson, B. S., D. L. Harris, R. D. Ludescher, A. Ruggiero, A. Cooney-Freed, and S. Cavalier, eds. *Fluorescence Probe Studies of Proteins and Membranes.* Applications of Fluorescence in the Biomedical Sciences, ed. D. L. Taylor. 1986, A. R. Liss: New York. 159–202.
12. Hofmann, K. and H. Bohn, *Studies on Polypeptides. XXXVI. The Effect of Pyrazole-Imidazole Replacements on the S-Protein Activating Potency of an S-Peptide Fragment.* J. Am. Chem. Soc., 1966. **88**: pp. 5914–5919.
13. Kaiser, E. T., *Synthetic Approaches to Biologically Active Peptides and Proteins including Enzymes.* Acc. Chem. Res., 1989. **22**: pp. 47–54.
14. Bain, J. D., C. G. Glabe, T. A. Dix, C. A. R., and E. S. Dalia, *Biosynthetic Site-Specific Incorporation of a Non-Natural Amino Acid into a Polypeptide.* J. Am. Chem. Soc., 1989. **111**: pp. 8013–8014.

15. Noren, C. J., S. J. Anthony-Cahill, M. C. Griffth, and P. G. Schultz, *A General Method for Site-Specific Incorporation of Unnatural Amino Acids into Proteins.* Science, 1989. **244**(182–188).
16. Cornish, V. W., D. R. Benson, C. A. Altenbach, K. Hideg, W. L. Hubbell, and P. G. Schultz, *Site-specific incorporation of biophysical probes into proteins.* Proc. Natl. Acad. Sci U S A, 1994. **91**(8): p. 2910.
17. Steward, L. E., C. S. Collins, M. A. Gilmore, J. E. Carlson, J. B. A. Ross, and A. R. Chamberlin, *In Vitro Site-Specific Incorporation of Fluorescent Probes into β-Galactosidase.* J. Am. Chem. Soc., 1997. **119**: pp. 6–11.
18. Mendel, D., V. W. Cornish, and P. G. Schultz, *Site-Directed Mutagenesis with an Expanded Genetic Code.* Ann. Rev. Biophys. Biomol. Struct., 1995. **24**: pp. 435–462.
19. Steward, L. E. and A. R. Chamberlin, *Protein Expression by Expansion of the Genetic Code*, in *Encyclopedia of Molecular Biology and Molecular Medicine*. 1996, VCH Publishers: New York.
20. Hogue, C. W., I. Rasquinha, A. G. Szabo, and J. P. MacManus, *A new intrinsic fluorescent probe for proteins. Biosynthetic incorporation of 5-hydroxytryptophan into oncomodulin.* FEBS Lett., 1992. **310**(3): pp. 269–272.
21. Ross, J. B. A., D. F. Senear, E. Waxman, B. B. Kombo, E. Rusinova, Y. T. Huang, W. R. Laws, and C. A. Hasselbacher, *Spectral enhancement of proteins: biological incorporation and fluorescence characterization of 5-hydroxytryptophan in bacteriophage lambda cI repressor.* Proc. Natl. Acad. Sci. U.S.A., 1992. **89**(24): pp. 12023–12027.
22. Ross, J. B. A., A. G. Szabo, and C. W. V. Hogue, *Enhancement of Protein Spectra with Tryptophan Analogs: Fluorescence Spectroscopy of Protein-Protein and Protein-Nucleic Acid Interactions*, in *Fluorescence Spectroscopy*, L. Brand and M. L. Johnson, Editors. 1997, Academic Press: New York. pp. 151–190.
23. Hogue, C. W. V., *Tryptophanyl-tRNA Synthetase and Its Role in the Incorporation of New Intrinsic Fluorescent Probes into Proteins.* 1994, Dissertation, University of Ottawa.
24. Hogue, C. W. V., S. Cyr, J. D. Brennen, T. L. Pauls, J. A. Cox, M. W. Berchtold, and A. G. Szabo, *Efficient Incorporation of Tryptophan Analogues in Recombinant Rat F102W Parvalbumin for Fluorescence and $^{19}F$ NMR Studies.* Biophys. J., 1995. **68**: p. A193.
25. Heyduk, E. and T. Heyduk, *Physical studies on interaction of transcription activator and RNA-polymerase: fluorescent derivatives of CRP and RNA polymerase.* Cell Mol. Biol. Res., 1993. **39**(4): pp. 401–407.
26. Heyduk, T. and S. Callaci, *Fluorescence Probes for Studying the Mechanisms of Transcription Activation.* Proc. SPIE-Int. Soc. Opt. Eng., 1994. **2137**: pp. 719–724.
27. Sambrook, J., E. F. Fritsch, and T. Maniatis, *Molecular Cloning, a Laboratory Manual.* 2nd ed. 1989, New York: Cold Spring Harbor Press.
28. Hasselbacher, C. A., R. Rusinova, E. Rusinova, and J. B. A. Ross, *Spectral Enhancement of Recombinant Proteins with Tryptophan Analogs: The Soluble Domain of Human Tissue Factor*, in *Techniques in Protein Chemistry*, J. W. Crabb, Editor. 1995, Academic Press: New York. pp. 349–356.
29. Soumillion, P., D. J. Sexton, and S. J. Benkovic, *Clamp subunit dissociation dictates bacteriophage T4 DNA polymerase holoenzyme disassembly.* Biochemistry, 1998. **37**(7): pp. 1819–1827.
30. Bronskill, P. M. and J. T. Wong, *Suppression of fluorescence of tryptophan residues in proteins by replacement with 4-fluorotryptophan.* Biochem. J., 1988. **249**(1): pp. 305–308.
31. Hott, J. L. and R. F. Borkman, *The non-fluorescence of 4-fluorotryptophan.* Biochem. J., 1989. **264**(1): pp. 297–299.
32. Farah, C. S. and F. C. Reinach, *Regulatory properties of recombinant tropomyosins containing 5-hydroxytryptophan: Ca2+-binding to troponin results in a conformational*

*change in a region of tropomyosin outside the troponin binding site.* Biochemistry, 1999. **38**(32): pp. 10543–10551.

33. Das, K., K. D. Ashby, A. V. Smirnov, F. C. Reinach, J. W. Petrich, and C. S. Farah, *Fluorescence properties of recombinant tropomyosin containing tryptophan, 5-hydroxytryptophan and 7-azatryptophan.* Photochem. Photobiol., 1999. **70**(5): pp. 719–730.
34. Studier, F. W. and B. A. Moffatt, *Use of bacteriophage T7 RNA polymerase to direct selective high-level expression of cloned genes.* J. Mol. Biol., 1986. **189**(1): pp. 113–130.
35. Hasselbacher, C. A., E. Rusinova, E. Waxman, W. Lam, A. Guha, R. Rusinova, Y. Nemerson, and J. B. A. Ross, *Probing the Structure of Human Tissue Factor by Site-Directed Mutagensis and in vivo Incorporation of Tryptophan Analogs.* Proc. SPIE, 1994. **2137**: pp. 312–323.
36. Zemsky, J., E. Rusinova, Y. Nemerson, L. A. Luck, and J. B. A. Ross, *Probing local environments of tryptophan residues in proteins: comparison of $^{19}F$ NMR results with the intrinsic fluorescence of soluble human Tissue Factor.* Proteins: Structure, Function and Genetics, 1999. **37**: pp. 709–716.
37. Waxman, E., E. Rusinova, C. A. Hasselbacher, G. P. Schwartz, W. R. Laws, and J. B. A. Ross, *Determination of the tryptophan:tyrosine ratio in proteins.* Anal. Biochem., 1993. **210**(2): pp. 425–428.
38. Edelhoch, H., *Spectroscopic determination of tryptophan and tyrosine in proteins.* Biochemistry, 1967. **6**(7): pp. 1948–1954.
39. Wetlaufer, D. B., *Ultraviolet spectra of proteins and amino acids*, in *Advances in Protein Chemistry*, C. B. Anfinsen, et al., Editors. 1962, Academic Press: New York. pp. 303–390.
40. Gill, S. C. and P. H. von Hippel, *Calculation of protein extinction coefficients from amino acid sequence data.* Anal. Biochem., 1989. **182**: pp. 319–326.
41. Mach, H., C. R. Middaugh, and R. V. Lewis, *Statistical determination of the average values of the extinction coefficients of tryptophan and tyrosine in native proteins.* Anal. Biochem., 1992. **200**: pp. 74–80.
42. Minks, C., R. Huber, L. Moroder, and N. Budisa, *Atomic mutations at the single tryptophan residue of human recombinant annexin V: effects on structure, stability, and activity.* Biochemistry, 1999. **38**(33): pp. 10649–10659.
43. Rich, R. L., M. Negrerie, J. Li, S. Elliott, R. W. Thornburg, and J. W. Petrich, *The photophysical probe, 7-azatryptophan, in synthetic peptides.* Photochem. Photobiol., 1993. **58**: pp. 28–30.
44. Brennan, J. D., C. W. V. Hogue, B. Rajendran, K. J. Willis, and A. G. Szabo, *Preparation of enantiomerically pure L-7-azatryptophan by an enzymatic method and its application to the development of a fluorimetric activity assay for tryptophanyl-tRNA synthetase.* Anal. Biochem., 1997. **252**(2): pp. 260–270.
45. Mendelson, R., *personal communication.*
46. Senear, D. F., J. B. A. Ross, and T. M. Laue, *Analysis of protein and DNA-mediated contributions to cooperative assembly of protein-DNA complexes.* Methods: A Companion to Methods in Enzymology, 1998. **16**(1): pp. 3–20.
47. Hasselbacher, C. A., E. Rusinova, E. Waxman, R. Rusinova, R. A. Kohanski, W. Lam, A. Guha, J. Du, T. C. Lin, I. Polikarpov, C. W. G. Boys, Y. Nemerson, W. H. Konigsberg, J. B. A. Ross, *Environments of the four tryptophans in the extracellular domain of human tissue factor: comparison of results from absorption and fluorescence difference spectra of tryptophan replacement mutants with the crystal structure of the wild-type protein.* Biophys. J., 1995. **69**(1): pp. 20–29.
48. Zemsky, J., *5-Fluoro-tryptophan as a probe for fluorescence and Flourine 19 NMR structure function studies: Analysis of 5-fluoro-tryptophan substituted soluble tissue factor,* 1998, Dissertation, City University of New York.

49. Sato, A. K., E. R. Bitten, D. F. Senear, J. B. A. Ross, and K. W. Rousslang, *Steady-State and Time-Resolved Phosphorescence of Wild-Type and Modified Bacteriophage λcI Repressors.* J. Fluorescence, 1994. **4**: pp. 195–201.
50. Cioni, P., L. Erijman, and G. B. Strambini, *Phosphorescence emission of 7-azatryptophan and 5-hydroxytryptophan in fluid solutions and in alpha2 RNA polymerase.* Biochem Biophys. Res. Commun., 1998. **248**(2): pp. 347–51.
51. Sato, A. K., E. R. Bitten, D. Lambert, and K. W. Rousslang, *Steady-State and Time-Resolved Phosphorescence of 5-hydroxy-L-tryptophan l cI Repressor Bound to DNA.* Proc. SPIE, 1994. **2137**: pp. 343–352.
52. Kwiram, A. L., *Optical Detection of Paramagnetic Resonance in Phosphorescent Triplet States.* Chem. Phys. Lett., 1967. **1**: pp. 272–275.
53. Schmidt, J., W. S. Veeman, and J. H. van der Waals, *Microwave Induced Delayed Phosphorescence.* Chem. Phys. Lett., 1969. **4**: pp. 341–346.
54. Ozarowski, A., J. Q. Wu, and A. H. Maki, *Global Analysis of Microwave-Induced Delayed Phosphorescence of Photoexcited Triplet States.* J. Magn. Reson., Ser. A, 1996. **121**: pp. 178–186.
55. Wong, C. Y. and M. R. Eftink, *Incorporation of tryptophan analogues into staphylococcal nuclease, its V66W mutant, and Delta 137–149 fragment: spectroscopic studies.* Biochemistry, 1998. **37**(25): pp. 8938–8946.
56. Ozarowski, A., J. Q. Wu, S. K. Davis, C. Y. Wong, M. R. Eftink, and A. H. Maki, *Phosphorescence and optically detected magnetic resonance characterization of the environments of tryptophan analogues in staphylococcal nuclease, its V66W mutant, and Delta 137–149 fragment.* Biochemistry, 1998. **37**(25): pp. 8954–8964.
57. Wong, C. Y. and M. R. Eftink, *Biosynthetic incorporation of tryptophan analogues into staphylococcal nuclease: effect of 5-hydroxytryptophan and 7-azatryptophan on structure and stability.* Protein Sci., 1997. **6**(3): pp. 689–697.
58. McCaul, C. and R. D. Ludescher, *Phosphorescence from tryptophan and tryptophan analogs in the solid state.* Proc. SPIE, 1998. **3256**: pp. 263–268.
59. Shen, F., S. J. Triezenberg, P. Hensley, D. Porter, and J. R. Knutson, *Transcriptional activation domain of the herpesvirus protein VP16 becomes conformationally constrained upon interaction with basal transcription factors.* J. Biol. Chem., 1996. **271**(9): pp. 4827–4837.
60. Callaci, S. and T. Heyduk, *Conformation and DNA binding properties of a single-stranded DNA binding region of sigma 70 subunit from Escherichia coli RNA polymerase are modulated by an interaction with the core enzyme.* Biochemistry, 1998. **37**(10): pp. 3312–3320.
61. Soumillion, P., L. Jespers, J. Vervoort, and J. Fastrez, *Biosynthetic incorporation of 7-azatryptophan into the phage lambda lysozyme: estimation of tryptophan accessibility, effect on enzymatic activity and protein stability.* Protein Eng., 1995. **8**(5): pp. 451–456.
62. Beckett, D., E. Kovaleva, and P. J. Schatz, *A minimal peptide substrate in biotin holoenzyme synthetase-catalyzed biotinylation.* Protein Sci., 1999. **8**(4): pp. 921–929.
63. Petra, P. H., P. C. Namkung, D. F. Senear, D. A. McCrae, K. W. Rousslang, D. C. Teller, and J. B. A. Ross, *Molecular characterization of the sex steroid binding protein (SBP) of plasma. Re-examination of rabbit SBP and comparison with the human, macaque and baboon proteins.* J. Steroid Biochem., 1986. **25**(2): pp. 191–200.

# 3

# Room Temperature Tryptophan Phosphorescence as a Probe of Structural and Dynamic Properties of Proteins

Vinod Subramaniam, Duncan G. Steel, and Ari Gafni

## 3.1. Introduction

Phosphorescence is defined as the emission from the first excited triplet state of an electronically excited molecular species, and is a versatile counterpart of the more commonly used singlet state emission, called fluorescence. While the fluorescence properties of protein tryptophan residues in solution have been long exploited in biophysical and biochemical studies, the triplet state emission of tryptophan at room temperature has been unequivocally demonstrated only relatively recently, when Saviotti and Galley observed Trp phosphorescence at room temperature from horse liver alcohol dehydrogenase (LADH) and *E. coli* alkaline phosphatase (AP).[1] The triplet state emission in solution is extremely sensitive to quenching by molecular oxygen, and thus it is necessary to reduce the oxygen content in solution to subnanomolar concentrations to effectively observe room temperature phosphorescence (RTP). In the absence of molecular oxygen, however, most proteins phosphoresce in solution at ambient temperature, with a triplet state lifetime

---

*Vinod Subramaniam* • Department of Molecular Biology, Max Planck Institute for Biophysical Chemistry, Am Fassberg 11, D-37077 Gottingen, Germany. email: vinod@mpc186.mpibpc.gwdg.de *Duncan G. Steel* • Departments of Physics and Electrical Engineering and Computer Science, Biophysics Research Division, and Institute of Gerontology, The University of Michigan, 300 N. Ingalls Building, Ann Arbor, MI 48109. email: dst@umich.edu *Ari Gafni* • Department of Biological Chemistry, Biophysics Research Division, and Institute of Gerontology, The University of Michigan, 300 N. Ingalls Building, Ann Arbor, MI 48109. email: arigafni@umich.edu

*Topics in Fluorescence Spectroscopy, Volume 6: Protein Fluorescence*, edited by Joseph R. Lakowicz. Kluwer Academic / Plenum Publishers, New York, 2000

in the ms range.[2–4] The detailed photophysics of Trp are complex and remain a subject of current investigation (see for example reference 5). Although the triplet to singlet transition is quantum mechanically forbidden for pure states, spin-orbit coupling relaxes this constraint. Since the triplet state $T_1$ is of lower energy than $S_1$, the phosphorescence emission is red-shifted with respect to fluorescence (Figure 3.1). In contrast to the fluorescence spectrum of Trp in proteins, which as a rule is broad and structureless as a consequence of the strong tendency of the excited singlet dipole to interact with the surrounding solvent, the triplet emission displays significant vibronic structure, representing a reduced interaction with the solvent of the smaller (relative to the singlet state) excited triplet dipole, and reflecting the fact that RTP is emitted only from highly buried Trps in rigid environments. The phosphorescence lifetime is sensitive to the local environment of the emitting residue, and is affected by factors such as solvent viscosity, proximity of charges and quenchers, and the "rigidity" of the residue. The RTP lifetime has thus been used as a sensitive probe of protein structure.

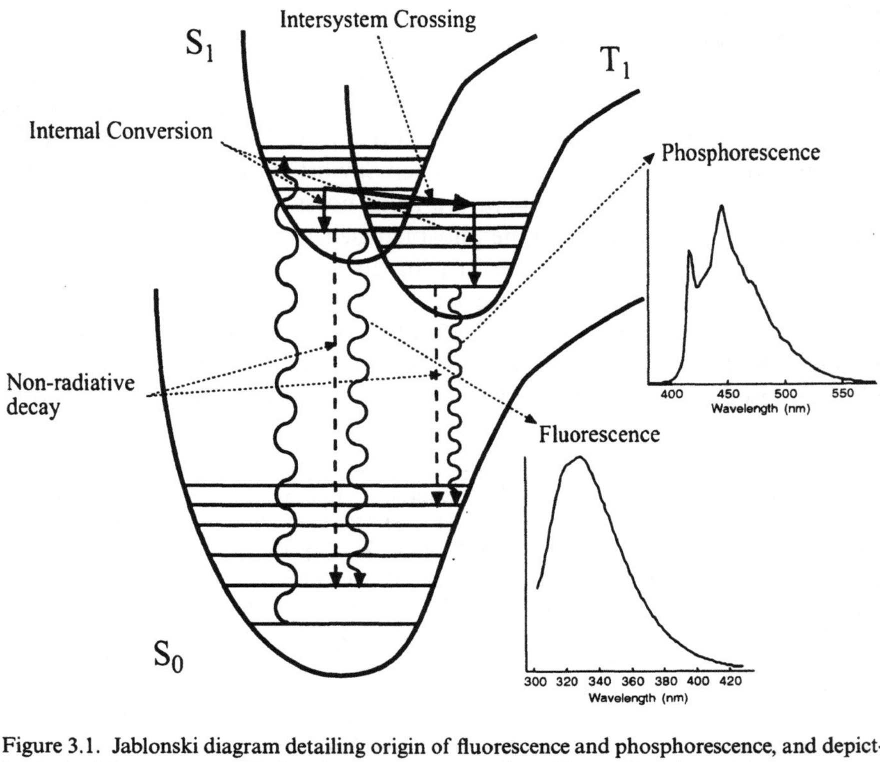


Figure 3.1. Jablonski diagram detailing origin of fluorescence and phosphorescence, and depicting typical fluorescence and phosphorescence spectra from Trp residues in proteins.

**Table 3.1. A Selection of Recent Applications Using Room Temperature Tryptophan Phosphorescence**

| Application | Reference |
|---|---|
| Trp RTP in Fluid Solution | 10 |
| Trp RTP in Proteins | 2 |
| Circularly Polarized Phosphorescence | 11 |
| Triplet State Energy Transfer | 12, 16 |
| Stopped flow RTP | 13, 14 |
| H-D Exchange | 15, 17 |
| Acrylamide quenching of RTP | 18, 19 |
| Refolding of Rnase T1 | 20 |
| Conformational dynamics in F1-ATPase | 21–25 |
| Ligand binding of Phosphorylase b | 26 |
| Unfolding of Alkaline Phosphatase | 27 |
| Refolding of Alkaline Phosphatase | 28 |
| Phosphate binding in Alkaline Phosphatase | 29 |
| Structure and Refolding of β-lactoglobulin | 30 |
| RTP from Trp analogues | 31, 32 |
| RTP from engineered Trp residues | 33 |

This contribution does not exhaustively review the origins and applications of RTP; these are discussed in earlier reviews.[4,6–8] A review of the methodologies and instrumentation used to detect RTP from proteins has been recently published by Schauerte *et al.*[9] Here we focus on recent work relating protein RTP to structural and dynamic properties of these macromolecules. Important recent contributions by Strambini and Gonelli have explored the factors affecting Trp phosphorescence in fluid solution[10] and have yielded new insights into the nature of RTP from proteins;[2] these results are summarized here. Other exciting new developments are also briefly described, including the use of the circularly polarized components of RTP to derive structural information,[11] the application of triplet-state energy transfer for distance determination,[12] the combination of RTP with stopped-flow techniques to study folding kinetics,[13,14] and the exploitation of H-D exchange methods in combination with RTP to extract detailed structural information.[15] A selection of the relevant recent literature is presented in Table 3.1.

## 3.2. Factors Influencing Tryptophan Phosphorescence in Fluid Solution and in Proteins

Population of the triplet state of Trp is usually achieved through excitation into the singlet manifold followed by intersystem crossing, a non-radiative transfer from an electronic state in the singlet manifold to an electronic state in the triplet manifold. The near-forbidden nature of the excited-triplet to ground-singlet transition yields exceptionally long Trp RTP lifetimes, reaching ≈ 2 seconds for *E. coli* alkaline phosphatase. Trp RTP lifetimes can vary about 3–4 orders of magnitude as a function of changes of the local environment of the indole ring, a significantly larger range than the factor of ≈ 10 associated with fluorescence lifetimes under the same conditions; this variation forms the basis of the sensitivity of RTP as a spectroscopic tool. Strambini and Gonnelli[34] initially showed that as the solvent viscosity was varied between $10^4$ and $10^9$ poise, the phosphorescence lifetime of various indole derivatives increased from 20–30 ms to 6 sec. These data have been recently extended to viscosities as low as $10^{-2}$ poise,[10] and are summarized in Figure 3.2. This work re-examined the intrinsic phosphorescence lifetime of indole in aqueous solution at room temperature using low chromophore concentrations ($3 \times 10^{-6}$ M), low excitation intensities, and rigorously cleaned and conditioned solutions and glassware. Under these conditions, a triplet state lifetime of 1.2 ms was observed, a factor of ≈ 60 larger than that reported using triplet-triplet absorption techniques.[35] While it is not certain that this is indeed the true intrinsic lifetime, indirect evidence from the dependence of lifetime on solvent viscosity suggests that it is likely to be the true intrinsic value. Fundamentally, the discrepancy between early work and this work is attributed to underestimation of triplet-triplet annihilation as a triplet deactivation mechanism in the earlier flash photolysis work. An immediate consequence of the more accurate determination of the intrinsic Trp lifetime is the reduction of the dependence of RTP lifetime on solvent

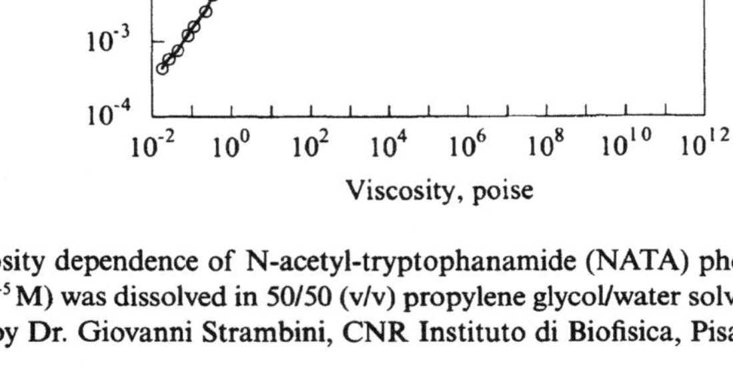


Figure 3.2. Viscosity dependence of N-acetyl-tryptophanamide (NATA) phosphorescence lifetimes. NATA ($10^{-5}$ M) was dissolved in 50/50 (v/v) propylene glycol/water solvent mixtures. Data kindlv provided by Dr. Giovanni Strambini, CNR Instituto di Biofisica, Pisa, Italy.

viscosity; the range of lifetimes in proteins should then span ≈ 1 ms to ≈ 2 s, a range of three orders of magnitude, rather than the 5 orders of magnitude previously accepted. However, the inability to detect RTP in some proteins suggests that the lifetime can be shorter than 1 ms, and thus must reflect quenching processes that contribute to the reduction of the lifetime. In their systematic re-examination of Trp RTP in proteins, Gonnelli and Strambini[2] determined that in addition to dynamic features of protein structure, intramolecular quenching reactions with His, Tyr, Trp and Cys side chains play an exceptionally important part in determining the RTP lifetime in proteins (see below).

Three related factors are thus involved in making Trp RTP a sensitive measure of protein flexibility:

(i) the long intrinsic lifetime of the excited triplet state,
(ii) as a consequence of (i), the RTP is exceptionally sensitive to quenching because it allows more time to interact with quenchers, and
(iii) the drastic dependence of RTP lifetime on solvent viscosity.

In the absence of quenching, non-radiative processes play an important role in determining the decay rate. For aromatic triplet states, such as indole, the major contribution is expected to be asymmetric out-of-plane vibrations which change the symmetry of the molecule and allow for greater mixing of triplet and singlet states.[36]

The long phosphorescence lifetime permits the monitoring of processes in proteins that occur on the msec-second range, which is very relevant to folding processes and which conventional fluorescence spectroscopy cannot access. Its long lifetime also makes RTP markedly more sensitive to quenching (than fluorescence) by short- and long-range processes, an attribute that can be used for studies of protein conformation and flexibility.

The RTP lifetime in proteins is thus affected by two broad classes of phenomena: (i) local environmental effects that influence rigidity, and (ii) effects due to specific quenching interactions such as energy transfer. The observation of a correlation between RTP lifetime and solvent viscosity for free indole in solution[34] has been confirmed in proteins. For example, chromophores that are in mobile sites, such as Trp residues that are substantially solvent exposed on surfaces, have extremely short RTP lifetimes. On the other hand, long-lived RTP is observed from Trp residues that occupy buried sites in protein interiors (such as Trp 109 in AP), exhibiting a much higher level of rigidity. The correlation between RTP lifetime and the effective local viscosity of the residue site has been explored further by inducing changes in structural flexibility of proteins by varying the temperature,[37] pressure,[38,39] cosolvent,[40,41] denaturant,[27,42] or upon ligand binding;[43,44] in all cases, the increase or decrease in the rigidity of the chromophore's environment was found to be reflected in the phosphorescence lifetime.

In addition to local environmental effects (the rigidity or effective local viscosity), both intrinsic and extrinsic quenching mechanisms play a very important role in determining the triplet state lifetime. The group of Vanderkooi has contributed extensively to understanding and applying quenching of Trp phosphorescence.[45–55] Molecular oxygen (dioxygen), whose ground electronic state is a triplet, is well known to be an effective quencher of phosphorescence,[3,45,51] and great care must be taken to reduce its concentrations to sub-nanomolar levels. Other intrinsic protein moieties are known to quench phosphorescence with varying efficiencies. Disulfide bonds in proteins are especially effective.[35,56–58] Recently, Gonelli and Strambini[2] have evaluated the quenching capabilities of various amino-acids. These studies revealed that cystine and cysteine are the most effective quenchers, with $k_q \approx 5.0 \times 10^8\,M^{-1}\,sec^{-1}$; protonated His residues and deprotonated Tyr residues are also very effective quenchers with $k_q \approx 2.0 \times 10^7\,M^{-1}\,sec^{-1}$, while the neutral residues are 20–50 fold less effective quenchers. In addition, Trp molecules in the ground state quench efficiently, with a quenching constant $\approx 1.0 \times 10^7\,M^{-1}\,sec^{-1}$.[10]

## 3.3. Protein Dynamics and Folding Studied Using RTP

The exquisite sensitivity of RTP to changes in the local environment of the emitting Trp residue, and the fact that the RTP lifetime is of the order of magnitude of the timescale of biologically relevant processes make RTP a useful technique in studying protein conformational dynamics. We summarize below results from some systems investigated in our laboratory.

### 3.3.1. Alkaline Phosphatase

*Escherichia coli* alkaline phosphatase (AP, E.C. 3.1.3.1), a phosphomonoesterase, is a dimer of approximately 94 kDa molecular weight exhibiting a very broad substrate specificity. AP is a metalloenzyme containing two zinc ions and one magnesium ion as well as two intramolecular disulfide bonds per subunit; the zinc ions are required for enzymatic activity,[59] while the magnesium ions have been shown to enhance the activity of the zinc containing enzyme.[60] AP has three tryptophan residues per monomer, in positions 109, 220 and 268, of which only Trp 109 phosphoresces at room temperature,[16,34] with a remarkably long RTP lifetime (~2 s). This has enabled its extensive use as a model system for RTP studies. Trp 109 is deeply buried in the hydrophobic core of the protein, situated close to the active site of the enzyme, thus providing a sensitive probe of the catalytic site.

*Unfolding and inactivation of Alkaline Phosphatase:* The inactivation of AP by ethylene diamine tetra-acetic acid (EDTA) and the denaturation of this protein by GuHCl, urea, and low pH were followed by monitoring changes in the enzyme activity and both its time-resolved room temperature phosphorescence intensity and lifetime.[27] The results indicated the existence of an enzymatically active, but structurally less rigid, intermediate protein conformation during unfolding, characterized by a shorter RTP lifetime. The time evolution of the denaturation curves showed that the pathways for denaturation of AP at low pH and in GuHCl were very different. While the denaturation by low concentrations of GuHCl was shown to be a single step process, the unfolding at low pH was more complex. During unfolding by low pH, two structural transitions were observed; in addition, clear evidence for an active intermediate state was seen. When AP was unfolded by GuHCl for short times, high concentrations (>4 M) of the denaturant induced the formation of an active unfolding intermediate with an RTP lifetime of ~800 ms. Upon further incubation, the protein unfolded extensively and the RTP signal was lost. RTP lifetimes for AP denatured by different methods also exhibited significantly broadened lifetime distributions, clearly demonstrating a heterogeneity in protein emitting species. The results suggested that the potential energy surface of the protein might be characterized by a distribution of substates separated by high energy barriers.

*Refolding and reactivation of AP:* The refolding of AP *in vitro*, following denaturation in GuHCl or acid, or inactivation by EDTA, have also been studied[28,61] using two experimental observables: RTP, probing the structural rigidity of the local environment of the luminescent tryptophan, and the lability of the protein to denaturation, reporting on the global protein status. These initial studies showed that when AP was refolded following extensive denaturation by GuHCl, the enzyme activity returned to the native state before the RTP lifetime indicating that structural changes continue after biological activity has been regained. Further work based on recovery of protein lability (a measure of the activation energy of unfolding) showed similar longer time scale structural events in the refolding of AP.[62] These studies also reveal that this slow phase in the postactivational conformational change is not due to proline isomerization, a common origin of slow events in protein folding, but in fact more likely due to conformational changes that accompany metal ion rearrangement (Dirnbach *et al.*, unpublished results). Long time-scale changes have been reported during metal-binding to demetalated (apo) AP,[63] and also upon refolding after thermal denaturation.[64] The structural changes associated with these reactions may be relevant to the refolding of the holoenzyme, and may provide further clues to the molecular mechanism behind the "annealing."

Site-directed mutagenesis provides a powerful means of engineering specific structural changes into a protein in an attempt to elucidate correlations between structure and function or other biophysical properties. Kantrowitz and coworkers have exploited this approach to explore the effect of mutation on the catalytic activity and metal-binding properties of *E. coli* AP.[65–72] Such approaches targeting residues in the neighbourhood of Trp109 provide a sensitive method of determining the effect of the Trp microenvironment on its RTP. Preliminary results using a series of mutant AP molecules suggested that a single-residue change in the vicinity of Trp109 can dramatically affect the RTP characterstics.[73] A number of mutations around Trp109 were created, both cavity-forming (Q320G, L159G, Y84G, see Figure 3.3) and those affecting the hydrogen-bonding within the core (Q320L), and the thermodynamic and RTP characteristics were measured. Significant changes were

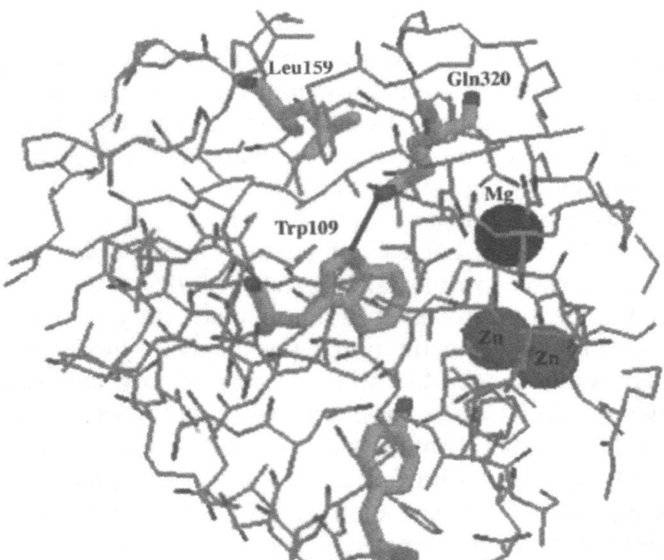


Figure 3.3. The environment of Trp109 in an *E. coli* alkaline phosphatase monomer showing the residues within 12.0 Å of this phosphorescent group. Trp109 and key residues involved in site-directed mutagenesis experiments are indicated in "stick" format. The two Zn and one Mg ion per monomer subunit of this metalloenzyme are indicated in spacefilling format. The hydrogen bond between the Trp109 enamine and Gln320 is indicated as a thick line. (Image rendered by RasMol.)

seen in response to the perturbation of the packing of the hydrophobic core of the protein; in contrast to the ~2 s RTP lifetime of WT AP, Q320L has a characteristic lifetime of ~1 s and Q320G, ~0.35 s.[74] Enzyme activity and thermodynamic stability of the mutant holo-proteins were less affected, although the demetalated (apo) monomers were significantly destabilized.[74] The altered subunit was also found to affect dimer interactions,[75] while the specific mutations significantly affected the hydrogen exchange kinetics (see below) of Trp109.[76] This mutational approach is also likely to shed more insight into whether the removal of a specific quenching interaction (for example, with Tyr 84 which is within 5 Å of Trp 109) may be implicated in the "annealing" phenomena.

### 3.3.2. Azurin

The *Pseudomonas aeruginosa* blue copper protein azurin contains a single copper atom and a single tryptophan residue (Trp 48) in a highly constrained and solvent-shielded environment. While the $Cu^{2+}$ ion strongly quenches all luminescence from the holo-azurin, the apoprotein exhibits a strong, long-lived RTP with a pH-dependent lifetime. As an initial approach to analysing the data in the pH range 4 to pH 8, Hansen *et al.*[77] assumed that the RTP decay could be well-fit by two fixed exponential components of 417 and 592 ms lifetime but with pH varying amplitudes. A theoretical fit of the fractional phosphoresence amplitudes of the 592 ms lifetime showed that the intensity of this component traced the deprotonation of a group with a $pK_a$ ~ 5.6. This correlates well with the deprotonation of His-35 in azurin, as previously determined. In general, multiexponential RTP decays from single emitting Trp residues in proteins have been attributed to ground state heterogeneity.[78] Here, the two lifetime components were interpreted to represent two different conformational states which are associated with the protonated/deprotonated states of His35. The protein apparently exhibits greater structural flexibility at lower pH. This result is of biological significance, suggesting that the active form of the protein is flexible enough to allow for efficient protein-protein interactions with the appropriate cytochrome ligand.

### 3.3.3. Beta-lactoglobulin

The bovine milk protein $\beta$-lactoglobulin A ($\beta$-LG) is a 36.8 KDa homodimer at neutral pH, with each subunit containing two tryptophan residues,

Trp 19 and Trp 61. Native $\beta$-LG has an RTP lifetime of ≈ 20 ms, attributed to Trp 19. Examination of the structure of $\beta$-LG determined by x-ray crystallography shows that Trp 61 is in a substantially solvent-exposed position and is also in close proximity to a disulfide bond formed between Cys 66 and Cys 160. Solvent-exposed tryptophans, as a rule, do not show long-lived RTP. Moreover, triplet states are known to be very effectively quenched by nearby disulfide bonds and the RTP from Trp 61 is thus expected to be extensively quenched. Trp 19, on the other hand, is buried in the calyx, and from its relative inaccessibility to the solvent is expected to be the sole phosphorescent Trp.

Using RTP and fluorescence-based lability approaches to monitor the *in-vitro* refolding of $\beta$-LG, we found that refolded $\beta$-LG adopted a non-native conformation with a shorter RTP lifetime (≈ 10 ms) than in the native state,[30] although the retinol-binding activity of the renatured protein was completely recovered. In contrast to the results obtained with *E. coli* AP, no structural "annealing" was observed and the refolded protein appeared permanently modified structurally. Similar results had earlier been observed by Hattori *et al.* using conformation specific monoclonal antibodies to probe for native-like structure.[79] It is interesting to note that the two monoclonal antibodies which detected a structural change in the refolded protein bind to epitopes around Trp 19, the putative phosphorescent residue, confirming that this domain does not recover during *in vitro* folding. Kinetic trapping of these non-native but biologically functional structures during the folding pathway is of considerable importance to understanding the protein folding process, and may have implications for the "aging" of proteins.

### 3.3.4. Ribonuclease T1

This is a small single domain protein with well-defined secondary and tertiary structures. The protein is stable both in the presence and absence of disulfide bonds and has recently received much attention as a model for studying the molecular aspects of the protein folding process.[80–82] Characterization of the folding kinetics of ribonuclease T1 has revealed complex behavior which has been attributed to slow cis-trans isomerization of proline residues.[80,81] The single Trp (Trp 59) in RNAse T1 from *Aspergillus oryzae* exhibits a measurable RTP signal (≈ 16 msec, 0.1 M NaOAc, pH 5.0, 10 °C) and provides a simple system to study the effects of proline isomerization on Trp luminescence from the protein.

*Unfolding and refolding of RNAse T1:* We have used Trp fluorescence and phosphorescence to follow the refolding of GdnHCl denatured RNAse T1.[61] The fluorescence recovery data is best fit to a sum of two exponentials, yielding rate constants of 0.252 $min^{-1}$ and 0.0198 $min^{-1}$ with relative amplitudes of 0.34 and 0.66 respectively, in reasonable agreement with those published previously.[80] In addition to an increase in the integrated fluorescence intensity, the spectra display a clear blue shift indicating that Trp 59 is being sequestered in a more hydrophobic environment. These slow changes in fluorescence have been previously assigned to cis-trans proline isomerization.

The results of RTP measurements during refolding of RNase T1 unexpectedly portray a different, and puzzling, picture. The RTP intensity increases quickly, and stabilizes at a relatively constant value within 10 minutes of refolding. A first order fit to the phosphorescence intensity recovery yields a rate constant of 1.075 $min^{-1}$, and does not exhibit a slow increase commensurate with the increase in fluorescence intensity. The cis-trans proline isomerization is thus surprisingly not reflected in the RTP data, despite the fact that the same chromophore is being interrogated. It is conceivable that the increase in Trp fluorescence reflects the increased Trp hydrophobicity, as demonstrated by the blue shift in the fluorescence spectrum, but the rigidity of the Trp environment (which RTP is sensitive to) does not change. The mechanisms responsible for these results still remain to be explained.

## 3.4. New Developments in RTP for Protein Studies

### 3.4.1. Distance Measurements Using RTP (Diffusion Enhanced Energy Transfer, Electron Transfer and Exchange Interactions)

As mentioned above, the long RTP lifetime makes this luminescence markedly more sensitive to quenching than fluorescence both by short- and long-range processes, an attribute that can be used for studies of protein conformation and flexibility. Since the degree of dipole-dipole interaction between acceptors and donors (Förster energy transfer) scales as $R^{-6}$, where R is the distance between the donor and acceptor (see references 84, 85 for reviews), energy transfer is a sensitive measure of distances on the molecular level. The ability to rapidly and accurately determine the RTP decay times makes it possible to use non-radiative luminescence energy transfer, with triplet Trp serving as the energy donor in combination with suitable acceptors.

The rate of energy transfer is enhanced by the diffusion of donor and acceptor (reviewed in reference 86). In the rapid diffusion limit, i.e. when the combined distance covered by the donor and acceptor during the excited state lifetime of the donor is much larger than their mean separation (or equivalently, when the donor excited state lifetime is long compared to the average diffusion time of the acceptors), the expressions describing the energy transfer are simplified considerably. The donor lifetime must be longer than the millisecond range for the rapid-diffusion limit to apply, a condition that is usually achieved using the long-lived luminscence of $Tb^{3+}$. This condition is also amply met by many phosphorescing Trp residues, even at low (sub-millimolar) concentrations of freely diffusing acceptors. Since Trp has the additional advantage of being an integral lumiphore in proteins, structural changes in the protein induced by the introduction of a donor group by chemical modification are avoided. This method offers an attractive approach to measuring the distance from core Trp residues to the protein surface, and has been exploited by Mersol *et al.* to measure the depth below the enzyme surface of Trp 109 in AP using various acceptors (small molecule quenchers and embedded heme groups in proteins).[16] The results were found to be in close agreement with structural data provided by X-ray crystallography. Traditional analysis of energy transfer assumes spherical symmetry of the interacting molecules, which is not necessarily the case for real systems, leading to significant errors in determining quenching rates. More sophisticated considerations of the geometrical and dipolar orientations of the donor and acceptor yielded analytical expressions accounting for the effects of non-spherical symmetry that improved the estimates of distance of closest approach between donor and acceptor.[87]

As the distance R between the donor and acceptor decreases, the energy transfer rate becomes dominated by the exchange term (Dexter exchange) and is characterized by an exponential dependence on donor-acceptor separation, $k^{ex}(R) = k_0^{ex} \exp(-2R/L)$, where L (0.8–1.0 Å) is an effective Bohr radius, and $k_0^{ex}$ is the quenching rate constant at the van der Waals contact distance between the donor-acceptor pair. This mechanism dominates in triplet state energy transfer from Trp109 to a Terbium ($Tb^{3+}$) atom substituted into the metal binding sites of *E. coli* AP[12] and changes in the distance between the donor and acceptor were monitored by the measurement of the sensitized $Tb^{3+}$ luminescence. The strong exponential dependence of the energy transfer coupled with the high accuracy of determination of RTP lifetimes can provide great accuracy that in principle can determine changes in the donor-acceptor distance ~0.1 Å. This technique can thus potentially monitor subtle structural changes in proteins in solution in real-time.

Additional modulation of the Dexter energy transfer rate may depend on relative orientations of the donor and acceptor. The original formalism of Dexter assumed hydrogenic wavefunctions, but a explicit consideration of structured electronic wavefunctions (such as in the $\pi - \pi^*$ transition in indole) will lead to an angular modulation of the effective orbital radius in the initial and final states, $L$. While the distance dependence is expected to be much more significant than the orientational dependence, for constant $R$, modulation of the orientation of the indole will be observable and used for conformation-determination. The applicability of this technique could in principle be tested by appropriate site directed mutagenesis to change the orientation of the indole ring.

### 3.4.2. H-D Exchange Studies

Hydrogen exchange studies have proven to be very powerful means of gaining insight into the dynamics of conformational changes in proteins. The dependence of the exchange kinetics on the experimental conditions provides information about the specific mechanisms of exchange, and in combination with site-directed mutagenesis, provides a powerful structural biological approach to understanding the details of the environment around specific residues. HD exchange detection by RTP spectroscopy was reported for the first time by Schlyer *et al.*[15] RTP was used to monitor hydrogen exchange within *E. coli* AP in solution by determining the change in the RTP decay rate. The phosphorescence lifetime of AP was seen to increase upon exchanging into $D_2O$ (see Figure 3.4), suggesting that conformational changes in the protein were reponsible for this effect. In this initial work, it was assumed that the HD exchange was characterized by a bimolecular (EX2) process. However, recent work has shown that the rate of the exchange process is not pH dependent, and thus is most likely to be of the EX1 type, i.e. rate-limited by "breathing" motions of the protein.[17] The exchange reaction was shown to be associated with replacement of a specific hydrogen, most likely the enamine group of Trp 109, which is hydrogen-bonded to a neighbouring residue, Gln 320 (see Figure 3.3). Site directed mutagenesis of this residue to remove the hydrogen bond (Q320L) yields changes in the exchange rates and in the activation energy of exchange,[76] as expected. This approach of combining hydrogen exchange with RTP has been extended to other proteins, including horse-liver alcohol dehydrogenase and glucose-6-phosphate dehydrogenase.[83]

### 3.4.3. Circularly Polarized Phosphorescence (CPP)

In contrast to circular dichroism, which yields information on the chirality of a chromophore's ground state, circularly polarized luminescence (CPL) reflects the chirality of the electronically excited chromophore, and in the particular case of CPP, of the excited triplet state. The existence of a

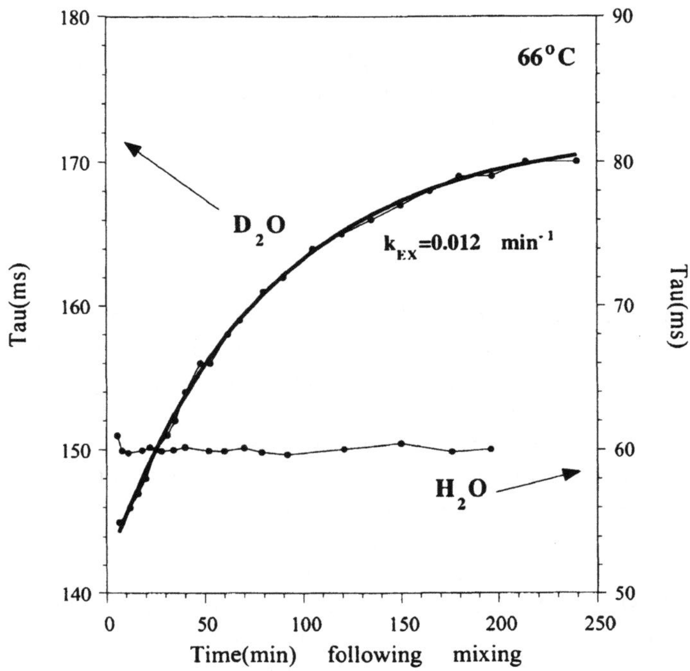


Figure 3.4. Change in average RTP lifetime of *E. coli* Alkaline Phosphatase upon transfer into deuterated and protonated solutions at 66 °C. The curve labeled $H_2O$ represent data for exchange into $H_2O$ buffer (10mM Tris, pH 8.0/$H_2O$) and that labeled $D_2O$ for exchange into $D_2O$ buffer (10mM Tris, pH 8.0/$H_2O$). For details, see reference 15. More recent work based on improved methodology shows that this process follows from a simple two state reaction (characterized by two time independent RTP lifetimes) between deuterated and protonated states (Fischer *et al.*, in press).

circularly polarized component in the RTP of proteins was demonstrated and applied to study several proteins including bacterial glucose 6-phosphate dehydrogenase,[11] which possesses several phosphorescent tryptophans. The great sensitivity of the RTP lifetime to the environment of the emitting chromophore frequently allows assignment of the different decay components to specific Trp residues in the protein; time-resolving the CPP therefore enables one to resolve the intrinsic excited-state chirality of each of the contributing Trps, thereby extracting additional structural information.

An advantage of the CPP method is that it frequently allows to distinguish, on the basis of the difference in their excited state chiralities, two or more phosphorescing moieties with similar lifetimes, the accurate resolution of which is limited by the Poisson noise inherent to the photon counting process. The time-resolved CPP instrument[11] uses a low repetition-rate laser system to excite the sample and a gated photon-counting photomultiplier to detect the phosphorescence light. An acousto-optic modulator is used to modulate the intensity of the circularly polarized component of the luminescence and a time delay gate generator which establishes two 6 microsecond gates centered about the maximum and minimum of the modulation sine wave. The signals collected during the opening period of each of these gates are routed respectively to one of two multichannel scalers (MCS). Since no correlation exists between the laser flashes and the modulation of emitted light, each of the MCS cards registers a continuous decay curve. The difference between the number of counts recorded in any given channel of the two MCS cards is proportional to the degree of circular polarization at that point in time. Evaluating the degree of circular polarization for all channels of the MCS cards yields a time-resolved CPP curve. The functional form of the time-resolved change in CPL is $f(t)=\frac{\sum_i \alpha_i g_{em,i}\exp(-t/\tau_i)}{\sum_i \alpha_i \exp(-t/\tau_i)}$, where the anisotropy factor $g_{em}=\frac{I_l - I_r}{\frac{1}{2}(I_l + I_r)}$, $I_l$ and $I_r$ are the intensities of left and right circularly polarized light in the emission respectively, and $\tau_i$ the decay lifetime of the *i*th lumiphore. Global analysis of the decay curves and the time-resolved CPL enables, for example, one to distinguish between the extremely similar metal binding sites in two closely related proteins, transferrin and conalbumin.[88] In this case bound Terbium served as the emitter, and the two decay components had a lifetime difference of 7% and a difference in emission anisotropy of $5 \times 10^{-2}$ (see Figure 3.5). Using these data along with the

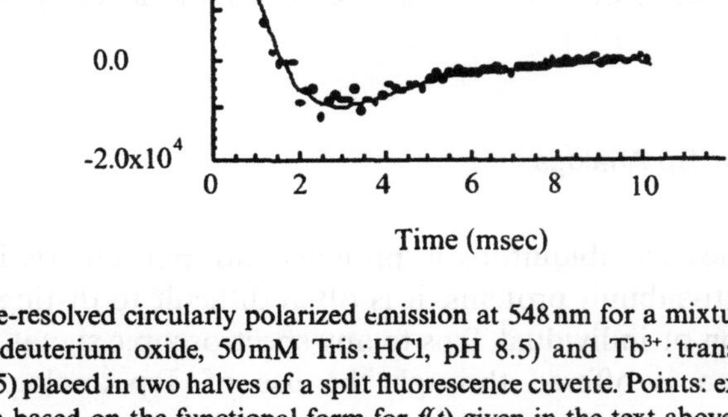


Figure 3.5. Time-resolved circularly polarized emission at 548 nm for a mixture of $Tb^{3+}$ : conalbumin (in 80% deuterium oxide, 50 mM Tris : HCl, pH 8.5) and $Tb^{3+}$ : transferrin (in 50 mM Tris : HCl, pH 8.5) placed in two halves of a split fluorescence cuvette. Points: experimental. Solid line—calculation based on the functional form for *f*(*t*) given in the text above.

values for $A_i$, $\tau_i$ determined from simultaneously recorded phosphorescence decay curves gave $g_1 = 0.015$ and $g_2 = -0.02$ at 548 nm. These numbers are in excellent agreement with control measurements made on separate complexes of $Tb^{3+}$ with each of the two proteins, thus demonstrating the ability of the instrument to resolve two very similar classes of sites.

### 3.4.4. Stopped Flow RTP

The requirement for thorough deoxygenation of protein solutions for observation of RTP has implied lengthy sample preparation procedures, limiting the use of RTP to systems exhibiting slow kinetics. Recent work has implemented RTP measurements in combination with a stopped-flow apparatus, yielding a dead time of ~10 ms.[14] This allows one to follow subtle changes in polypeptide conformation using RTP, which may not be reflected in the more often used stopped-flow fluorescence or circular dichroism methods. Stopped-flow RTP has been applied to study the denaturation of LADH by urea and guanidinium hydrochloride, revealing details of the different unfolding mechanisms associated with these denaturants and the heterogeneous nature of the unfolding kinetics.[13] Specifically, for denaturation in up to 8 M urea, there is little change in the phosphorescence lifetime, indicating that there is only a single phosphorescing species during the course of denaturation, in which the environment of the phosphorescing Trp 314 is native-like. The phosphorescence intensity, in contrast, decreased steadily, reflecting the fact that denaturation yields a non-phosphorescent species. The denaturation in GuHCl revealed a different behavior where the RTP lifetime of LADH was reduced drastically within the dead-time (~10 ms) of the mixing, and exhibited significant heterogeneity at concentrations of GuHCl up to 4.5 M. These data suggest that denaturation in GuHCl proceeds from a partially unfolded intermediate state. The heterogeneity of the phosphorescence decay and denaturation kinetics suggest the existence of multiple stable conformations and multiple unfolding pathways.

### 3.4.5. RTP from Trp Analogs

Trp residues are ubiquitous in proteins and, particularly in multi-tryptophan or multi-subunit proteins, it is often difficult to distinguish between the contribution of individual Trps to any spectroscopic signal. One solution to this problem is offered through the use of Trp analogs which have spectral characteristics distinct from Trp, but whose structure is similar, thus avoiding steric problems. In addition, these Trp analogs can be specifically incorporated into proteins by using site-directed mutagenesis techniques. The fluorescence from Trp analogs has been exploited by several groups (reviewed in ref. 89), but until recently there have been no reports regarding the phosphoresence characteristics of these analogs. While McCaul and Ludescher studied the RTP properties of Trp analogs in amorphous sucrose,[31] Cioni *et al.* have characterized RTP from the analogs 7-Azatryptophan (7AW) and 5-Hydroxytryptophan (5OHW) in solution and incorporated into the $\alpha_2$ subunit of RNA polymerase,[32] and have reported that the triplet emission from these analogs is strongly quenched by very efficient non-radiative processes.

### 3.4.6. Concluding Remarks and Prospects for the Future

Fluorescence from proteins containing aromatic amino acids is a well-established phenomenon and has been exploited for structural and dynamic studies for several decades. Unlike fluorescence, RTP is only observed in the absence of oxygen, and has thus only relatively recently received more wide-spread attention. RTP offers some advantages over fluorescence, including:

(i) RTP originates from deeply buried Trps and thus can be used to selectively probe a particular residue in a multi-Trp protein,
(ii) Unlike fluorescence which reflects contributions from all Trps in a protein, long lived RTP arises only from the very few Trps which are deeply buried, thus providing a more local site specific probe for structural studies,
(iii) The long lifetime makes Trp RTP susceptible to many quenching processes, a feature which can be exploited for structural studies,
(iv) The effect of chirality, leading to circularly polarized luminescence, is an order of magnitude larger for RTP than in fluorescence, and
(v) RTP is very susceptible to changes in the local environment and to interaction with quenchers, as reflected in the large dynamic range in RTP lifetimes, and thus can be a very sensitive monitor of protein conformation and flexibility.

RTP can be used in combination with other biophysical techniques, such as hydrogen exchange, stopped-flow methodologies, polarization sensitive detection, and energy transfer, to enhance the utility of this spectroscopy and to enable it to yield high-resolution structural information as well as real-time dynamic information on the appropriate timescales. In conjuction with molecular biological approaches allowing site-directed mutagenesis, replacement and engineering of Trp residues and Trp analogs, phosphorescent "beacons" placed into specific regions in proteins enable one to answer specific structural questions. We have reported initial results on RTP from a Trp residue engineered into the micrococcal nuclease from *Staphylococcus aureus*.[33] Other directions for the future include using RTP as a reporter *in vivo*, first demonstrated by Horie and Vanderkooi,[90] and recently used in our laboratory to follow the folding of AP (Dirnbach *et al.*, unpublished results). For specific situations, RTP has the potential of providing a reporter signal free of the autofluoresence from other proteins in the cellular environment.

## Acknowledgments

We thank Prof. Giovanni Strambini for the data used to construct Figure 3.2 and for sharing manuscripts prior to publication, and Prof. Richard Ludescher for providing unpublished manuscripts. Research at the University of Michigan was supported by the National Institute on Aging (Grant AG09761), Office of Naval Research (Grant N00014-91-J-1938), and a National Institutes of Health Molecular Biophysics Training Grant (Grant GM08270). VS was the recipient of postdoctoral fellowships from the Human Frontiers Science Program Organization and the Max Planck Society.

## References

1. M. L. Saviotti and W. C. Galley, Room temperature phosphorescence and the dynamic aspects of protein structure, *Proc. Natl. Acad. Sci. USA 71*, 4154–4158, (1974).
2. M. Gonnelli and G. B. Strambini, Phosphorescence lifetime of tryptophan in proteins, *Biochemistry 34*, 13847–13857, (1995).
3. J. M. Vanderkooi, D. B. Calhoun and W. S. Englander, On the prevalence of room-temperature protein phosphorescence, *Science 236*, 568–569, (1987).
4. J. M. Vanderkooi, Tryptophan phosphorescence from proteins at room temperature, in: *Topics in Fluorescence Spectroscopy*, Vol. 3, Lakowicz, J. R., Ed., pp. 113–136, Plenum Press, New York, (1992).
5. P. R. Callis, $^1L_a$ and $^1L_b$ transitions of Tryptophan: applications of theory and experimental observations to fluorescence of proteins, in: *Methods in Enzymology*, Vol. 278, Brand, L. and Johnson, M. L., Eds., pp. 113–150, Academic Press, New York, (1997).
6. S. V. Konev, *Fluorescence and phosphorescence of proteins and nucleic acids*, Plenum Press, New York, (1967).

7. S. Papp and J. M. Vanderkooi, Tryptophan Phosphorescence At Room-Temperature As a Tool to Study Protein-Structure and Dynamics, *Photochem. Photobiol. 49*, 775–784, (1989).
8. J. M. Vanderkooi and J. W. Berger, Excited Triplet-States Used to Study Biological Macromolecules At Room-Temperature, *Biochim. Biophys. Acta 976*, 1–27, (1989).
9. J. A. Schauerte, D. G. Steel and A. Gafni, Time-resolved room temperature tryptophan phosphorescence in proteins, in: *Methods in Enzymology*, Vol. 278, pp. 49–71, (1997).
10. G. B. Strambini and M. Gonnelli, Tryptophan phosphorescence in fluid solution, *J. Am. Chem. Soc. 117*, 7646–7651, (1995).
11. J. A. Schauerte, D. G. Steel and A. Gafni, Time-resolved circularly polarized protein phosphorescence, *Proc. Natl. Acad. Sci. USA 89*, 10154–10158, (1992).
12. B. D. Schlyer, D. G. Steel and A. Gafni, Direct kinetic evidence for triplet state energy transfer from *Escherichia coli* alkaline phosphatase tryptophan 109 to bound terbium, *J. Biol. Chem. 270*, 22890–22894, (1995).
13. M. Gonnelli and G. B. Strambini, Time-resolved protein phosphorescence in the stopped-flow: Denaturation of horse liver alcohol dehydrogenase by urea and guanidine hydrochloride, *Biochemistry 36*, 16212–16220, (1997).
14. G. B. Strambini, A. Puntoni and M. Gonnelli, A modified stopped-flow apparatus for time-resolved protein phosphorescence, *Rev. Sci. Instrum. 68*, 4583–4587, (1997).
15. B. D. Schlyer, D. G. Steel and A. Gafni, Long time-scale probing of the protein globular core using hydrogen-exchange and room temperature phosphorescence, *Biochem. Biophys. Res. Commun. 223*, 670–674, (1996).
16. J. V. Mersol, D. G. Steel and A. Gafni, Quenching of tryptophan phosphorescence in *Escherichia coli* alkaline phosphatase by long-range transfer mechanisms to external agents in the rapid-diffusion limit, *Biochemistry 30*, 668–675, (1991).
17. C. J. Fischer, J. A. Schauerte, K. C. Wisser, A. Gafni and D. G. Steel, Hydrogen exchange at the enamine nitrogen of Trp 109 in *E. coli* alkaline phosphatase studied by room-temperature trp phosphorescence, *Biochemistry in press*, (1999).
18. P. Cioni and G. B. Strambini, Acrylamide quenching of protein phosphorescence as a monitor of structural fluctuations in the globular fold, *J. Am. Chem. Soc. 120*, 11749–11757, (1998).
19. P. Cioni and G. Strambini, Pressure/Temperature effects on protein flexibility from acrylamide quenching of protein phosphorescence, *J. Mol. Biol. 291*, 955–964, (1999).
20. Y. Kai and T. Maeda, Room temperature phosphorescence study of refolding of disulfide reduced RNase T-1, *J. Phys. Soc. Japan 67*, 1486–1491, (1998).
21. G. Solaini, A. Baracca, E. Gabellieri and G. Lenaz, Modification of the mitochondrial F-1-ATPase epsilon subunit, enhancement of the ATPase activity of the IF1-F-1 complex and IF1-binding dependence of the conformation of the epsilon subunit, *Biochem. J. 327*, 443–448, (1997).
22. G. Solaini, A. Baracca, G. P. Castelli and G. B. Strambini, Tryptophan phosphorescence as a structural probe of mitochondrial F1-ATPase epsilon subunit, *Eur. J. Biochem. 214*, 729–734, (1993).
23. E. Gabellieri, G. B. Strambini, A. Baracca and G. Solaini, Structural mapping of the epsilon-subunit of mitochondrial H+-ATPase complex (F-1), *Biophys. J. 72*, 1818–1827, (1997).
24. A. Baracca, S. Barogi, E. Gabellieri, G. Lenaz and G. Solaini, A study of the mitochondrial F1-ATPase tryptophan phosphorescence at 273 K, *Biochem. Biophys. Res. Commun. 207*, 369–374, (1995).
25. A. Baracca, E. Gabellieri, S. Barogi and G. Solaini, Conformational changes of the mitochondrial F1-ATPase epsilon subunit induced by nucleotide binding as observed by phosphorescence spectroscopy, *J. Biol. Chem. 270*, 21845–21851, (1995).

26. A. Szarka, E. Gabellieri, M. Gonnelli, P. Cioni, Z. Lakos and B. Somogyi, Alteration of the intramolecular dynamics of glycogen phosphorylase b by allosteric ligands, *J. Photochem. Photobiol. B. 42*, 52–56, (1998).
27. J. V. Mersol, D. G. Steel and A. Gafni, Detection of intermediate protein conformations by room temperature tryptophan phosphorescence spectroscopy during denaturation of *Escherichia coli* alkaline phosphatase, *Biophys. Chem. 48*, 281–291, (1993).
28. V. Subramaniam, N. C. H. Bergenhem, A. Gafni and D. G. Steel, Phosphorescence reveals a continued slow annealing of the protein core following reactivation of *Escherichia coli* alkaline phosphatase, *Biochemistry 34*, 1133–1136, (1995).
29. L. Sun, E. R. Kantrowitz and W. C. Galley, Room temperature phosphorescence study of phosphate binding in Escherichia coli alkaline phosphatase, *Eur. J. Biochem. 245*, 32–39, (1997).
30. V. Subramaniam, D. G. Steel and A. Gafni, In vitro renaturation of bovine beta-lactoglobulin A leads to a biologically active but incompletely refolded state, *Prot. Sci. 5*, 2089–2094, (1996).
31. C. P. McCaul and R. D. Ludescher, Room temperature phosphorescence from tryptophan and halogenated tryptophan analogs in amorphous sucrose, *Photochem. Photobiol. 70*, 166–171, (1999).
32. P. Cioni, L. Erijman and G. B. Strambini, Phosphorescence emission of 7-azatryptophan and 5-hydroxytryptophan in fluid solutions and in alpha(2) RNA polymerase, *Biochem. Biophys. Res. Commun. 248*, 347–351, (1998).
33. V. Subramaniam, A. Gafni and D. G. Steel, Time-resolved tryptophan phosphorescence spectroscopy: A sensitive probe of protein folding and structure, *IEEE J. Spec. Top. Quant. Elec. 2*, 1107–1114, (1996).
34. G. B. Strambini and M. Gonnelli, The indole nucleus triplet-state lifetime and its dependence on solvent microviscosity, *Chem. Phys. Lett. 115*, 196–200, (1985).
35. D. V. Bent and E. Hayon, Excited state chemistry of aromatic amino acids and related peptides. III. Tryptophan, *J. Am. Chem. Soc. 97*, 2612–2619, (1975).
36. S. K. Lower and M. A. El-Sayed, The triplet state and molecular electronic processes in organic molecules, *Chemical Reviews 66*, 199–241, (1966).
37. G. B. Strambini and E. Gabellieri, Temperature dependence of tryptophan phosphorescence in proteins, *Photochem. Photobiol. 51*, 643–648, (1990).
38. P. Cioni and G. B. Strambini, Pressure effects on protein flexibility monomeric proteins, *J. Mol. Biol. 242*, 291–301, (1994).
39. P. Cioni and G. B. Strambini, Pressure effects on the structure of oligomeric proteins prior to subunit dissociation, *J. Mol. Biol. 263*, 789–799, (1996).
40. M. Gonnelli and G. B. Strambini, Glycerol effects on protein flexibility—a tryptophan phosphorescence study, *Biophys. J. 65*, 131–137, (1993).
41. S. Hogiu, M. Enescu and M. L. Pascu, Dynamic and thermodynamic effects of glycerol on bovine serum albumin in aqueous solution: a tryptophan phosphorescence study, *J. Photochem. Photobiol. B. 40*, 55–60, (1997).
42. G. B. Strambini and M. Gonnelli, Effects of urea and guanidine hydrochloride on the activity and dynamical structure of equine liver alcohol dehydrogenase, *Biochemistry 25*, 2471–2476, (1986).
43. G. B. Strambini, P. Cioni, A. Peracchi and A. Mozzarelli, Conformational changes and subunit communication in tryptophan synthase—Effect of substrates and substrate analogs, *Biochemistry 31*, 7535–7542, (1992).
44. G. B. Strambini and M. Gonnelli, Tryptophan luminescence from liver alcohol dehydrogenase in its complexes with coenzyme. A comparative study of protein conformation in solution, *Biochemistry 29*, 196–203, (1990).

45. D. B. Calhoun, J. M. Vanderkooi, G. V. Woodrow III and S. W. Englander, Penetration of dioxygen into proteins studied by quenching of phosphorescence and fluorescence, *Biochemistry 22*, 1526–1532, (1983).
46. D. B. Calhoun, J. M. Vanderkooi and S. W. Englander, Penetration of Small Molecules Into Proteins Studied By Quenching of Phosphorescence and Fluorescence, *Biochemistry 22*, 1533–1539, (1983).
47. D. B. Calhoun, J. M. Vanderkooi and S. W. Englander, Study of Internal Protein Motion By Fluorescence and Phosphorescence Quenching, *Biophys. J. 47*, A209–A209, (1985).
48. D. B. Calhoun, S. W. Englander, W. W. Wright and J. M. Vanderkooi, Quenching of Room-Temperature Protein Phosphorescence By Added Small Molecules, *Biochemistry 27*, 8466–8474, (1988).
49. V. Dadak, J. M. Vanderkooi and W. W. Wright, Electron-Transfer From Excited Tryptophan to Cytochrome-C—Mechanism of Phosphorescence Quenching, *Biochim. Biophys. Acta 1100*, 33–39, (1992).
50. S. Papp, W. W. Wright, S. W. Englander, J. M. Vanderkooi and C. S. Owen, Phosphorescence Quenching of Proteins At Room-Temperature By Added Small Molecules, *Biophys. J. 57*, A77–A77, (1990).
51. J. M. Vanderkooi, G. Maniara, T. J. Green and D. F. Wilson, An optical method for measurement of dioxygen concentration based upon quenching of phosphorescence, *J. Biol. Chem. 262*, 5476–5482, (1987).
52. J. M. Vanderkooi, S. W. Englander, S. Papp, W. W. Wright and C. S. Owen, Long-Range Electron Exchange Measured in Proteins By Quenching of Tryptophan Phosphorescence, *Proc. Natl. Acad. Sci. USA 87*, 5099–5103, (1990).
53. W. W. Wright, C. S. Owen and J. M. Vanderkooi, Penetration of Analogs of H2o and Co2 in Proteins Studied By Room-Temperature Phosphorescence of Tryptophan, *Biochemistry 31*, 6538–6544, (1992).
54. J. W. Berger and J. M. Vanderkooi, Characterization of Lens Alpha-Crystallin Tryptophan Microenvironments By Room-Temperature Phosphorescence Spectroscopy, *Biochemistry 28*, 5501–5508, (1989).
55. S. Papp, T. E. King and J. M. Vanderkooi, Intrinsic Tryptophan Phosphorescence As a Marker of Conformation and Oxygen Diffusion in Purified Cytochrome-Oxidase, *FEBS Lett. 283*, 113–116, (1991).
56. Z. Li, W. E. Lee and W. C. Galley, Distance dependence of the tryptophan-disulfide interaction at the triplet level from pulsed phosphorescence studieson a model system, *Biophys. J. 56*, 361–367, (1989).
57. Z. Li and W. C. Galley, Evidence for ligand-induced conformational changes in proteins from phosphorescence spectroscopy, *Biophys. J. 56*, 353–560, (1989).
58. Z. Li, A. Bruce and W. C. Galley, Temperature dependence of the disulfide perturbation to the triplet state of tryptophan, *Biophys. J. 61*, 1364–1371, (1992).
59. D. J. Plocke, Alkaline phosphatase of *Escherichia coli*: A zinc metalloenzyme, *Biochemistry 1*, 373–378, (1962).
60. W. F. Bosron, R. A. Anderson, M. C. Falk, F. S. Kennedy and B. L. Vallee, Effect of magnesium on the properties of zinc alkaline phosphatase, *Biochemistry 16*, 610–614, (1977).
61. V. Subramaniam, *Application of time-resolved tryptophan phosphorescence spectroscopy to protein folding studies*, University of Michigan, Ann Arbor, (1996).
62. E. Dirnbach, D. G. Steel and A. Gafni, Proline isomerization is unlikely to be the cause of slow annealing and reactivation during the folding of alkaline phosphatase, *J. Biol. Chem. 274*, 4532–4536, (1999).
63. P. Gettins and J. Coleman, 113Cd nuclear magnetic resonance of Cd(II) alkaline phosphatases, *J. Biol. Chem. 258*, 396–407, (1983).

64. J. F. Chlebowski and S. Mabrey, Differential scanning calorimetry of apo-, apophosphoryl, and metalloalkaline phosphatases, *J. Biol. Chem. 252*, 7042–7052, (1977).
65. A. Chaidaroglou and E. R. Kantrowitz, Alteration of aspartate 101 in the active site of *Escherichia Coli* alkaline phosphatase enhances the catalytic activity, *Prot. Eng. 3*, 127–132, (1989).
66. L. Ma, T. T. Tibbitts and E. R. Kantrowitz, *Escherichia coli* alkaline phosphatase—X-ray structural studies of a mutant enzyme (His-412-Asn) at one of the catalytically important Zinc-binding sites, *Prot. Sci. 4*, 1498–1506, (1995).
67. L. Ma and E. R. Kantrowitz, Mutations at histidine-412 alter zinc-binding and eliminate transferase-activity in *Escherichia coli* alkaline phosphatase, *J. Biol. Chem. 269*, 31614–31619, (1994).
68. L. Ma and E. R. Kantrowitz, Kinetic and X-ray structural studies of a mutant *Escherichia coli* alkaline phosphatase (His-412 → Gln) at one of the zinc binding sites, *Biochemistry 35*, 2394–2402, (1996).
69. J. E. Murphy, X. Xu and E. R. Kantrowitz, Conversion of a magnesium binding site into a zinc binding site by a single amino acid substitution in *Escherichia coli* alkaline phosphatase, *J. Biol. Chem. 268*, 21497–21500, (1993).
70. J. E. Murphy, T. T. Tibbitts and E. R. Kantrowitz, Mutations at position 153 and position 328 in *Escherichia coli* alkaline phosphatase provide insight towards the structure and function of mammalian and yeast alkaline phosphatases, *J. Mol. Biol. 253*, 604–617, (1995).
71. B. Stec, M. J. Hehir, C. Brennan, M. Nolte and E. R. Kantrowitz, Kinetic and X-ray structural studies of three mutant E-coli alkaline phosphatases: Insights into the catalytic mechanism without the nucleophile Ser102, *J. Mol. Biol. 277*, 647–662, (1998).
72. L. Sun, D. C. Martin and E. R. Kantrowitz, Rate-determining step of Escherichia coli alkaline phosphatase altered by the removal of a positive charge at the active center, *Biochemistry 38*, 2842–2848, (1999).
73. N. C. Bergenhem, K. C. Wisser, J. A. Schauerte, V. Subramaniam, D. G. Steel and A. Gafni, Effects of cavity forming mutations on the phosphorescence of tryptophan-109 in the core of alkaline phosphatase, *Biophys. J. 70*, SU202–SU202, (1996).
74. J. A. Schauerte, K. C. Wisser, N. C. Bergenhem, D. G. Steel and A. Gafni, Room temperature phosphorescence studies of the structural stability and folding pathways of single point mutants of alkaline phosphatase, *Biophys. J. 74*, A168–A168, (1998).
75. J. A. Schauerte, K. C. Wisser, D. G. Steel and A. Gafni, Identification of intermediates in the folding of native and mutant alkaline phosphatase: Effects of altered thermodynamic stability of subunits on dimer interactions, *Biophys. J. 76*, A173, (1999).
76. C. J. Fischer, J. A. Schauerte, A. Gafni and D. G. Steel, Effects of site-directed mutations in *E. coli* alkaline phosphatase on the activation energy for hydrogen exchange at the tryptophan 109 enamine studied by tryptophan phosphorescence, *Biophys. J. 76*, A119, (1999).
77. J. E. Hansen, D. G. Steel and A. Gafni, Detection of a pH-dependent conformational change in azurin by time-resolved phosphorescence, *Biophys. J. 71*, 2138–2143, (1996).
78. B. D. Schlyer, J. A. Schauerte, D. G. Steel and A. Gafni, Time-resolved room temperature protein phosphorescence—nonexponential decay from single emitting tryptophans, *Biophys. J. 67*, 1192–1202, (1994).
79. M. Hattori, A. Ametani, Y. Katakura, M. Shimizu and S. Kaminogawa, Unfolding/refolding studies on bovine $\beta$-lactoglobulin with monoclonal antibodies as probes, *J. Biol. Chem. 268*, 22414–22419, (1993).

80. T. Kiefhaber, R. Quaas, U. Hahn and F. X. Schmid, Folding of ribonuclease $T_1$. 1. Existence of multiple unfolded states created by proline isomerization, *Biochemistry 29*, 3053–3061, (1990).
81. T. Kiefhaber, R. Quaas, U. Hahn and F. X. Schmid, Folding of Ribonuclease $T_1$. 2. Kinetic models for the folding and unfolding reactions, *Biochemistry 29*, 3061–3070, (1990).
82. M. Mücke and F. X. Schmid, Intact disulfide bonds decelerate the folding of ribonuclease $T_1$, *JMB 239*, 713–725, (1994).
83. P. Wolanin, J. A. Schauerte, A. Gafni and D. G. Steel, Hydrogen exchange kinetics of proteins monitored by time-resolved room temperature phosphorescence, *Biophys. J. 76*, A167, (1999).
84. L. Stryer, Fluorescence energy transfer as a spectroscopic ruler, *Ann. Rev. Biochem. 47*, 819–846, (1978).
85. P. R. Selvin, Fluorescence resonance energy transfer, *Methods Enzymol. 246*, 300–334, (1995).
86. L. Stryer, D. D. Thomas and C. F. Meares, Diffusion-enhanced fluorescence energy transfer, *Annual Review of Biophysics & Bioengineering 11*, 203–222, (1982).
87. J. V. Mersol, H. Wang, A. Gafni and D. G. Steel, Consideration of dipole orientation angles yields accurate rate equations for energy transfer in the rapid diffusion limit, *Biophys. J. 61*, 1647–1655, (1992).
88. J. A. Schauerte, A. Gafni and D. G. Steel, Improved differentiation between luminescence decay components by use of time-resolved optical activity measurements and selective lifetime modulation, *Biophys. J. 70*, 1996–2000, (1996).
89. J. B. Ross, A. G. Szabo and C. W. Hogue, Enhancement of protein spectra with tryptophan analogs: fluorescence spectroscopy of protein-protein and protein-nucleic acid interactions, *Methods Enzymol. 278*, 151–190, (1997).
90. T. Horie and J. M. Vanderkooi, Phosphorescence of alkaline phosphatase of E. coli in vitro and in situ, *Biochim. Biophys. Acta 670*, 294–297, (1981).

# 4

# Azurins and Their Site-Directed Mutants

Giampiero Mei, Nicola Rosato, and
Alessandro Finazzi Agrò*

## 4.1. A Brief Overview on Azurin and Its Dynamic Fluorescence Properties

Azurin is a blue copper-containing protein which functions as a redox mediator in the electron transfer system of denitrifying bacteria.[1] From the structural point of view azurin is a globular monomer of about 14.6 kDa, containing one copper atom per molecule. As revealed by X-ray crystallography and NMR measurements, the protein tertiary structure is characterized by a $\beta$-barrel arrangement of eight strands, plus a short helix of $\approx 20$ residues.[2–4] Despite its small size and the presence of a single tryptophan residue, Trp48, azurin exhibits a lot of unique spectroscopical features, in the visible and in the near UV. In particular, the peculiar emission spectrum, centered at 308 nm, is the lowest-wavelength protein fluorescence spectrum known so far. Its unusual vibrational fine structure and the absorption and circular dichroism (CD) signals around 292 nm, reveal the extremely hydrophobic nature of Trp48 microenvironment, which in fact has been found by X-ray crystallography to be constituted by a group of apolar side chains.[3] On the other hand the copper coordination geometry gives rise to a rather intense, broad absorption around 627 nm and to a complex CD spectrum in the range 400–700 nm, with different bands assigned to specific transition through ligand field calculations.[5] The metal binding site has also been extensively studied by Raman, epr and spectrochemical techniques,[6,7] in

---

*Giampiero Mei, Nicola Rosato, and Alessandro Finazzi Agrò* • Dipartimento di Medicina Sperimentale, e Scienze Biochimiche, Universita' di Roma "Tor Vergata", Via di Tor Vergata, 135, 00133 Roma, ITALY. *To whom all correspondence should be addressed at: Dipartimento di Medicina Sperimentale e Scienze Biochimiche, Universita' di Roma, "Tor Vergata", Via di Tor Vergata, 135, 00133 Roma, Italy, Tel. +39-06-72596460, Fax. +39-06-72596468, email: mei@med.uniroma2.it

*Topics in Fluorescence Spectroscopy, Volume 6: Protein Fluorescence*, edited by Joseph R. Lakowicz. Kluwer Academic / Plenum Publishers, New York, 2000

order to better characterize the relationship between the structure of the active site and its spectroscopical and redox properties.

The fluorescence decay kinetics of azurin has been also measured with different techniques, at various pHs, temperatures and emission wavelengths.[8–15] Since the first measurement by Grinvald and coworkers in 1975, it was clear that at least two discrete lifetimes are required to satisfactorily fit the data (Table 4.1): a short component, $\tau_1$, of hundreds of picoseconds, and a longer one, with a $\tau_2 = 4.5$ ns, which however accounts only for a small percentage of the total signal. As the dynamic fluorescence data of the copper-free protein (apo-azurin) may be fitted with a single exponential function (Table 4.1), it follows that the short component is essentially dependent on the presence of the metal ion. Despite this finding, the exact origin of a double exponential decay in the case of holo-azurin is not yet clear. The similarity between the fluorescence lifetime of the apo-protein and the long component of the copper-containing azurin, $\tau_2$, suggested the presence of an "apo-like" contaminant in the holo-azurin sample.[11] Alternatively, Szabo and coworkers proposed that this fluorescence heterogeneity might arise from at least two[10] or three[12] different protein conformations, in the proximity of Trp48. These conformers should correspond to different geometries of the

Table 4.1. Fluorescence Lifetimes of Holo- and Apo-azurin

| Holo | $\tau_1$ | $\tau_2$ | $\tau_3$ | $\alpha_1$[a] | $\alpha_2$ |
|---|---|---|---|---|---|
| ref. (8) | 0.8 | 4.5 | — | 0.65 | 0.35 |
| ref. (9) | 0.75 | 4.15 | — | 0.50 | 0.50 |
| ref. (10) | 0.18 | 4.78 | — | 0.80 | 0.20 |
| ref. (11) | 1.02 | 4.15 | — | 0.97 | 0.03 |
| ref. (12) | 0.097 | 0.36 | 4.80 | 0.92 | 0.05 |
| ref. (13) | 0.10 | 4.23 | — | 0.97 | 0.03 |
| ref. (14) | 0.22 | 4.51 | — | 0.93 | 0.07 |
| ref. (15) | 0.06 | 0.14 | — | 0.80 | 0.20 |
| **Apo** | $\tau_1$ | $\tau_2$ | $\tau_3$ | $\alpha_1$ | $\alpha_2$ |
| ref. (8) | 4.7 | — | — | 1.00 | — |
| ref. (9) | 4.79 | 0.88 | — | 0.60 | 0.40 |
| ref. (10) | 4.86 | — | — | 1.00 | — |
| ref. (11) | 5.16 | — | — | 1.00 | — |
| ref. (12) | 5.08 | — | — | 1.00 | — |
| ref. (13) | 4.94 | — | — | 1.00 | — |
| ref. (14) | 4.70 | — | — | 1.00 | — |
| ref. (15) | 0.13 | 1.31 | 4.71 | 0.18 | 0.13 |

[a] $\alpha_i$ are the pre-exponential factor ($\Sigma_i \alpha_i = 1$).

copper-ligand field, one of which, having a stronger quenching effect on azurin fluorescence, could explain the small value of $\tau_1$. This quenching mechanism has been attributed to an electron transfer from Trp48 to the metal site[11] or in terms of a non-radiative energy transfer process.[13,16] At variance with $\tau_1$, longer lifetimes have been found to depend on pH,[12] indicating the possible involvement of an histidine residue. It has also been suggested that conformational changes might be induced by protonation of His35. In a previous paper,[17] we have demonstrated that indeed substitution of His35 with different residues decreased the $\tau_2$ value from 4.51 ns to ≈ 3.9 ns, without any effect on the apo-azurin decay, ruling out at least in that case, the presence of an "apo-like" impurity. Differences in the long component of copper-free and copper-containing azurin have been found also in the case of two core mutants, namely Phe110Ser (F110S) and Ile7Ser (I7S), again demonstrating that the longer lifetimes of holo-azurins do not trivially originate from some copper-free molecules.[14]

The anisotropy decay of azurin has also been studied in detail, in order to evaluate the rotational correlation lifetimes associated to global and local motions. Both the holo- and apo-proteins display a longer component (≥6.5 ns), associated with the tumbling of the whole molecule, and a shorter one (≤0.5 ns), which might be ascribed to the intrinsic dynamic of Trp48 (Table 4.2). In some cases the spatial amplitude of this fast rotation was extimated by the "wobbling-cone" model,[18] yielding a semi-angle, $\theta$, of about ≈ 30°–40° (Table 4.2). As already pointed out in earlier studies,[9] this result is quite interesting as it shows that proteins may have an internal

**Table 4.2. Rotational Correlation Lifetimes of Holo- and Apo-azurin**

| Holo | $\Phi_1$ | $\Phi_2$ | $\alpha_1$ | $r_0$[a] | $\theta$[b] |
|---|---|---|---|---|---|
| ref. (9) | 0.51 | 11.8 | 0.101 | 0.233 | 34° |
| ref. (17) | 0.19 | 6.71 | 0.165 | 0.270 | 43° |
| ref. (15) | 0.70 | 7.00 | 0.08 | 0.14 | — |
| **Apo** | $\Phi_1$ | $\Phi_2$ | $\alpha_1$ | $r_0$ | $\theta$ |
| ref. (9) | 0.49 | 6.84 | 0.139 | 0.231 | — |
| ref. (11) | — | 4.94 | — | 0.26 | — |
| ref(17) | 0.14 | 7.01 | 0.180 | 0.268 | 47° |
| ref. (15) | 0.30 | 6.70 | 0.05 | 0.13 | — |

[a] $r_0$ is the total anisotropy value at time $t = 0$.
[b] $\theta$ is the semi-angle of Trp48 movement estimated by the "wobbling-cone model".[18]

conformational fluidity in the subnanosecond time range. This local, structural heterogeneity reflects the large number of conformational substates found in the azurin molecule studying ligand binding equilibria.[19] It has been found that core mutations, creating cavities inside the protein molecule,[20] increase the mobility of both the Trp48[15,17] and the mutated residues.[20]

In this study we extend the previous spectroscopical characterization of these mutants, reporting the effects that an increased flexibility and a reduced hydrophobicity of Trp48 microenvironment have on azurin stability.

## 4.2. Experimental Procedures

Recombinant wild-type azurin, I7S and F110S mutant, dissolved in TrisHCl 50 mM, pH = 7.2, were expressed, and purified as previously described.[14] Equilibrium unfolding measurements were performed using stock solution of ultrapure guanidinium hydrochloride (GdHCl) and incubating the samples at 10 °C for at least 12 h in the presence of different amounts of denaturant. Reversibility of the denaturation transition was checked by diluting fully unfolded samples.

The apo-proteins were prepared by a 30 minutes dialysis at 4 °C against a K-phosphate buffer (80 mM, pH = 6) containing 20 molar excess ascorbate with respect to the protein. Then, a second dialysis in the same buffer was performed in presence of 50 mM KCN for 45 minutes, at 4 °C, followed by at least two subsequent dialyses against a Tris-HCl buffer (50 mM, pH = 7.2). This procedure, slightly different from that of previous experiments,[14,17] resulted to be more correct, since it preserves intact the secondary structure of the apo-proteins, which in fact showed circular dichroism spectra perfectly superimposable to those of the respective holo-samples.

Circular dichroism spectra were recorded on a Jasco J700 spectropolarimeter, using 0.1 cm quartz cuvettes. The protein optical density at 280 nm, measured on a Perkin Elmer Lambda 18 spectrophotometer, was always 0.09, using an optical path of 1 cm.

Steady-state fluorescence spectra were recorded with a ISS-K2 fluorometer (ISS, Champaign, IL, USA) with an excitation wavelength $\lambda$ = 280 nm. High pressure experiments were performed using the ISS pressure cell, as described by Paladini and Weber.[21] Dynamic fluorescence experiments were carried out at LASP facility (Laboratorio di Spettroscopia al Picosecondo, University of Rome, "Tor Vergata", Italy), using a Nd-Yag-pumped, frequency-doubled, Rhodamine 6G dye laser, and the phase-shift/demodulation technique as elsewhere described.[22] The dynamic fluorescence and depolarization data were fitted using the GLOBAL

Unlimited software,[23] based on a Marquardt minimization of the reduced chi-squared value.[24]

## 4.3. Copper-containing Azurins

The unfolding of wt-azurin, I7S and F110S mutants by GdHCl has been studied by steady-state fuorescence and circular dichroism. It was previously found[25] that fully unfolded azurin has a "normal" tryptophan fluorescence at ≈ 355 nm. The red-shift in the fluorescence spectrum peak, observed at increasing GdHCl concentrations and diagnostic of a progressive exposure of Trp48 to solvent, is accompanied by a decrease in the CD signal at 220 nm, indicating the simultaneous loss of tertiary and secondary structure (Figure 4.1).

The sigmoidal shape of the transition suggested to interpolate the data according to a simple two-state process:

$$N \xleftrightarrow{K} U$$

in which the only species involved are the native and the unfolded protein, so that K can be easily expressed as the ratio between the fractional populations of the two states:

$$K = f_U / f_N$$

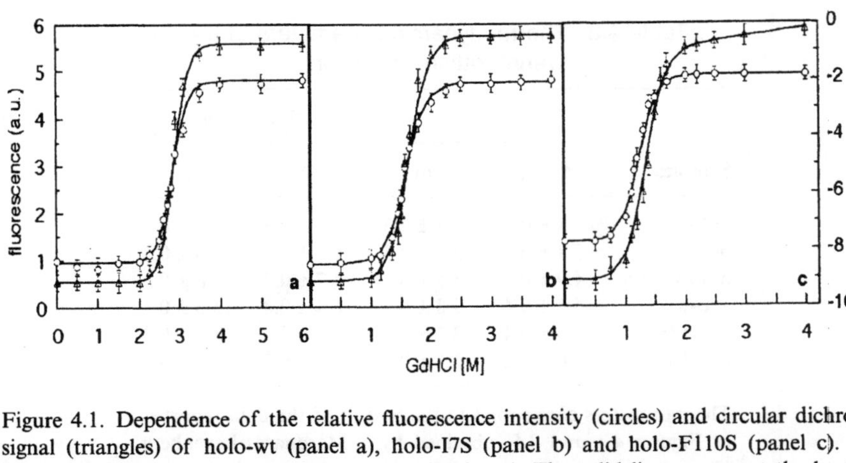


Figure 4.1. Dependence of the relative fluorescence intensity (circles) and circular dichroism signal (triangles) of holo-wt (panel a), holo-I7S (panel b) and holo-F110S (panel c). The quantum yield of the wt sample was normalized to 1. The solid lines represent the best fits obtained.

The parameters corresponding to the best fit are reported in Table 4.3, assuming a linear dependence of the unfolding free energy, $\Delta G = -RT\ln K$, on denaturant concentration:[26]

$$\Delta G = \Delta G_{H_2O} - m[GdHCl]$$

The results demonstrate that the substitution of a single apolar residue with a serine is sufficient to dramatically affect the protein stability, decreasing the $\Delta G_{H_2O}$ value by several kcal/mol. The analysis of the crystallographic structure,[20] anisotropy decay measurements[17] and red-edge excitation spectroscopy[14] revealed that for both mutants no structural modification occurs but the mobility and average dielectric constant of Trp48 microenvironment. This demonstrates that the hydrophobic interactions within the core are crucial for azurin stability. The case of F110S, which exhibits the lowest denaturation free energy value, deserves some additional comments. Dynamic fluorescence measurements have shown that the emission decay of this sample is more heterogeneous than that of the wild-type protein, requiring a double distribution of fluorescence lifetimes rather than simply two lifetimes.[14] This greater heterogeneity may be interpreted in terms of the presence of solvent in the hydrophobic core of the protein,[17] since the substitution of Phe110 with a serine residue creates a cavity of about 100 Å$^3$. This hypothesis has been confirmed by X-ray crystallography which showed that F110S has two or three water molecules near Trp48.[20] The presence of solvent inside the hydrophobic core is important because it enhances the local mobility and decreases its packing density, thus deacreasing the protein stability.[27–29] The

Table 4.3. Thermodynamic Parameters of the Denaturation Transition

| | CD | | fluorescence | |
|---|---|---|---|---|
| Samples | $\Delta G_{H_2O}$ | m | $\Delta G_{H_2O}$ | m |
| holo-wt | 9.8 ± 0.4 | 3.4 ± 0.3 | 9.1 ± 0.3 | 3.2 ± 0.2 |
| holo-I7S | 5.9 ± 0.4 | 3.6 ± 0.3 | 5.7 ± 0.2 | 3.6 ± 0.2 |
| holo-F110S | 4.9 ± 0.3 | 3.5 ± 0.2 | 4.7 ± 0.2 | 3.8 ± 0.3 |
| apo-wt | 6.5 ± 0.4 | 3.8 ± 0.3 | 6.4 ± 0.4 | 3.6 ± 0.3 |
| apo-I7S | 3.4 ± 0.4 | 3.3 ± 0.3 | 3.3 ± 0.3 | 3.6 ± 0.2 |
| apo-F110S | 2.6 ± 0.2 | 3.1 ± 0.1 | 2.8 ± 0.1 | 3.3 ± 0.2 |

The parameters correspond to the best fit of the data reported in Figures 4.1 and 4.5, obtained by a Marquardt-Levenberg algorithm, using a two-state equilibrium scheme and assuming a linear dependence of $\Delta G$ on GdHCl concentration.

interior of most globular proteins is indeed characterized by the presence of non-polar residues, well packed together, to form a dense and tough structure, almost inaccessible to solvent. The creation of a cavity is therefore another source of destabilization, whose contribution has been evaluated[30] to be in the range 24–33 kcal/mol Å$^3$, that for F110S azurin would represent a decrease in the $\Delta G_{H_2O}$ value of about ≈ (2.4–3.3) kcal/mol.

The increased flexibility at the hydrophobic core of mutants suggests a greater compressibility with respect to wt-azurin. It is well known that small globular proteins exhibit a high stability below 4 kbar,[31] because their compact structure prevents the penetration of solvent in the hydrophobic regions. In line with this result we have found that wt-azurin is practically unaffected by hydrostatic pressure (Figure 4.2a). Previous experiments by Cioni and Strambini[32] gave a similar result, where the only observable change in the pre-denaturational pressure range (≤3 kbar) was the phosphorescence lifetime of the metal-depleted enzyme. Instead Figures 4.2b and 4.2c show that the fluorescence spectrum of I7S and F110S was significantly modified under pressure.

Dynamic fluorescence measurements performed on these samples (Figures 4.3a and 3b) showed that $\tau_2$, the long component of the decay, was more sensitive than $\tau_1$, being slightly shorter in both proteins (Table 4.4). This effect is accompanied by the narrowing of both lifetime distributions in the case of F110S, indicating a lower conformational heterogeneity experienced by the tryptophylic residue.

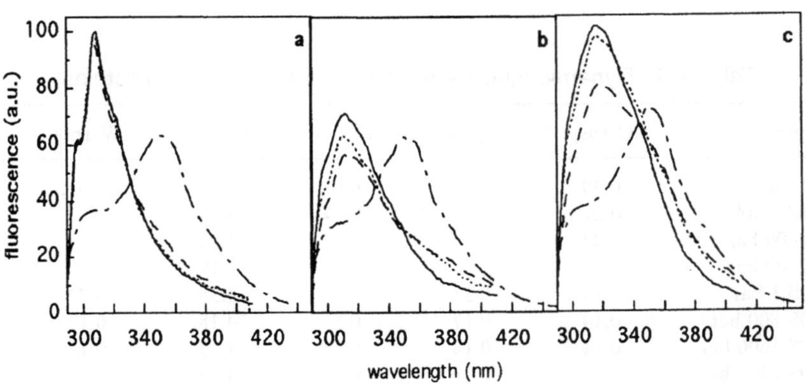


Figure 4.2. Relative steady-state fluorescence spectra of holo-wt (panel a), holo-I7S (panel b) and holo-F110S (panel c) at 1 bar (solid line), 2600 bar (dashed line) and 1 bar again (dotted line) after 2600 bar. The wt sample was normalized to 100. The spectra of the proteins unfolded by GdHCl 4 M (long dashes) have been reported for comparison, and reduced in size by a factor of 4.

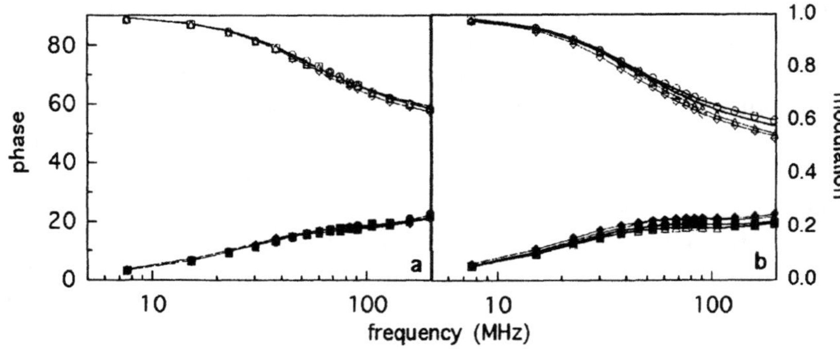


Figure 4.3. Phase shift and demodulation data of the holo-I7S (panel a) and holo-F110S (panel b) azurins. Phase (filled symbols) and modulation (open symbols) data are recorded at 1 bar (diamonds), 800 bar (triangles), 1600 bar (squares) and 2400 bar (circles).

This larger compressibility, which induces the thightening of F110S hydrophobic core, might be accounted by the empty space created by the substitution of the bulkier Phe with the smaller Ser, in agreement with the data obtained for other globular proteins.[33] Interestingly the fluorescence spectra of F110S and I7S collected at atmospheric pressure before and after compression at 2600 bar, show an evident hysteresis of the process (Figure 4.2).

In order to investigate the main characteristic of this persistent structural modification, we have compared the absorption and CD spectrum of

Table 4.4. Dynamic Fluorescence Parameters of Holo Proteins[a]

| Samples | $C_1$ (ns) | $W_1$ (ns) | $F_1$ (%) | $C_2$ (ns) | $W_2$ (ns) | $\chi^2$ |
|---|---|---|---|---|---|---|
| I7S 1 bar | 0.20 | — | 0.61 | 3.41 | — | 1.1 |
| I7S 800 bar | 0.21 | — | 0.64 | 3.34 | — | 0.9 |
| I7S 1600 bar | 0.22 | — | 0.62 | 3.29 | — | 1.2 |
| I7S 2400 bar | 0.23 | — | 0.63 | 3.18 | — | 1.3 |
| F110S 1 bar | 0.15 | 0.21 | 0.55 | 4.38 | 0.22 | 1.0 |
| F110S 800 bar | 0.14 | 0.17 | 0.55 | 4.18 | 0.19 | 1.2 |
| F110S 1600 bar | 0.14 | 0.14 | 0.58 | 4.13 | 0.18 | 1.0 |
| F110S 2400 bar | 0.16 | 0.09 | 0.60 | 4.02 | 0.06 | 0.9 |

[a]$C_{1,2}$—center of lorentzian distribution and/or discrete lifetime component ($\Delta C_1 \approx 20$ ps, $\Delta C_2 \approx 50$ ps).

$W_{1,2}$—width of lorentzian distribution ($\Delta W_{1,2} \approx 30$ ps).

$F_1$—fractional intensity relative to $C_1$ ($F_1 + F_2 = 1$; $\Delta F \approx 0.01$).

$\chi^2$—reduced chi-squared values.

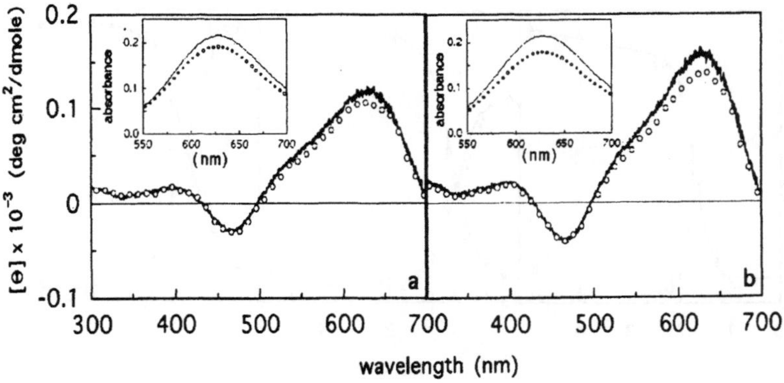


Figure 4.4. CD spectra of holo-I7S (panels a and b) and holo-F110S (panels c and d) at 1 bar before (solid lines) and after (circles) 2600 bar. Insets: absorption spectra of the same samples at 1 bar (solid lines) and after recovery of atmospheric pressure (circles).

the holo proteins at 1 bar with that recorded after returning back from high pressure. No difference at all was observed in the spectra between 200 and 300 nm (data not shown) demonstrating that both the secondary and the tertiary structure are fully recovered. However small changes occurred in the visible band (Figure 4.4) indicating some modification at the copper binding site. It should be noted that these changes cannot be ascribed to a denaturation process, nor simply to the loss of some copper from the protein molecules. In fact both unfolding and copper removal result in a large increase in the fluorescence intensity of holo-azurins[25,34] while the data obtained at high pressure or after returning at 1 bar show a lower florescence (Figure 4.2). Since the band at 627 nm has been assigned to a charge tranfer transition from Cys112 to copper,[5] the small permanent distortion of the ligand field induced by pressure could be attributed to an increased distance between the metal and the cysteine residue. It has been shown that Trp48 may be involved in one possible electron transfer pathway to copper,[35] so that detailed investigation on the correlation between the Cys112-copper distance and a decrease in the fluorescence quantum yield of mutants, could give in future new insights on the fluorescence quenching mechanism in azurin.

## 4.4. The Apo-proteins

The GdHCl-induced unfolding of copper-free azurin samples are reported in Figure 4.5.

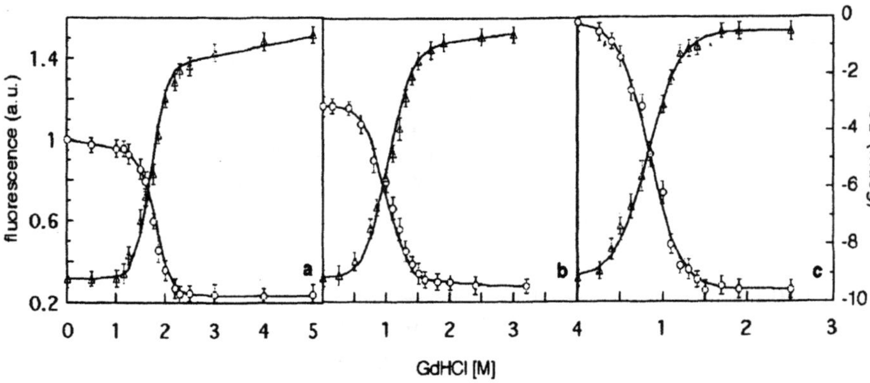


Figure 4.5. Dependence of the relative fluorescence intensity (circles) and circular dichroism signal (triangles) of apo-wt (panel a), apo-I7S (panel b) and apo-F110S (panel c) azurin. The quantum yield of the wt sample was normalized to 1. The same experimental conditions of the holo-proteins (Figure 4.1) were used.

As shown by the midpoint of the unfolding transition all apo-azurins are less stable than the respective holo-forms, confirming also for the two mutants the important structural role of copper found in the case of wt-azurin.[36] In particular a comparison among the unfolding parameters for the holo- and apo-forms (Table 4.3) allows to estimate the contribution of copper to the overall protein stability, which is of the order of $\Delta\Delta\, G_{holo\text{-}apo} \approx 2\text{–}3\,\text{kcal/mol}$. Despite this result, it is known that copper removal does not decrease the stability of azurin at high pressure.[32] Indeed, as reported in Figure 4.6a only negligible effects may be detected in the range 1–3 kbar for apo-wt. Instead the fluorescence spectra of the apo-mutants are considerably affected by hydrostatic pressure, showing a large quenching effect, associated to a shift of the center of mass towards longer wavelengths (Figures 4.6b and 6c). As previously reported[14,17] the fluorescence decay of apo-I7S and apo-F110S is more heterogeneous than the corresponding single lifetime of apo-wt. Dynamic fluorescence measurements demonstrate that this heterogeneity progressively increases from 1 to 2400 bar (Figure 4.7 and Table 4.5).

This finding, together with the steady-state fluorescence results, point out that, at variance with the holo-samples, the apo-mutants may be unfolded well below 3 kbar. In order to better characterize this effect, we have measured also the anisotropy decay of these samples as a function of hydrostatic pressure. As in the case of apo-wt (Table 4.2) interpolation of the phase and demodulation data collected at 1 bar yielded two rotational correlation times (Figure 4.8). The longer one, which is similar for the two

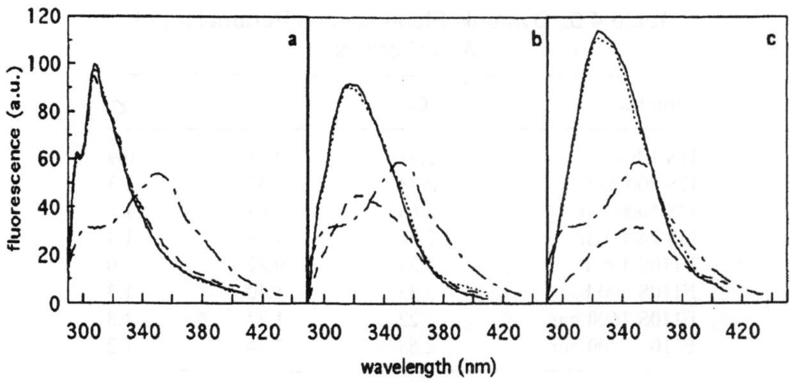


Figure 4.6. Relative steady-state fluorescence spectra of apo-wt (panel a), apo-I7S (panel b) and apo-F110S (panel c) at 1 bar (solid line), 2600 bar (dashed line) and 1 bar again (dotted line) after 2600 bar. The wt sample was normalized to 100. The spectra of the proteins unfolded by GdHCl 4 M (long dashes) have been reported for comparison, and enhanced in size by a factor of 4.

samples ($\Phi_2 \approx 6$ ns), is compatible with the rotational motion of the whole azurin molecule. The shorter component, $\Phi_1$, varied from 0.06 ns (apo-F110S) to 0.20 ns (apo-I7S) and may be therefore assigned to the local movement of the Trp 48 residue. These results are slightly different from the data already published[17] which indicated a partial loosening of the secondary and tertiary structure of both I7S and F110S upon copper removal. In contrast, here the smoother method of preparing the copper-free samples (see Section 2) allows the preservation of the native structure (same CD

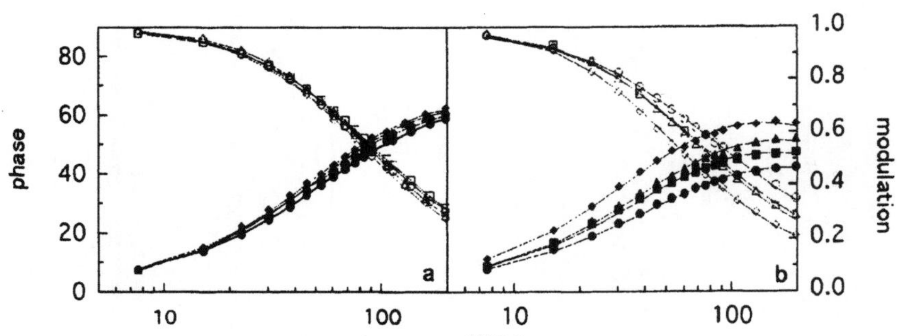


Figure 4.7. Phase shift and demodulation data of the apo-I7S (panel a) and apo-F110S (panel b) azurins. Phase (filled symbols) and modulation (open symbols) data are recorded at 1 bar (diamonds), 800 bar (triangles), 1600 bar (squares) and 2400 bar (circles).

Table 4.5. Dynamic Fluorescence Parameters of Apo-Proteins[a]

| Samples | $C_1$ | $W_1$ | $\chi^2$ |
|---|---|---|---|
| I7S 1 bar | 2.91 | 0.66 | 0.9 |
| I7S 800 bar | 2.78 | 0.53 | 1.3 |
| I7S 1600 bar | 2.68 | 1.03 | 1.1 |
| I7S 2400 bar | 2.51 | 1.18 | 1.3 |
| F110S 1 bar | 4.34 | 0.82 | 1.0 |
| F110S 800 bar | 3.35 | 1.40 | 1.3 |
| F110S 1600 bar | 3.22 | 1.77 | 1.3 |
| F110S 2400 bar | 2.58 | 2.34 | 1.2 |

[a]see Table 4.4.

spectrum and $\Phi_2$ value of holo-azurin). The rotational correlation lifetimes of apo-I7S and apo-F110S are dramatically affected by hydrostatic pressure. In particular, already at 1500 bar, an evident decrease of the $\Phi_1$ and $\Phi_2$ values (Figure 4.8) indicated that a faster dynamic is taking place. As shown in Figure 4.8, a fairly good reversibility is achieved, recovering the initial atmospheric pressure. This result demonstrate the larger flexibility of the apo-structures, while the presence of copper in the holo-proteins increases their stability, providing a stiffer tridimensional arrangement which in that case does not allow reversibility.

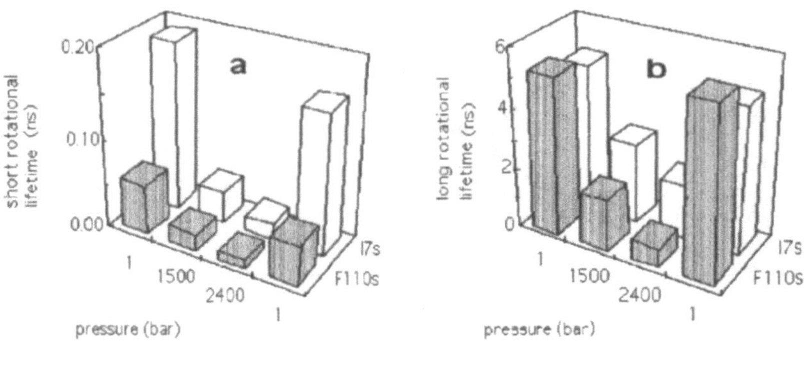


Figure 4.8. Rotational correlation lifetimes as a function of pressure of apo-I7S (light bars) and apo-F110S (dark bars). $\phi_1$ and $\phi_2$ represent the short (panel a) and the long (panel b) components.

## 4.5. Conclusions

The detailed knowledge of azurin structure and the new possibilities offered by site-directed mutagenesis make it a convenient model for studies on the stability of small globular proteins. In particular the importance of a very stable hydrophobic core for maintaining the native, biologically active conformation appears evident. The peculiar location of the single tryptophan, just at heart of this core, has two important consequences. First, the spectroscopic features of this tryptophan are similar to those of indole in non-aqueous solutions and very low temperatures, even though, as demonstrated by the anisotropy decay, it has a considerable freedom of rotation. Second, it represents a very useful, built-in probe not only of the native-denatured transition, but also of subtler modifications of the structure which may preceed its collapse toward a disordered state.

## References


1. S. R. Parr, D. Barber, C. Greenwood, B. W. Phillips, and J. Melling, A purification procedure for the soluble cytochrome oxidase and some other respiratory proteins from Pseudomonas aeruginosa, *Biochem. J. 157*, 423–430 (1976).
2. E. T. Adman, R. E. Stenkamp, L. C. Sieker, and L. H. Jensen, A crystallographic model for azurin at 3 Å resolution, *J. Mol. Biol. 123*, 35–47 (1978).
3. H. Nar, A. Messerschmidt, R. Huber, M. van de Kamp, and G. W. Canters, Crystal structure analysis of oxidized Pseudomonas aeruginosa azurin at pH 5.5 and pH 9.0, *J. Mol. Biol. 221*, 765–772 (1991).
4. M. van de Kamp, G. W. Canters, S. S. Wijmenga, A. Lommen, C. W. Hilbers, H. Nar, A. Messerschmidt, and R. Huber, Complete sequential $^{1}H$ and $^{15}N$ nuclear magneic resonance assignments and solution secondary structure of the blue copper protein azurin from pseudomonas aeruginosa, *Biochemistry 31*, 10194–10207 (1992).
5. E. I. Solomon, J. W. Hare, D. M. Dooley, J. H. Dawson, P. J. Stephens, and H. B. Gray, Spectroscopic studies of stellacyanin, plastocyanin and azurin. Electronic structure of the blue copper sites, *J. Am. Chem. Soc. 102*, 168–178 (1980).
6. T. J. Thamann, P. Frank, L. J. Willis, and T. M. Loehr, Normal coordinate analysis of the copper center of azurin and the assignment of its resonance raman spectrum, *Proc. Natl. Acad. Sci. 79*, 6396–6400 (1982).
7. E. W. Ainscough, A. G. Bingham, A. M. Brodie, W. R. Ellis, H. B. Gray, T. M. Loehr, J. E. Plowman, G. E. Norris, and E. N. Baker, Spectrochemical studies on the blue copper protein azurin from alcaligenes denitrificans, *Biochemistry 26*, 71–82 (1987).
8. Grinvald, J. Schlessinger, I. Pecht, and I. Z. Steinberg, Homogeneity and variability in the structure of azurin molecules studied by fluorscence decay and circular polarization, *Biochemistry 14*, 1921–1929 (1975).
9. Munro, I. Pecth, and L. Stryer, Subnanosecond motion of tryptophan residues in proteins, *Proc. Natl. Acad. Sci. 76*, 56–60 (1979).

10. G. Szabo, T. M. Stepanik, D. M. Wayner, and N. M. Young, Conformational heterogeneity of the copper binding site in azurin, A time-resolved fluorescence study, *Biophys. J. 41*, 233–242 (1983).
11. J. W. Petrich, J. W. Longworth, and G. R. Fleming, Internal motions and electron transfer in proteins: a picosecond fluorescence study of three homologous azurins, *Biochemistry 26*, 2711–2722 (1987).
12. M. Hutnik, and A. G. Szabo, Confirmation that multiexponential fluorescence decay behaviour of holoazurin originates from conformational heterogeneity, *Biochemistry 28*, 3923–3934 (1989).
13. J. E. Hansen, J. M. Longworth, and G. R. Fleming, Photophysics of metalloazurins, *Biochemistry 29*, 7329–7338 (1990).
14. G. Gilardi, G. Mei, N. Rosato, G. W. Canters, and A. Finazzi Agrò, Unique environment of Trp48 in pseudomonas aeruginosa azurin as probed by site-directed mutagenesis and dynamic fluorescence spectroscopy, *Biochemistry 33*, 1425–1432 (1994).
15. S. J. Kroes, G. W. Canters, G. Gilardi, A. van Hoek, and A. J. W. G. Visser, Time resolved fluorescence study of azurin variants: conformational heterogeneity and tryptophan mobility, *Biophys. J. 75*, 2441–2450 (1998).
16. J. A. Sweeney, P. A. Harmon, S. A. Asher, C. M. Hutnik, and A. G. Szabo, UV resonance raman examination of the azurin tryptophan environment and energy relaxation pathways, *J. Am. Chem. Soc. 113*, 7531–7537 (1991).
17. G. Mei, G. Gilardi, M. Venanzi, N. Rosato, G. W. Canters, and A. Finazzi Agrò, Probing the structure and mobility of Pseudomonas aeruginosa azurin by circular dichroism and dynamic fluorescence anisotropy, *Protein Sci. 5*, 2248–2254 (1996).
18. G. Lipari, and A. Szabo, Effect of librational motion on fluorescence depolarization and nuclear magnetic resonance relaxation in macromolecules and membranes, *Biophys J. 30*, 489–506 (1980).
19. Ehrenstein, and G. U. Nienhaus, Conformational substates in azurin, *Proc. Natl. Acad. Sci. 89*, 9681–9685 (1992).
20. Hammann, A. Messerschmidt, R. Huber, H. Nar, G. Gilardi, and G. W. Canters, X-ray crystal structure of the two site-specific mutants Ile7Ser and Phe110Ser of azurin from pseudomonas aeruginosa, *J. Mol. Biol. 255*, 362–366 (1996).
21. Paladini, and G. Weber, Absolute measurements of fluorescence polarization at high pressure, *Rev. Sci. Instrum. 52*, 419–427 (1981).
22. J. R. Lakowicz, and I. Gryczynski, Frequency-domain fluorescence spectroscopy, in: *Topics in fluorescence spectroscopy* (J. R. Lakowicz, ed.), Vol. 1, pp. 293–335, Plenum Press, New York (1991).
23. J. M. Beechem, and E. Gratton, Fluorescence spectroscopy data analysis environment: a second generation global analysis program, *Proc. SPIE-Int. Soc. Opt. Eng. 909*, 70–81 (1988).
24. M. Straume, S. G. Frasier-Cadoret, and M. L. Johnson, Least-squares analysis of fluorescence data, in: *Topics in fluorescence spectroscopy* (J. R. Lakowicz, ed.), Vol. 2, pp. 177–240, Plenum Press, New York (1991).
25. A. Finazzi Agrò, G. Rotilio, L. Avigliano, P. Guerrieri, V. Boffi, and B. Mondovi', Environment of copper in pseudomonas fluorescens azurin: fluorimetric approach, *Biochemistry 9*, 2009–2014 (1970).
26. N. Pace, B. A. Shirley, and J. A. Thomson, Measuring the conformational stability of a protein, in: *Protein Structure, A Practical Approach* (T. E. Creighton, ed.), pp. 311–330, IRL Press, New York (1989).
27. J. R. Desjarlais, and T. M. Handel, De novo design of the hydrophobic cores of proteins, *Protein Sci. 4*, 2006–2018 (1995).

28. M. Munson, S. Balasubramanian, K. G. Fleming, A. D. Nagi, R. O'Brian, J. M. Sturtevant, and L. Regan, What makes a protein a protein? Hydrophobic core designs that specify stability and structural properties, *Protein Sci. 5*, 1584–1593 (1996).
29. Dahiyat, and S. L. Mayo, Probing the role of packing specificity in protein design, *Proc. Natl. Acad. Sci. USA 94*, 10172–10177 (1997).
30. E. Eriksson, W. A. Baas, X. J. Zhang, D. W. Heinz, M. Blaber, E. P. Baldwin, and B. W. Matthews, Response of a protein structure to cavity-creating mutations and its relation to the hydrophobic effect, *Science 255*, 178–183 (1992).
31. M. Gross, and R. Jaenicke, Proteins under pressure. The influence of high hydrostatic pressure on structure, function and assembly of proteins and protein complexes, *Eur. J. Biochem. 221*, 617–630 (1994).
32. P. Cioni, and G. B. Strambini, Pressure effects on protein flexibility monomeric protein, *J. Mol. Biol. 242*, 291–301 (1994).
33. K. Gekko, and Y. Hasegawa, Compressibility-structure relationship of globular proteins, *Biochemistry 25*, 6563–6571 (1986).
34. P. Guptasarma, Resolving multiple protein conformers in equilibrium unfolding reactions: a time-resolved emission spectroscopic (TRES) study of azurin, *Biophys. Chem. 65*, 221–228 (1996).
35. O. Farver, L. K. Skov, G. Gilardi, G. van Pouderoyen, G. W. Canters, S. Wherland, and I. Pecht, Structure-function correlation of intramolecular electron transfer in wild type and single-site mutated azurins, *Chem. Phys. 204*, 271–277 (1996).
36. J. Leckner, N. Bonander, P. Wittung-Staffshede, B. G. Malmström, and B. G. Karlsson, The effect of the metal ion on the folding energetics of azurin: a comparison of the native, zinc and apoprotein, *Biochim. Biophys. Acta 1342*, 19–27 (1997).

# 5

# Barnase: Fluorescence Analysis of A Three Tryptophan Protein

Yves Engelborghs and Alan Fersht

## 5.1. Introduction

Barnase is an extracellular ribonuclease that is produced by the prokaryote *Bacillus amyloliquefaciens*. It is a small (110 residues, Mw = 12382) single domain enzyme, the structure of which is characterized by a twisted, five stranded antiparallel $\beta$-sheet and two $\alpha$-helices, the first of which packs against the $\beta$-sheet.[1] It is an enzyme that has been extensively used as a model for studying the principles that rule protein stability and protein folding,[2,3] as well as protein-protein interactions,[4,5,6] substrate binding,[7,8] and electrostatics.[9,10] In many of these studies the fluorescence of the protein is used as a tool. The protein fluorescence is governed by the contributions of the tryptophan residues, especially when the protein is excited at 295 nm. In barnase, three tryptophan residues are present and are found at positions 35, 71 and 94 (Figure 5.1). W35 is near the C-terminal end of the second $\alpha$-helix and relatively far away from the other two (22–25 Å) tryptophan residues. W71 is located in a hydrophobic region at the beginning of the second strand of the $\beta$-sheet and only 11 Å away from W94. W94 is situated at the beginning of the fourth strand of the $\beta$-sheet and is in close contact with the imidazole ring of H18, that lies at the C-terminal end of the first $\alpha$-helix. Tryptophan residues 35 and 71 are almost completely shielded from the solvent, while W94 shows a pronounced exposure. The close contact between H18 and W94 explains the pH-dependence of the protein fluorescence, since protonated His is known to be a fluorescence quencher.[11,12] The short distance between W94 and W71 suggests the possibility of energy transfer.

*Yves Engelborghs* • Laboratory of Biomolecular Dynamics, University of Leuven, Celestijnenlaan 200D, B-3001 Heverlee, Belgium. *Alan Fersht* • Cambridge Center for Protein Engineering, Cambridge University, Lensfield Road, Cambridge CB2 1EW, United Kingdom.

*Topics in Fluorescence Spectroscopy, Volume 6: Protein Fluorescence*, edited by Joseph R. Lakowicz. Kluwer Academic / Plenum Publishers, New York, 2000

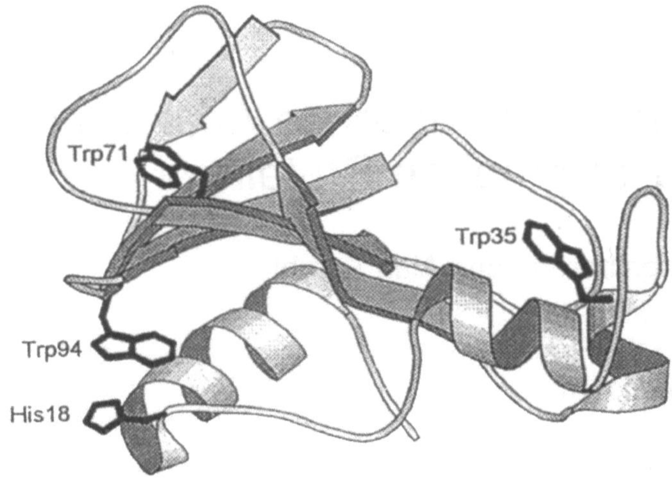


Figure 5.1. Schematic representation of the structure of barnase showing the positions of the three tryptopan residues and His 18.

A lot of information about the fluorescence properties of individual tryptophan residues can be obtained by the method of subtraction: the tryptophan is removed by site directed mutagenesis, and a fluorescence difference spectrum is determined between the spectrum of the WT and the mutant protein. In a similar way lifetimes can be assigned, provided that the removal of a tryptophan residue results in the clear cut disappearance of one lifetime component. This technique was extensively applied to the study of the fluorescence of barnase.[13,14]

It is clear that the method of subtraction has its limitations: only lifetimes that disappear are unambiguously attributed to the Trp-residue that has been removed. However, it can not be excluded that the removed Trp has more lifetimes, in common with and therefore masked by the remaining Trp-residues.

Therefore one-tryptophan-containing mutants were also produced and their steady-state and time-resolved fluorescence and phosphorescence parameters were analysed in order to explore in more detail the luminescence properties of the individual tryptophan residues.[15] The experimental results obtained in this way were compared with the results previously calculated by subtraction.

To probe the mobility of the tryptophan environment, fluorescence anisotropy measurements were also performed. In addition to this, the room-temperature phosphorescence properties of barnase were examined, as a probe for local structure and dynamics. In the concentration dependence of both the fluorescence anisotropy and the phosphorescence, indications for protein-protein interactions were found.

## 5.2. Results Obtained by the Method of Subtraction

### 5.2.1. pH-Dependency of the Fluorescence

In a first series of studies the fluorescence properties of wild-type barnase and of single tryptophan mutants (W35F, W71Y and W94F) were determined.[13] The tryptophan residues were replaced by the amino acids phenylalanine and tyrosine that do not contribute to fluorescence when excited at 295 nm. In order to probe the role of H18 additional mutants were made: H18G and W94L.

As expected from the analysis of the structure, showing the vicinity of W94 and H18, the fluorescence of the wild-type showed strong pH-dependency: the fluorescence is quenched especially at low pH. The pH-dependency fits perfectly the Henderson-Hasselbalch equation and a pKa of $7.75 \pm 0.02$ was calculated.[13] The same type of curve and the same pKa values were found for the mutant proteins W35F and W71Y. However, the fluorescence of the mutants W94F, W94L and H18G was pH-independent. These results clearly indicate that H18 in its protonated state is responsible for the quenching of W94 and therefore for the pH-dependence of the protein fluorescence. It is interesting to note that the titration curve of barnase can be fitted with the Henderson-Hasselbalch equation for a single acid, while one would expect to have to use the Linderstrom-Lang equation taking the overall charge of the protein into account. The pH-dependent fluorescence change linked to the ionisation of H18 can be used very fruitfully to study electrostatic interactions in proteins.[9] Unfortunately, electrostatic effects at the active site have to be studied on the basis of the pH-dependence of catalysis because the active site is too far away from any of the three tryptophan residues.[10]

### 5.2.2. The Effect of Removing W35

The fluorescence spectra of the different proteins were obtained at low (5.5) and high pH (9.4). At these pH values (using the pKa of 7.75) the protonation of H18 is 95% and 5% respectively. The properties of the individual tryptophan residues can be obtained by subtraction. When W35 is mutated the fluorescence decreases by 70% at low pH and 45% at high pH when compared to the wild-type protein at identical concentration. This shows that W35 is the major contributor to the protein fluorescence in both pH regions. Mutation of W35 results in a red shift relative to the wild-type protein at high pH. This proves that the spectrum of W35 is responsible for the more blue emission of the wild-type protein, and that the other tryptophan residues have a more red-shifted emission.

### 5.2.3. The Effect of Removing W71

When W71 is mutated to Tyr there is only a small decrease of about 20% of the fluorescence intensity both at low and high pH. Therefore W71 contributes the least to the emission intensity of the wild-type protein. This is also suggested by the fact that this mutation is not accompanied by a shift of the wavelength of maximum emission. On mutation of W71 there was probably no change in the environment of the other tryptophan residues. This is proven for W94, since the pKa of H18 did not change in mutant protein W71Y. W71 is strongly buried and therefore its low fluorescence intensity is puzzling. The reason for its small contribution to the fluorescence of the wild-type protein is probably energy transfer to W94 (see below).

### 5.2.4. The Effect of Removing W94

The fluorescence intensity of mutant proteins W94F and W94L is higher than the intensity of the wild-type protein. This indicates either that W94 in wild-type protein behaves as a sink for the fluorescence of the other residues or, alternatively, that there is a change in the environment of the other tryptophan residues on mutation at position 94. This latter hypothesis is unlikely since the two mutant proteins, one with a rather conservative mutation (W94F) and one with a much less conservative mutation (W95L) show the same fluorescence properties. Furthermore the two proteins remain fully active. The more plausible explanation is therefore that W94, which is itself rather strongly quenched by H18, is a sink of fluorescence energy by transfer from the other tryptophans. The emission spectra of the buried residues W71 and W35 (as indicated by the blue shift of the mutants W94F and W94L) is blue-shifted relative to W94. The blue-shifted emission spectra of W35 and W71 provide an effective overlap with the absorption spectrum of W94 in the UV-region. The mutant H18G shows a 150% increase of fluorescence intensity at low pH (55% at high pH) relative to the wild-type protein at the same pH-values. This indicates the quenching effect of H18, especially in the protonated state. Finally the higher exposure to the solvent of W94 in the mutant H18G and its increased contribution to the fluorescence spectrum relative to the other tryptophan residues explain why this mutant exhibits the most red shifted spectrum of all.

### 5.2.5. Calculation of the Absorption and Fluorescence Emission Spectra of the Individual Tryptophans

The calculation of the absorption and emission spectra of the individual tryptophan residues by subtraction, has to be done very carefully, taking into account the possibility of energy transfer, in this case between W71 and W94. The absorption spectrum of W35 was obtained by subtracting the spectrum of the mutant W35F from that of the wild-type protein (the contribution of F35 was neglected). In this way energy transfer between W71 and W94 was present in both proteins. The spectrum of W71 was obtained by subtracting the spectrum of W35 from that of mutant W94F. The spectrum of W94 was obtained by subtracting the spectrum of W35 from that of the mutant W71Y, and after correction for the contribution of Y.[13]

The calculated absorption spectra of the three tryptophan residues in barnase show the typical three peak structure of Trp absorption spectra.[16] The spectrum of W94 is red-shifted with respect to the spectra of the two other tryptophan residues.

The emission spectra of W71 and W94 (Figure 5.2) have been calculated from the emission spectra of proteins in which only one of the two residues was present and, therefore, represent the fluorescence emission spectra of these tryptophan residues in the absence of energy-transfer between them (see below). W71 shows the highest quantum yield and the most blue shifted emission spectrum of the three tryptophan residues. W94 shows the lowest quantum yield and the most red-shifted spectrum at both low and high pH. Both the quantum yield and the wavelengths of maximum emission of W71 and W35 are practically pH-independent. This does not apply to W94 which shows, at low pH, a lower quantum yield and a less red-shifted spectrum than at high pH.

The spectral properties of the three tryptophan residues can be rationalized in terms of their environment in the barnase molecule. W35 and W71 are buried residues. The solvent accessible areas for the indole rings are 10 and 7 Å$^2$ respectively. W94 is a more exposed residue (58 Å$^2$ of exposed area).[14] Accordingly, the maxima of emission of the three tryptophan residues (-ET) is progressively shifted to the red following the increase in solvent exposed area in the series W71, W35, W94 (Figure 5.2).

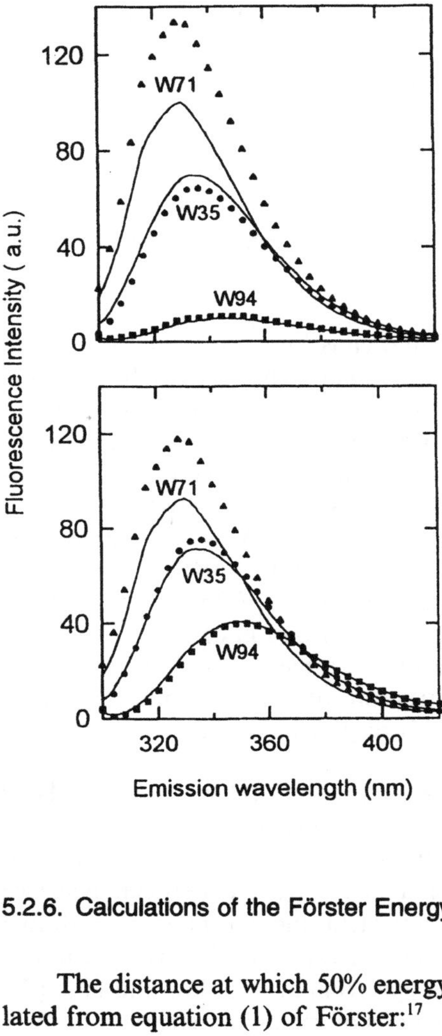


Figure 5.2. Fluorescence emission spectra of the one-tryptophan-containing mutants at low pH (top) and high pH (bottom). Spectra are recorded at the same protein concentration of 20 μM (lines), or calculated from the spectra of WT and the two-tryptophan-containing mutants (symbols: W35 (●), W71 (▲), W94(■); see text).

### 5.2.6. Calculations of the Förster Energy-Transfer on the Basis of Spectral Data

The distance at which 50% energy-transfer occurs ($R_0$ in cm) was calculated from equation (1) of Förster:[17]

$$R_0^{\,6} = 8.8 \times 10^{-25} \times \left(J_{AD}.n^{-4}.\kappa^2.\phi_D\right)^6 \tag{5.1}$$

Where $J_{AD}$ (in $cm^6 mmole^{-1}$) is the overlap integral, calculated from the absorption and fluorescence emission spectra of the individual tryptophan groups according to the classical equation. The refractive index of the medium (n) was taken as 1.5.[14] The geometric orientation factor ($\kappa$) has been calculated from equation (5.2):

$$\kappa^2 = \left[\cos\theta_T - 3.\cos\theta_D.\cos\theta_A\right]^2 \tag{5.2}$$

where $\theta_T$ is the angle between the emission dipole of the donor and the absorption dipole of the acceptor, $\theta_D$ and $\theta_A$ are the angle between these dipoles and the vector joining the midpoints of the CE2/CD2 bond of the donor and the acceptor respectively. Indole has two excited states termed $^1L_a$ and $^1L_b$ as shown by Valeur and Weber.[16] Since the absorption of indole in the region of 295 nm, where overlap with the emission spectra occurs, is mainly due to the $^1L_a$ state, the $^1L_b$ state was ignored in the calculation of the geometric orientation factors. The direction of transition moment of the $^1L_a$ state was defined, as the line linking NE1 and a point one-fifth of the distance along the bond between the midpoints CE3 and CZ3.[14] All the angles were calculated on the basis of the X-ray structure.[1]

The fluorescence quantum yield of the donor in the absence of acceptor $\phi_D$, was calculated from the determined lifetimes of the donors (in the absence of energy-transfer) and the average natural lifetime of 24 ± 8 ns, obtained from 15 known pairs of quantum yields and lifetimes for tryptophan.[18] Finally, the efficiency of energy-transfer ($E^a$) was calculated from equation (5.3):

$$E^a = \frac{R_0^{\ 6}}{R_0^{\ 6} + r^6} \tag{5.3}$$

The values of r were obtained from the X-ray structure.[1] Upper and lower limits of the overlap integrals were calculated by assuming an absolute error of ±1% in the molar absorptivity at the maximum in the absorbance spectra (Table 5.1). The transfer efficiencies ($E^a$) were calculated using these overlap integrals and the other parameters shown in Table 5.1.

Our results indicate that there is energy-transfer between W71 and W94. This energy-transfer process occurs in both directions, though it is greater from W71 to W94. The calculated one way energy-transfer efficiencies are similar at high and low pH, except for the reverse transfer from W94 to W71 which is of lower efficiency at low pH. W35, however, is a lone tryptophan residue, not involved in energy-transfer with the other two tryptophans (the transfer efficiencies $E^b$ were calculated from the lifetimes and are discussed further on).

### 5.2.7. The Fluorescence Lifetimes

#### *5.2.7.1. Measured and Calculated Lifetimes*

The fluorescence lifetimes of the different proteins were determined by automatic multi frequency phase fluorometry.[19] For WT barnase a triple

Table 5.1. Calculated Distances (Å), Overlap Integrals $J_{AD}$ ($\times 10^{-16}$ cm$^6$ mmol$^{-1}$), Donor Quantum Yields ($Q_D$) $R_0$ Values, and Calculated Transfer Energies (a) Based on Spectral Data, and (b) Based on Lifetime Data., at Low (c) and High (d) pH

| | Tryptophan | Pairs | | |
|---|---|---|---|---|
| | 35 → 71 | 35 → 94 | 71 → 94 | 94 → 71 |
| Dist. (Å) | 22.4 | 24.6 | 10.8 | 10.8 |
| $\kappa^2$ | 0.17 | 1.21 | 1.73 | 1.73 |
| $J_{AD}$ (c) | 0.2–0.98 | 0.5–1.2 | 0.6–1.5 | 0.004–0.6 |
| $Q_D$ (c) | 0.18 | 0.18 | 0.2 | 0.034 |
| $R_0$(Å) (c) | 6.9–8.9 | 11–13 | 12.5–14.5 | 4.4–9.3 |
| $E^a$(%) (c) | 0.09–0.4 | 0.5–2.0 | 70–85 | 0.5–29 |
| $E^b$ (%) (c) | — | — | 86 ± 2 | 4 ± 2 |
| $J_{AD}$ (d) | 0.2–0.98 | 0.5–1.2 | 0.7–1.6 | 0.06–0.6 |
| $Q_D$ (d) | 0.18 | 0.18 | 0.2 | 0.065 |
| $R_0$ (Å) (d) | 6.9–8.9 | 11–13 | 13–15 | 7–10 |
| $E^a$ (%) (d) | 0.09–0.4 | 0.5–2.0 | 73–86 | 7–41 |
| $E^b$ (%) (d) | — | — | 71 ± 2 | 36 ± 2 |

exponential decay fits best to the frequency dependence of all the phase measurements.[14] The calculated theoretical curves follow closely the experimentally measured phase shifts. The weighted residuals do not show any systematic deviation. The autocorrelation function falls quickly to zero and remains close to it. However, the standard error estimates on the lifetimes are rather large. Therefore a global analysis was performed on the emission data, giving good values for the reduced chi square and a better definition of the parameters.[20] Table 5.2 shows the measured lifetimes and amplitude fractions.

Although the time-dependent fluorescence emission of the wild-type enzyme can be described by a sum of three exponentials, they cannot be assigned to the different residues without reference to the mutant proteins. The assignment is further complicated by the presence of two way energy transfer, since both residues contribute to the two lifetimes, as shown in the model of Porter.[21]

The calculated lifetimes and amplitudes of the mutant proteins W35F, W94F and W71Y, at both low and high pH, are also shown in Table 5.2. W94 F could be fitted with one exponential decay. By looking for lifetime components that disappear upon the removal of a tryptophan residue, lifetimes were assigned to single residues, as shown in Table 5.2.

Table 5.2.A. Lifetimes and Amplitude Fractions ($a_i$) Observed at Low pH and 25 °C. # Lifetime Estimated by Subtraction, $\langle\tau\rangle$ Amplitude Average Lifetime

| | $\tau_1$ ($a_1$) | $\tau_2$ ($a_2$) | $\tau_3$ ($a_3$) | $\langle\tau\rangle$ |
|---|---|---|---|---|
| WT | 4.48 (0.27) | 0.89 (0.58) | 0.50 (0.14) | 1.8 |
| W35F | — | 0.89 (0.61) | 0.65 (0.39) | 0.8 |
| W71Y | 4.34 (0.40) | 0.82 (0.60) | — | 2.2 |
| W94F | 4.7 (1.0) | — | — | 4.7 |
| W35 # | 4.34–4.7 (1.0) | — | — | 4.3–4.7 |
| W71 (-ET)# | 4.7 (1.0) | — | — | 4.7 |
| W94 (-ET)# | — | 0.82 (1.0) | — | 0.82 |

Table 5.2.B. Lifetimes and Amplitude Fractions ($a_i$) Observed at High pH and 25 °C. # Lifetime Estimated by Subtraction, $\langle\tau\rangle$ Amplitude Average Lifetime

| | $\tau_1$ ($a_1$) | $\tau_2$ ($a_1$) | $\tau_3$ ($a_3$) | $\langle\tau\rangle$ |
|---|---|---|---|---|
| WT | 4.79 (0.32) | 2.44 (0.43) | 0.77 (0.45) | 2.93 |
| W35F | 5.05 (0.12) | 2.42 (0.42) | 0.74 (0.45) | 1.95 |
| W71Y | 4.48 (0.58) | 1.57 (0.41) | — | 3.24 |
| W94F | 4.73 (1.0) | — | — | 4.73 |
| W35 # | 4.48–4.79 (1.0) | — | — | 4.48–4.79 |
| W71 (-ET)# | 4.73 (1.0) | — | — | 4.73 |
| W94 (-ET)# | — | 1.57 (1.0) | — | 1.57 |

### *5.2.7.2. Energy Transfer Calculations Using Lifetime Data*

When the acceptor is a fluorescent group identical or related to the donor, reverse transfer can occur. Porter[21] worked out the coupled differential equations for this system and showed that the fluorescence decay is described by two lifetimes to which both the donor and the acceptor contribute. A similar calculation was done by Woolley *et al.*[22] for intramolecular two way energy transfer. The efficiencies of energy transfer in both directions can be calculated from the lifetimes in the presence and absence of energy transfer, using the formulae (5.4) and (5.5) from Porter:[21]

$$k_{12} = \frac{(\lambda_1 - k_1)(k_1 - \lambda_2)}{(k_1 - k_2)} \tag{5.4}$$

and $$k_{21} = \frac{(\lambda_1 - k_2)(\lambda_2 - k_2)}{(k_1 - k_2)} \tag{5.5}$$

where $\lambda_1$ ($\lambda_2$) is the inverse of the shortest (longest) lifetime observed in the presence of two way energy transfer, $k_1$ and $k_2$ are the inverse lifetimes obtained in the absence of energy transfer, $k_{12}$ and $k_{21}$ are the rate constants for forward and backward energy transfer. Efficiencies can then be calculated as follows:

$$E_1^{\,b} = k_{12}/(k_1 + k_{12}) \quad \text{and} \quad E_2^{\,b} = k_{21}/(k_2 + k_{21}) \tag{5.6}$$

Whether $k_1$ is assigned the largest or smallest value, the calculated one way efficiency is always the corresponding one. The calculated efficiencies will not be substantially different for the new data, if average lifetimes are taken for the single tryptophan residues.

### 5.2.8. Discussion of Data Obtained From Single Tryptophan Mutants

Energy-transfer between W71 and W94 is favoured by the close distance of the two residues and their relative orientation ($\kappa^2$) and is suggested to occur from the steady -state emission spectra of wild-type barnase and mutant proteins. The lifetimes for the pair were determined independently from two proteins: W35F and the wild-type protein. At low pH, the 0.89 ns and 0.65 ns lifetimes of the mutant W35F were assigned to the energy-transfer couple W71/W94, since the same two lifetimes are recognized in the data obtained for wild-type protein. At high pH the corresponding lifetimes are 2.42 ns and 0.74 ns. At high pH an additional lifetime of 5.05 ns appears in the mutant W35F and cannot be unambiguously assigned to a single residue. It could originate from W71, W94 or both and could arise from a fraction of the protein in a conformation locally ordered in such a way as to prevent energy-transfer.

The lifetimes of the two residues, when not involved in energy-transfer, i.e. W71(-ET) and W94 (-ET), were determined from the mutant W94F and W71Y respectively. W71 (-ET) has a long lifetime of 4.7 ns at low pH and 4.73 at high pH. W94 (-ET) has a short lifetime of 0.82 ns at low pH and 1.57 at high pH.

The W71/W94 couple was analysed according to Porter.[21] The energy-transfer in both directions can be estimated, on the basis of Eqs. (5.4)–(5.6) using the empirically determined lifetimes (without energy-transfer) for W71(-ET) (4.7 ns) and for W94(-ET) (1.57 ns at high pH and 0.82 ns at low pH) and the lifetimes observed in the wild-type. The calculated values are 71% for the forward transfer (from W71 to W94) and 36% for reverse transfer at high pH and 86% for the forward transfer and 4% for the reverse transfer at low pH (all at ±2%). These values lie within or close to the limits calculated from the spectra (Table 5.1).

The lifetime of W35 was determined from wild-type barnase and the mutants W71Y and W94F. At low pH the lifetime of 4.48 ns in the wild-type protein is attributed to W35 since the two other lifetimes have already been assigned to the two other tryptophans. A similar value of 4.34 ns in the mutant W71F can be attributed to W35. The corresponding value for W35 at high pH is 4.79 ns. W35, as expected, behaves as a lone tryptophan; mutation of the other two tryptophan residues hardly alters its lifetime, nor does mutation of W35 alter the lifetime of the two other tryptophan residues.

The lifetimes of W35 and W71 (when this residue is not involved in energy-tansfer) of about 4–5 ns are within the range of lifetimes observed for tryptophan residues in proteins.[18] The lifetime of W94 is shorter, 0.8–1.6 ns, and is dependent on pH, indicating that this is a strongly quenched residue. The trajectory of a 120 ps molecular dynamics simulation of barnase in water shows that W94 and H18 are often in close contact.[14] The observations that the lifetime of W94 is halved at low pH while the lifetimes of the other two tryptophan residues are hardly changed strongly suggest that H18 is responsible for the short lifetime of W94. This is similar to other systems in which indole fluorescence is quenched by a neighbouring histidine in a pH dependent way.[23]

Despite the fact that W35 and W71 (-ET) have a very similar lifetime, the fluorescence intensity of W71 (-ET) is much higher than that of W35, indicating that the latter may be decreased by static quenching.

## 5.3. Characterization of the Double Mutant Protein

### 5.3.1. Steady-State Fluorescence Parameters

The emission of the individual tryptophan residues as calculated by subtraction is compared with the experimentally observed fluorescence emission of single tryptophan containing mutants (Figure 5.2). Apart from the fluorescence intensity of W71, the calculated curves coincide with the measured curves with regard to the position of the maximum of the emission wavelength and the fluorescence intensity. The deviation of the spectrum of W71 will be discussed further on. The spectral properties of the individual tryptophans are reflected in the other proteins. W35 and W71 display blue-shifted emission (330–335 nm), the wavelength position and the intensity of the maximum being essentially unaffected by pH. W94 is characterized by a pronounced red-shifted emission, which is pH-dependent. Upon increasing the pH from 5.8 to 8.9, a red-shift of 5 nm is accompanied with a fourfold fluorescence increase.

### 5.3.2. Fluorescence Lifetimes

Since the first lifetime studies[14] on barnase, the phase fluorimeter has been improved. The dye laser has been replaced by a solid state laser (Tsunami, Spectra Physics) which considerably improves the stability of the system and reduces the noice of the phase measurements. Also the bandwidth of the detection system has been increased and we are currently measuring phases up to 1 GHz. Moreover, the theory of the analysis of multi tryptophan proteins has been more elaborated.[24]

The lifetimes of WT barnase were measured again and an excellent agreement is obtained at low pH.[15] At high pH an additional component of 1% amplitude of 9.53 ns was observed. The phase data for the single tryptophan mutants had to be fitted with a sum of three exponentials, indicating that for these proteins additional components (compared to the previous measurements) were resolved but again with very small amplitudes.

The fluorescence decay parameters of the single tryptophan residues were determined at pH 5.8 and pH 8.9, and at emission wavelengths ranging from 330 nm to 380 nm with 10 nm intervals. The data reported in Table 5.3 are the result of a global analysis of the measurements at these wavelengths. At low pH, the best fit (lowest $\chi^2_R$) was obtained when using a triple-exponential decay. The only exception to this is W35, which displays two

**Table 5.3.A.** Measured and Predicted Lifetime(s) for the Single Tryptophan Residues at Low pH

| | predicted | $\langle\tau\rangle$ | $\tau_1$ ($a_1$) | $\tau_2$ ($a_2$) | $\tau_3$ ($a_3$) |
|---|---|---|---|---|---|
| W35 | 4.34–4.7 | 4.5 | 4.54 (0.99) | 0.88 (0.01) | — |
| W71 | 4.7 | 5.1 | 5.06 (0.97) | 2.52 (0.06) | 0.40 (−0.03) |
| W94 | 0.82 | 0.6 | 4.39 (0.01) | 0.78 (0.59) | 0.23 (0.39) |

**Table 5.3.B.** Measured and Predicted Lifetime(s) for the Single Tryptophan Residues at High pH

| | predicted | $\langle\tau\rangle$ | $\tau_1$ ($a_1$) | $\tau_2$ ($a_2$) | $\tau_3$ ($a_3$) | $\tau_4$ ($a_4$) |
|---|---|---|---|---|---|---|
| W35 | 4.48–4.8 | 4.7 | 4.70 (1.0) | — | — | — |
| W71 | 4.73 | 4.9 | 4.99 (0.98) | 1.92 (0.015) | 0.22 (0.05) | — |
| W94 | 1.57 | 2.9 | 3.93 (0.48) | 1.91 (0.32) | 0.38 (0.16) | 7.51 (0.04) |

lifetimes. At high pH, a small contribution of a long lifetime emerges in all mutants containing W94. Although amplitude fractions at 340 nm for $\tau_1$ are sometimes very small, the corresponding lifetimes make a considerable improvement on the fittings.

The lifetimes for the single tryptophan residues calculated by subtraction correspond very good with the major component of the direct measurements and with the average lifetime (except for W94 at high pH).

For W94 rising the pH causes not only the appearance of an extra lifetime and an increase in quantum yield, but also the lengthening of the lifetimes $\tau_2$ and $\tau_3$, and a red-shift (~5 nm) of the emission maxima of the DAS spectra. If we interpret the decrease of $\tau_2$ and $\tau_3$ upon decreasing the pH as due to quenching by H18, an intramolecular collisional frequency of 5 to $2\,ns^{-1}$ can be calculated, using a quenching efficiency of 0.32 for Trp-His quenching previously determined.[12] This reflects the internal dynamics of the protein. The pH-dependent changes in the fluorescence parameters of the isolated W94 are transferred to the multiple-tryptophan proteins.

### 5.3.3. Calculation of the Fluorescence Decay Parameters of Multi-Tryptophan Proteins from the Emission of Single-Tryptophan Proteins

The lifetimes $\langle\tau_i\rangle$ and amplitude fractions $\langle a_i\rangle$ of multi-tryptophan proteins can be calculated from the linear combination of the lifetimes and amplitude fractions of the individual emitting tryptophans or tryptophan pairs by making combinations within lifetime groups (short, middle, long) using the following equations:[24]

$$\langle\tau_i\rangle = \frac{\sum_j \frac{a_{ij}\varepsilon_j}{\langle\tau_R\rangle_j}\tau_{ij}}{\sum_j \frac{a_{ij}\varepsilon_j}{\langle\tau_R\rangle_j}} \tag{5.7}$$

$$\langle a_i\rangle = \frac{\sum_j \frac{a_{ij}\varepsilon_j}{\langle\tau_R\rangle_j}}{\sum_i \sum_j \frac{a_{ij}\varepsilon_j}{\langle\tau_R\rangle_j}} \tag{5.8}$$

where i is the index of the lifetime group (short, middle, long = 1, 2, 3) and j is a single tryptophan or a tryptophan pair. The pre-exponential factors are weighted by $\varepsilon/\langle\tau_R\rangle$, with $\varepsilon$ being the absorbance of the respective tryptophan and $\langle\tau_R\rangle$ its radiative lifetime.

When applied to barnase, the overall lifetime of the WT protein cannot be calculated by the summation of the lifetimes of the individual trptophan residues.[15] However, combining the data of W35 with the data of the single mutant W35F gives a nice fit, indicating again energy transfer within the W71/W94 pair. Also the average lifetime of the single mutant W94F can be calculated by combining data from W35 and W71 indicating the absence of interactions among these residues.

## 5.4. Fluorescence Anisotropy

Time-resolved fluorescence anisotropy was obtained by differential phase measurements performed on the one-tryptophan-containing mutants in order to gain information on the mobility of the tryptophan environments.[25] The best fits were obtained with a double-exponential decay giving two rotational correlation times.[15] For WT barnase and all mutant proteins, the anisotropy is dominated by a large $\phi_2$, which can be attributed to the global rotation of the protein. For a spherical protein in water, with a molecular mass of 12.4 kDa and a $1\,\mathrm{cm}^3/\mathrm{g}^{-1}$ hydration, a rotational correlation time of 5.1 ns is calculated using the Stokes-Einstein equation ($\phi = \eta V/kT$, where $\eta$ is the viscosity of water and V is the hydrated volume). The calculated value is considerably smaller than the experimental average (8 ns ± 1 ns). This phenomenon is observed for a number of proteins and can have multiple causes. A deviation from spherical symmetry as well as increasing hydration will result in an elevated rotational correlation time of the protein. In this case, however, the longest rotational correlation time corresponding to the rotation of the whole protein shows a concentration dependence that can be described by an overall trimerisation process with the following equation:

$$\phi_{prot} = \frac{\phi_2 . c_1 + 3\phi_2 3K^2 c_1{}^3}{c_1 + 3K^2 c_1{}^3} \tag{5.9}$$

Where $c_1$ is the concentration of the monomer, $\phi_2$ its rotational correlation time, $3\phi_2$ is asumed to be that of the trimer and $3K^2c_1{}^3$ is the mass concentration of trimers. The data can be simulated very well with an association constant of $K = 0.1 \pm 0.05\,\mu\mathrm{M}^{-1}$ and a correlation time $\phi_2$ of $4\,\mathrm{ns}^{-1}$ (Figure 5.3).

Small contributions of a shorter component to the anisotropy decay arise from the segmental motion of the tryptophan environment.[26] At both pH-values, the movement of W71 is most restrained (highest $\phi_1$).[15] The mobility of W35 is increased with increasing pH. The rotational correlation time of W94 is very small at low pH, indicating a highly flexible environment.

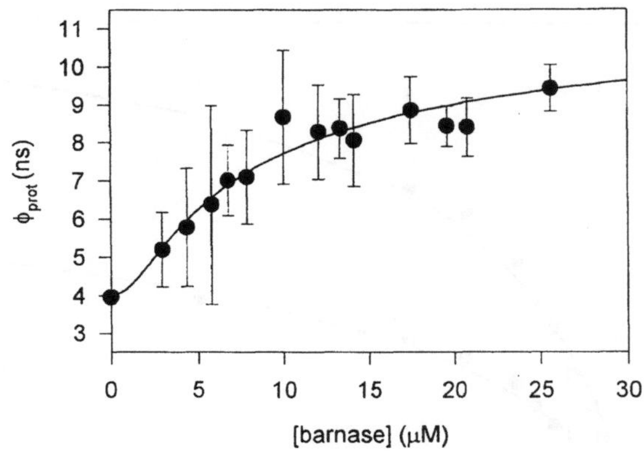


Figure 5.3. Protein concentration dependence of the large rotational correlation time of barnase. The continuous line is the best simulation for the weight average correlation time in case of a trimerisation using equation 9.

## 5.5. Steady-State Phosphorescence

The phosphorescence spectra of WT barnase and the different mutants were determined at 21 °C and at pH 7, after removing oxygen from the solution.[27] The obtained spectra are very typical of tryptophan emission spectra reported for proteins at room-temperature. They are characterised by a 0–0 transition near 420 nm and a emission maximum at 441–445 nm. Substitution of W71 by tyrosine has no significant change on the phosphorescence emission of the protein. In the mutant W94Y, the phosphorescence intensity is increased by more than 200% relative to WT. When W35 is replaced by phenyl-alanine, there is a decrease of about 70%. From the spectra of the one-tryptophan-containing mutants, it turns out that W94 does not show any detectable phosphorescence. W71 shows the highest intensity and W35 exhibits an intermediate phosphorescence intensity. The phosphorescence of W71 is also strongly reduced in the presence of W94.

## 5.6. Concentration Dependence of Phosphorescence Intensity

Since the phosphorescence quantum yield of most proteins is very low (about $10^{-6}$), measurement are often performed at high protein concentration (1 to 2 mg/mL). Under these conditions, the phosphorescence intensity of barnase is no longer linearly concentration dependent. To investigate the origin of this effect, the phosphorescence emission of WT barnase was

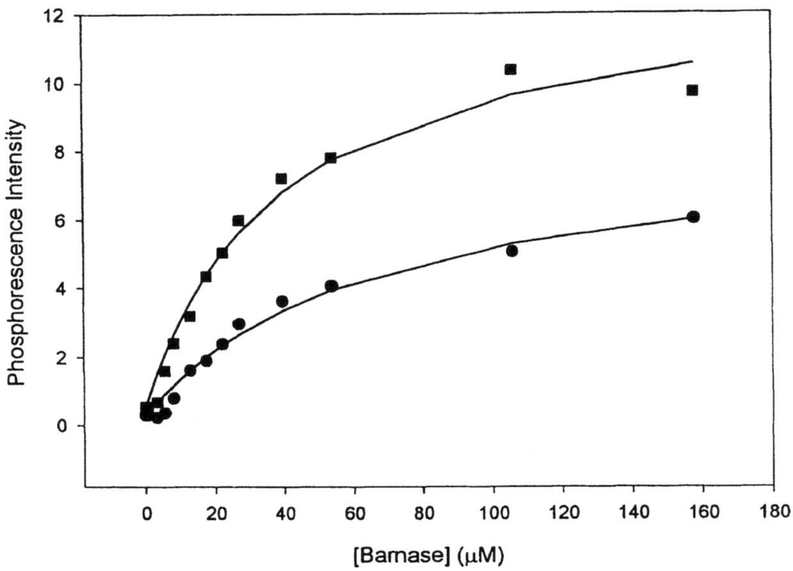


Figure 5.4. Phosphorescence intensity of WT barnase versus protein concentration at excitation wavelengths 280 nm (upper) and 295 nm (lower) and emission wavelength 441 nm at pH 7.0. Data were fitted to equation (10).

determined over a concentration range from 0.6 to 160 μM (0.05 to 2 mg/mL), and this for excitation wavelengths 280 nm and 295 nm. Data are shown in Figure 5.4.

Deviation from linearity start at 25 μM of barnase, irrespective of the optical density of the sample and is not due to an inner filter effect. The data can be fairly fitted to the equation of dynamic quenching, although the slight sigmoidality indicates that they are influenced by trimer formation as well:

$$I = A_0 \cdot \frac{[protein]}{k_0 + k_q \cdot [protein]} + I_d \tag{5.10}$$

Where I is the measured phosphorescence intensity corrected for inner filter effects, $I_d$ is the dark current, $A_0$ the phosphorescence amplitude, $k_0$ is the inverse phosphorescence lifetime and $k_q$ the collisional quenching constant. Using $1\,ms^{-1}$ for $k_0$ a value of $5 \times 10^6\,M^{-1}\,s^{-1}$ for $k_q$ is found, indicating that only a limited number of collisions lead to phosphorescence quenching.

## 5.7. Conclusions

The construction of the double mutants has allowed us to experimentally determine the fluorescence properties of the individual tryptophan residues of barnase, and to compare them with the values previously predicted on the basis of subtraction (fluorescence data of WT—fluorescence data of single tryptophan mutants).

The spectra of the individual tryptophan residues compare very well with the ones obtained by subtraction. Only for W71 a deviation is observed (Figure 5.2). Since the spectrum of W71 is calculated by a double subtraction: (spectrum W71 = spectrum W94F—spectrum W35; while spectrum W35 = spectrum WT—spectrum W35F) it is possible that deviations in the spectrum of W94F are responsible for this. In many respects the W94F mutant behaves differently: it has a high quantum yield and high $k_r$ as compared to the other proteins which cannot be explained.[15]

The fluorescence lifetimes of the single tryptophan residues were predicted in the same way. In contrast to these predictions, the single tryptophan residues (in the double mutants) show several lifetimes, but their amplitude average lifetime corresponds very well with the single lifetime that was predicted (except for W94). The agreement between the amplitude average lifetimes of the mutants from the previous and the new measurements is very good, especially in acid medium. In basic medium a very small fraction of a long lifetime component has been observed, which seems to be linked to the presence of W94. These results indicate that a very broad frequency range has to be scanned for good lifetime resolution.

Data on single tryptophans can be used to check additivity or interactions in the WT. The best parameter to be used for this purpose in our experience turned out to be the amplitude average lifetime. This is again confirmed here. The data clearly shows that the combination of the lifetimes of all three tryptophans do not reproduce the average lifetime of the WT. However, combination where the single Trp-mutant W35F is combined with the lifetime data of W35 gives a very good approximation, indicating that energy transfer occurs between W71 and W94. The average lifetime of mutant W94F can neatly be obtained by combining the individual data of W71 and W35 indicating additivity and no interactions between W35 and W71.

The individual tryptophan mutants allow us also to determine the ratio of $\varepsilon_{295}/\varepsilon_{280}$. This ratio does not correspond with the value expected, taking into account the number of tryptophan and tyrosine residues present in the proteins.[15] This proves that, although the extinction coefficient at 280 nm of a protein can generally accurately be calculated from the amino acid composition,[28] this is not true for the extinction coefficient at 295 nm, due to changes in the bandwith.

The phoshorescence properties of the individual tryptophans complete the picture obtained from the lifetime and anisotropy analysis. The most mobile tryptophan residues, as deduced from the correlation times, also show the lowest phosphorescence intensity and lifetime. The most surprizing result obtained, however, is the concentration dependence of the phosphorescence intensity and of the fluorescence anisotropy. The concentration dependence of the phosphorescence intensity cannot be explained by the inner filter effect. Gabellieri and Strambini[29] observed a decrease in the phosphorescence lifetime of alcohol dehydrogenase (LADH) and glyceralaldehyde-3-phosphate dehydrogenase (Ga*P*DH) with increasing concentration of unrelated proteins. The authors interpret these findings as association reactions which cause temporary structure fluctuations.

The concentration dependence of the fluorescence anisotropy also indicates the presence of protein-protein interactions. A nice fit between the measured rotational correlation time and simulation is obtained for a mechanism of trimer formation. Evidence for trimer formation by domain swapping in the crystalline state has recently been published by Zegers *et al.*[30]

## References


1. Y. Mauguen, R. W. Hartley, G. G. Dodson, E. J. Dodson, G. Bricogne, C. Chothia, and A. Jack. Molecular Structure of a new family of ribonucleases. Nature 297, 162–164, 1982.
2. A. R. Fersht. Protein folding and stability: the pathway of folding of barnase. FEBS Letters 325, 5–16, 1993.
3. J. F. Corrales and A. R. Fersht. The folding of GroEL-bound barnase as a model for chaperonin-mediated protein folding. Proc. Natl. Acad. Sci. USA 92, 5326–5330, 1995.
4. R. W. Hartley. Barnase and barstar: two small proteins to fold and fit together. TIBS 14, 450–454, 1989.
5. A. M. Buckle, G. Schreiber, and A. R. Fersht. Protein-protein recognition: crystal structural analysis of a barnase-barstar complex at 2.0-Å resolution. Biochemistry 33, 8878–8889, 1994.
6. G. Schreiber and A. R. Fersht. Energetics of protein-protein interactions: analysis of barnase-barstar interface by single mutations and double mutant cycles. J. Mol. Biol. 248, 478–486, 1995.
7. D. E. Mossakowska, K. Nyberg, and A. R. Fersht. Kinetic characterisation of the recombinant ribonuclease from Bacillus amyloliquefaciens (barnase) and investigation of key residues in catalysis by site-directed mutagenesis. Biochemistry 28, 3843–3850, 1989.
8. A. M. Buckle and A. R. Fersht. Substrate binding in an RNase: Structure of a barnase-tetranucleotide complex at 1.76 Å resolution. Biochemistry 33, 1644–1653, 1994.
9. R. Loewenthal, J. Sancho, T. Reinikainen, and A. R. Fersht. Long-Range Surface Charge-Charge Interactions in Proteins. Comparison of Experimental Results with Calculations from a Theoretical Model. J. Mol. Biol. 232, 574–583, 1993.
10. K. Bastyns, M. Froeyen, J. F. Diaz, G. Volckaert, and Y. Engelborghs. Experimental and Theoretical Study of Electrostatic Effects on the Isoelectric pH and the pKa of the Cat-

alytic Residue His-102 of the Recombinant Ribonuclease From Bacillus amyloliquefaciens (Barnase). Proteins, Struc., Func., Gen. 24, 370–378, 1996.
11. T. L. Bushueva, E. P. Busel, V. N. Bushueva, and E. A. Burstein. The interaction of protein functional groups with indole chromophore. I. Imidazole group. Stud. Biophys. 44, 129–139, 1974.
12. R. Vos and Y. Engelborghs. A Fluorescence Study of Tryptophan-Histidine Interactions in the Peptide Anantin and in Solution. Photochem. Photobiol. 60, 24–32, 1994.
13. R. Loewenthal, J. Sancho, and A. R. Fersht. Fluorescence spectrum of barnase: contribution of three tryptophan residues and a histidine-related pH dependence. Biochemistry 30, 6775–6779, 1991.
14. K. Willaert, R. Loewenthal, J. Sancho, M. Froeyen, A. R. Fersht, and Y. Engelborghs. Determination of the excited-state lifetimes of the tryptophan residues in barnase, via multifrequency phase fluorometry of tryptophan mutants. Biochemistry 31, 711–716, 1992.
15. K. De Beuckeleer, G. Volckaert, and Y. Engelborghs. Time Resolved Fluorescence and Phosphorescence Properties of the Individual Tryptophan Residues of Barnase: Evidence for Protein-portein Interactions. Proteins, Struc. Function and Genetics 36, 42–53, 1999.
16. B. Valeur and G. Weber. Resolution of the fluorescence excitation spectrum of indole into the 1La and 1Lb excitation bands. Photochem. Photobiol. 25, 441–444, 1977.
17. Th. Förster. Intermolecular Energy Migration and Fluorescence. Ann. Phys. (Leipzig) 2, 55–75, 1948.
18. E. A. Burstein, N. S. Vedenka, and M. N. Ivkova. Fluorescence and the location of tryptophan residues in protein molecules. Photochem. Photobiol. 18, 263–279, 1973.
19. G. Weber. Resolution of the fluorescent lifetimes in a heterogeneous system by phase and modulation measurements. J. Phys. Chem. 85, 949–953, 1981.
20. J. M. Beechem, J. R. Knutson, J. B. A. Ross, B. W. Turner, and L. Brand (1983) Global resolution of heterogeneuos decay by phase modulation fluorometry: mixtures and proteins. Biochemistry 22, 6054–6058, 1983.
21. G. B. Porter. Reversible Energy Transfer. Theor. Chim. Acta (Berlin) 24, 265–270, 1971.
22. P. Woolley, K. G. Steinhauser, and B. Epe. Forster-type Energy transfer. Simultaneous "forward" and "reverse" transfer between unlike fluorophores. Biophys. Chem. 26, 367–374, 1987.
23. M. Shinitzky and R. Goldman. Fluorometric detection of histidine-tryptophan complexes in peptides and proteins. Eur. J. Biochem. 3, 139–144, 1967.
24. A. Sillen and Y. Engelborghs. The Correct Use of "Average" Fluorescence Parameters. Photochem. Photobiol. 67, 475–486, 1998.
25. G. Weber. Theory of differential phase fluorometry: detection of anisotropic molecular rotations. J. Phys. Chem. 66, 4081–4091, 1977.
26. J. R. Lakowicz, B. P. Maliwal, H. Cherek, and A. Balter. Rotational Freedom of Tryptophan Resiudes in Proteins and Peptides. Biochemistry 22, 1741–1752, 1983.
27. S. W. Englander, D. B. Calhoun, and J. J. Englander. Biochemistry without oxygen. Anal. Biochem. 161, 300–306, 1987.
28. H. Mach, R. Middaugh, and R. V. Lewis. Statitical determination of the average values of the extinction coefficients of tryptophan and tyrosin in native proteins. Anal. Biochem. 200, 74–80, 1992.
29. E. Gabellieri and G. B. Strambini. Conformational changes in proteins induced by dynamic associations. A tryp tophan phosphorescence studie. Eur. J. Biochem. 221, 77–85, 1994.
30. I. Zegers, J. Deswarte, and L. Wyns. Trimeric domain-swapped barnase. Proc. Natl. Acad. Sci. USA 96, 818–822, 1999.

# 6

# Fluorescence Study of the DsbA Protein from Escherichia Coli

## *Energy Transfer, Quenching by the Catalytic Disulfide and Microstate Reshuffling*

Alain Sillen, Jens Hennecke, Rudi Glockshuber, and Yves Engelborghs

## 6.1. Introduction

Enzymes of the thiol-disulphide oxidoreductase (TDOR) family are involved in numerous processes in prokaryotic and eukaryotic cells, including protein folding, DNA synthesis, cytochrome biogenesis, and photosynthesis [for reviews, see Gilbert,[1] Bardwell and Beckwith,[2] and Loferer and Hennecke[3]]. All TDORs catalyse the formation, isomerization or reduction of structural, regulatory, or catalytic disulphide bridges in target proteins by disulphide exchange reactions with their substrates. The C-X-X-C motif of the active-site disulphide is characteristic for all TDORs. Reduction of the catalytic disulphide bridge in thioredoxin, DsbA, and TlpA has been shown to cause a strong increase in tryptophan fluorescence.[4–6] The fluorescence properties of DsbA from Escherichia coli have been studied in detail. DsbA is a monomeric, periplasmic 21.1 kDa protein (189 aa) that is required for efficient disulphide bond formation in secretory proteins in the bacterial periplasm.[7,8] The enzyme contains a single, catalytic disulphide with the active-site sequence C30-P31-H32-C33. The X-ray structure of oxidized DsbA[9] as well as the X-ray[10] and NMR-structure[11] of reduced DsbA has

*Alain Sillen and Yves Engelborghs* • Laboratory of Biomolecular Dynamics, University of Leuven, Celestijnenlaan 200D, B-3001 Leuven, Belgium. *Jens Hennecke and Rudi Glockshuber* • Institut für Molekularbiologie und Biophysik, Eidgenössische Technische Hochschule Hönggerberg, CH-8093 Zürich, Switzerland.

*Topics in Fluorescence Spectroscopy, Volume 6: Protein Fluorescence*, edited by Joseph R. Lakowicz. Kluwer Academic / Plenum Publishers, New York, 2000

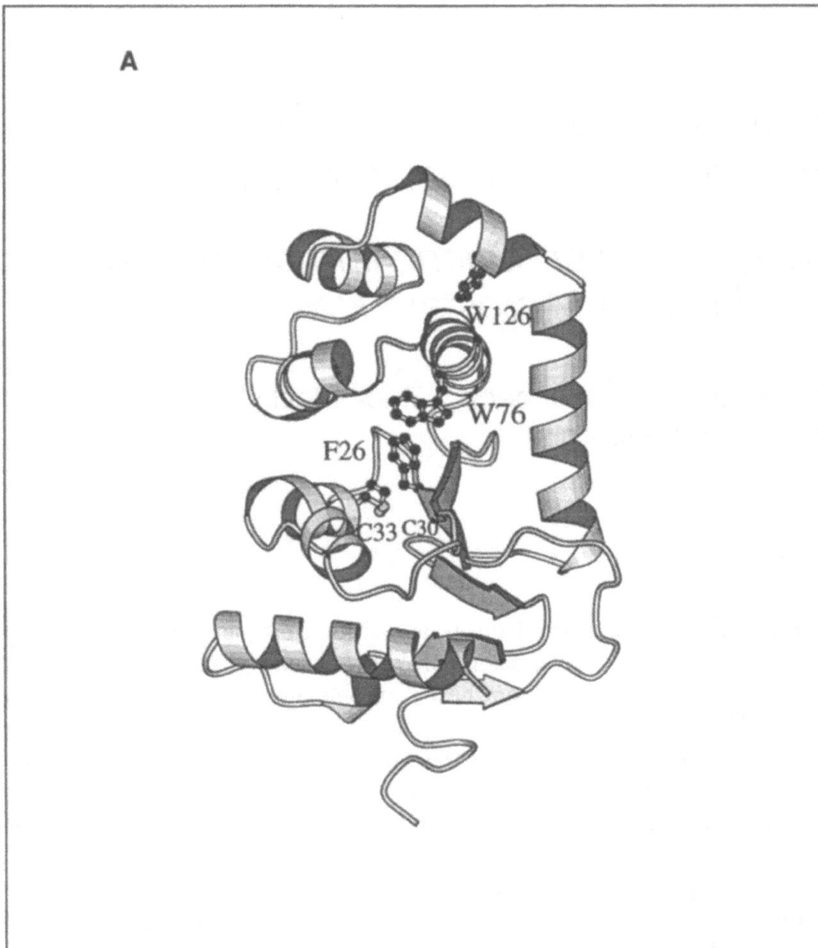


Figure 6.1A. Molscript representation[36] of the X-ray structure of oxidized DsbA from E. Coli. The disulfide, the two tryptophan residues and the side chain F26 are shown.

revealed that the enzyme possesses a thioredoxin-like domain (residues 1–62 and 139–189), a motif found in all known structures of disulphide oxidoreductases.[12] The sequence of the thioredoxin-like domain of DsbA is, however, only 10% identical with E. coli thioredoxin. DsbA possesses a second domain (residues 63–138) of unknown function, which is inserted into the thioredoxin motif and exclusively consists of $\alpha$-helices (Figure 6.1). In contrast to thioredoxin, DsbA does not contain a tryptophan residue amino-terminal to the catalytic disulphide. Nevertheless, a strong (about 3-fold) increase of

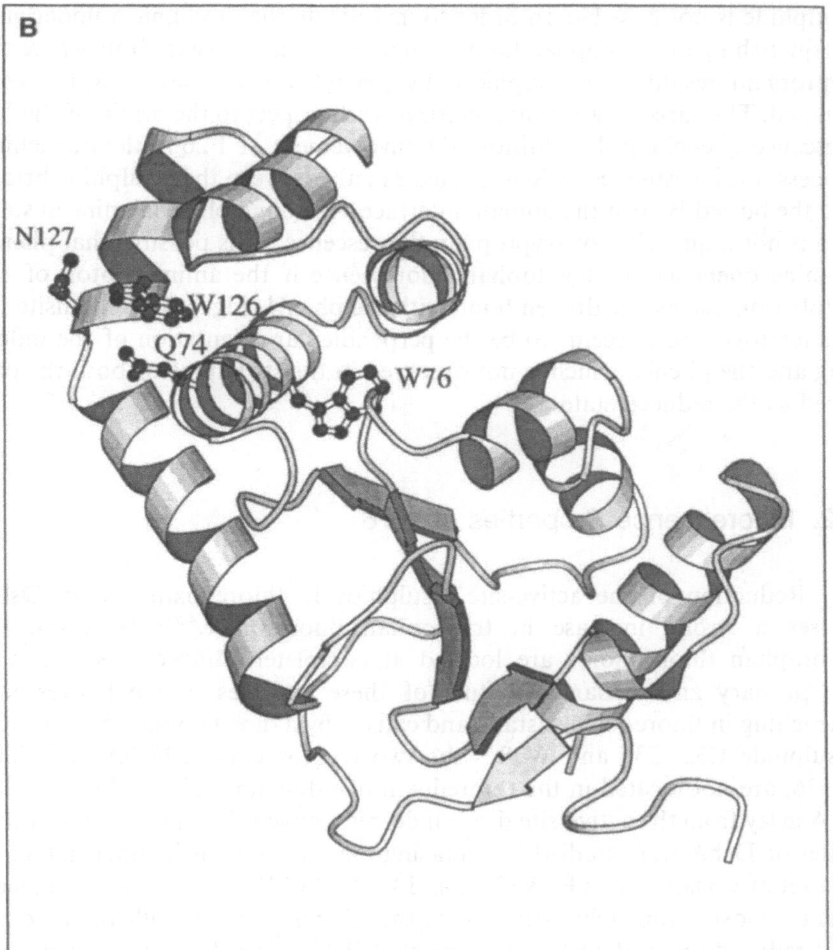


Figure 6.1B. Molscript representation[36] of the X-ray structure of oxidized DsbA from E. Coli. The tryptophan residues, and the side chains of Q74 and N127 are shown.

tryptophan fluorescence is observed upon reduction of its disulphide.[5,13] This was used to measure the redox potential of the protein and to monitor its interaction with substrate proteins.[5,13-15] Interestingly, both tryptophans of DsbA, W76 and W126, are not contained in the thioredoxin domain and are located in the $\alpha$-helical domain (Figure 6.1). W76 is buried and about 12 Å apart from the disulphide, whereas W126 is even further away from the disulphide bridge (about 20 Å) and partially solvent-accessible. Hence, quenching of the tryptophan fluorescence by the direct contact between W76 and the disulphide is not possible. In order to investigate the mechanism underlying the quenching of tryptophan fluorescence in detail, DsbA variants where the tryptophan residues were replaced by phenylalanine residues where constructed. The variants were characterized with respect to the origin of the fluorescence quenching. In addition, the involvement of F26 in the quenching process was investigated. F26 is located exactly between the disulphide bridge and the buried W76 at the domain interface. Although phenylalanine in solution is not a quencher of tryptophan fluorescence[16] it is possible that phenylalanine quenches the tryptophan fluorescence if the amine proton of the indole ring makes a hydrogen bond with the phenyl ring.[17] A prerequisite for this hydrogen bond seems to be the perpendicular orientation of the indole ring and the phenyl, which is not observed in the structures of both the oxidized as the reduced state.

## 6.2. Fluorescence Properties of W76

Reduction of the active-site disulphide in thioredoxin and in DsbA causes a strong increase in tryptophan fluorescence.[4,5,13] However, the tryptophan fluorophores are located at completely different positions in the primary and tertiary structure of these enzymes. While fluorescence quenching in thioredoxin is static and caused by a direct contact between the disulphide C32–C35 and W28,[18] the two tryptophans in DsbA, W76 and W126, are not located in the thioredoxin-like domain and are about 12 and 20 Å away from the active-site disulphide, respectively. The fluorescence properties of DsbA were studied by replacing the tryptophans by phenylalanines in a set of variants (W76F, W126F, and W76F/W126F).[19] The W76F replacement almost completely extinguishes the fluorescence of both the oxidized and reduced form of DsbA showing that W126 must be heavily quenched, while W76 is identified as the most prominent active tryptophan fluorophore. W76 is buried in the hydrophobic domain interface of the protein. Consistently, the fluorescence emission maximum of DsbA WT (326 nm) is blue-shifted compared to that of E. coli thioredoxin (341 nm), where the critical fluorophore W28 is significantly solvent-exposed.[20,21] Despite the removal of a fluorophore, the reduced variant W126F shows a higher tryptophan fluorescence compared to reduced DsbA WT. A comparable observation was made for E. coli thioredoxin[22,23] and barnase[24,25] upon removal of one of the tryptophans. In case of DsbA, this phenomenon can be explained by the presence of energy transfer from W76 to the heavily quenched W126 in the wild type protein, and the absence of this phenomenon in the variant W126F.

Disulfides are known to be effective quenchers of tryptophan fluorescence.[26] However, in the case of DsbA the question arises how the disulfide is able to quench the fluorescence of W76 although it is more than 12 Å away. Since F26 makes a van der Waals contact with both W76 and C33 of the catalytic disulfide in oxidized DsbA, it is possible that F26 is involved in the quenching process. Indeed, exchange of the aromatic residue F26 against leucine diminishes disulphide-dependent quenching of W76, while the steady-state fluorescence properties of the reduced F26L variant remain essentially unchanged. As the fluorescence of oxidized F26L is still 1.7-fold lower than that of the reduced variant, a limited, redox-state-dependent quenching of W76 still occurs in F26L. The fluorescence intensity of W126 is extremely quenched. Therefore energy transfer from W76 to W126 makes W126 an energy sink for W76.

We can conclude from the fluorescence intensity (Table 6.1) and lifetime measurements (Table 6.2) that two distinct quenching processes can be operative in DsbA: unidirectional nonradiative energy transfer from W76 to W126, and dynamic quenching by the disulphide bond or the –SH groups. In our previous paper[19] we have assumed that the long lifetime of 3.6 ns ($a$ = 0.66) in WTred was reduced to 1.0 ns ($a$ = 0.67) in the WTox, due to dynamic quenching by the disulfidebridge. However, a more detailed analysis is possible which is presented here.

Different lifetimes for a single tryptophan are usually explained in terms of the existence of different conformers of that residue.[27] Upon changing the redox state or changing a residue in the vicinity of tryptophan it is possible that tryptophan changes the relative population of its different conformations. This phenomenon, in case of residue replacement, is described in more detail in the section about the fluorescence properties of W126. The redox state-dependent accessibility of W76 was analyzed by measuring the dynamic

Table 6.1. Molar Absorption Coefficients ($\varepsilon_{280}$), Quantum Yields (Q), Amplitude Average Lifetimes $\langle\tau\rangle$ and Average Radiative Rate Constant $\langle k_r\rangle$ of DsbA WT and DsbA Variants

| | $\varepsilon_{280}$ ($M^{-1}cm^{-1}$) | Q | $\langle\tau\rangle$ (ns) | $\langle k_r\rangle$ ($ns^{-1}$) |
|---|---|---|---|---|
| $WT_{ox}$ | 23 322 | 0.03 | 0.84 | 0.036 |
| $WT_{red}$ | ″ | 0.10 | 2.85 | 0.035 |
| $F26L_{ox}$ | 23 919 | 0.06 | 1.69 | 0.036 |
| $F26L_{red}$ | ″ | 0.09 | 2.40 | 0.038 |
| $W126F_{ox}$ | 17 680 | 0.05 | 1.31 | 0.038 |
| $W126F_{red}$ | ″ | 0.20 | 3.70 | 0.054 |

Table 6.2. Lifetimes ($\tau$) and Amplitude Fractions (a) at 340 nm and $\chi_R^2$ as Obtained by Global Analysis of the Fluorescence Decay of DsbA and its Variants F26L and W126F in the Oxidized and Reduced States (Excitation at 295 nm)

| | $a_1$ | $\tau_1$ (ns) | $a_2$ | $\tau_2$ (ns) | $a_3$ | $\tau_3$ (ns) | $\chi_R^2$ |
|---|---|---|---|---|---|---|---|
| $WT_{ox}$ | 0.29 | 0.14 | 0.67 | 1.00 | 0.04 | 3.0 | 2.2 |
| | ± 0.01 | ± 0.02 | ± 0.02 | ± 0.03 | ± 0.02 | ± 0.2 | |
| $WT_{red}$ | 0.19 | 0.13 | 0.15 | 2.78 | 0.66 | 3.6 | 0.9 |
| | ± 0.03 | ± 0.03 | ± 0.01 | ± 0.09 | ± 0.03 | ± 0.1 | |
| $F26L_{ox}$ | 0.25 | 0.34 | 0.71 | 2.03 | 0.04 | 4.5 | 2.4 |
| | ± 0.02 | ± 0.06 | ± 0.02 | ± 0.08 | ± 0.028 | ± 1.1 | |
| $F26L_{red}$ | 0.21 | 0.16 | 0.28 | 2.17 | 0.51 | 3.4 | 1.4 |
| | ± 0.01 | ± 0.02 | ± 0.02 | ± 0.06 | ± 0.01 | ± 0.1 | |
| $W126F_{ox}$ | 0.14 | 0.38 | 0.77 | 1.26 | 0.09 | 3.1 | 1.3 |
| | ± 0.01 | ± 0.04 | ± 0.03 | ± 0.09 | ± 0.03 | ± 0.4 | |
| $W126F_{red}$ | 0.01 | 0.45 | 0.38 | 1.9 | 0.61 | 4.90 | 0.8 |
| | ± 0.02 | ± 0.06 | ± 0.02 | ± 0.1 | ± 0.02 | ± 0.02 | |

Table adapted from ref 19.

quenching by acrylamide.[19] Interestingly, acrylamide quenching of tryptophan fluorescence is slightly higher for reduced DsbA than for the oxidized protein. This indicates that there is a small increase in the accessibility of W76. The following analysis suggest that this could be linked to a change of the relative population of its microconformations. Indeed inspection of the amplitude fractions of oxidized and reduced WT shows that the second lifetime is the most populated one in the oxidized state while the longest lifetime is the most populated one in the reduced state. We suggest to analyze this phenomenon in the following way: the ratio of the quantum yields of different variants can be split into a factor ($f_{kr}$) representing the change in $k_r$ or homogenous static quenching (i.e. static quenching that does not alter the ratio of the amplitude fractions), a factor ($f_{PR}$) reflecting population reshuffling and/or heterogenous static quenching and a factor ($f_{DQ}$) representing pure dynamic quenching[28]:

$$\frac{Q}{Q_0} = \frac{\langle k_r \rangle \sum \alpha_i \tau_i}{\langle k_{r0} \rangle \sum \alpha_{0i} \tau_{0i}} = \frac{\langle k_r \rangle}{\langle k_{r0} \rangle} \frac{\sum \alpha_i \tau_{0i}}{\sum \alpha_{0i} \tau_{0i}} \frac{\sum \alpha_i \tau_i}{\sum \alpha_i \tau_{0i}}$$
$$= f_{kr} \times f_{PR} \times f_{DQ} \tag{6.1}$$

The factor $f_{PR}$ is affected by static quenching only if the static quenching is heterogenous. If there is static quenching and an increase of the fluorescence due to changes in the populations of the different conformers, then $f_{PR}$ is the minimum factor by which the fluorescence intensity increases due to a change in the balance of the micro conformations. The calculation of the factor $\Sigma\alpha_i\ \tau_{0i}$ is somewhat arbitrary: which new amplitude has to be combined with which old lifetime? For limited modifications it seems logical to make the combination within the classes of short, middle and long lifetimes. Only for WT protein there exist an ambiguity, and two combinations are possible. The data summarized in Table 6.3 clearly show that upon the transition from the reduced to the oxidized state the quantum yield drops to 30%. This 70% decrease in quantum yield of W76 in WT is due to population reshuffling (28% or 10%) and also due to both disulphide quenching and energy transfer to W126 (59% or 68%). In the F26L variant only the decrease in fluorescence due to population reshuffling remains, while there is little or no dynamic quenching due to disulphide bond quenching or energy transfer to W126. The results of this variant strongly support the combination with $f_{PR} = 0.7$ for the WT protein. Oxidation in the DsbA variant W126F causes a 29% decrease of $k_r$, which is difficult to explain. Due to this decrease of $k_r$, the factor $f_{PR}$ is not only population reshuffling but the product of both population reshuffling and heterogeneous static quenching. Only the 34% decrease in fluorescence due to dynamic quenching in this variant is attributed to disulphide quenching.

In order to calculate the rate constant for dynamic quenching (either by collisional quenching or by energy transfer) from the observed average lifetimes of the different proteins, we also have to correct for the possibility of

Table 6.3. Quenching Analyses: The Ratio of the Quantum Yields ($Q/Q_0$) and the Quenching Factors Due to the Change in Radiative Rate Constant ($f_{kr}$), Population Reshuffling ($f_{PR}$) and Dynamic Quenching ($f_{DQ}$)

| | $Q/Q_0$ | $f_{kr}$ | $f_{PR}$ | $f_{DQ}$ |
|---|---|---|---|---|
| WT red → ox | 0.30 | 1.01 | 0.72[b]/0.90[c] | 0.41[b]/0.32[c] |
| F26L red → ox | 0.67 | 0.95 | 0.71 | 0.99 |
| W126F red → ox | 0.25 | 0.71[a] | 0.53 | 0.66 |

[a]Because $f_{PR}$ significantly differs from 1, $f_{PR}$ is not purely due to population reshuffling, but also contains information about heterogeneous static quenching.
[b]Calculated by the amplitudes and lifetimes as in Table 6.1.
[c]Calculated by combining lifetime 3.6 ns (a = 0.66) in WTred with 1.0 ns (a = 0.67) in WTox and 2.78 (a = 0.15) in WTred with 3.0 (a = 0.04) in WTox.

the reshuffling of micro conformations of tryptophan induced by either a change in the oxidation state or a mutation. Therefore the amplitude fractions are kept constant in calculating the amplitude average lifetime of two different variants or states. Out of the ratio of these amplitude average lifetimes an average rate constant for dynamic quenching $\langle k_{dq}\rangle$ can be calculated. (For an extensive derivation of these equations, see ref. 28.):

$$\langle k_{dq}\rangle = \frac{\sum a_{0i}\tau_{0i}\tau_i k_{dqi}}{\sum a_{0i}\tau_{0i}\tau_i} \tag{6.2}$$

The simplest situation is found in the reduced variant W126F, where no energy transfer from W76 to W126 and no quenching by the disulphide bridge can occur. Since the average lifetime of W76 is relatively long (3.7 ns), the possible quenching by the –SH groups must be very small. We therefore made the simplifying assumption that the lifetime of W76 in the reduced variant W126F equals the intrinsic lifetime of W76 in all oxidized and reduced DsbA proteins. In the variant W126F, the decreased fluorescence of W76 in the oxidized protein is exclusively caused by quenching by the disulphide. The dynamic part of quenching can be described by

$$\frac{\langle \tau_0\rangle}{\langle \tau\rangle} = 1 + \frac{\sum a_{0i}\tau_{0i}\tau_i}{\langle \tau\rangle}\langle k_Q\rangle \tag{6.3}$$

where $\langle \tau_0\rangle$ and $\langle \tau\rangle$ are the average lifetimes in the absence and presence of quenching respectively, and $\langle k_Q\rangle$ is the average rate constant of quenching. If we assume that the conformational effects of oxidation on the intrinsic lifetimes are negligible, we can use $\langle \tau_0\rangle$ from the reduced variant W126F and estimate $\langle k_Q\rangle$. Applying this to the variant W126F gives a $k_Q$ of $0.13\,\mathrm{ns}^{-1}$ (Table 6.4). The differences between the lifetimes of the reduced variants W126F and F26L can only be due to nonradiative energy transfer from W76 to W126, as described by

$$\frac{\langle \tau_0\rangle}{\langle \tau\rangle} = 1 + \frac{\sum a_{0i}\tau_{0i}\tau_i}{\langle \tau\rangle}\langle k_{ET}\rangle \tag{6.4}$$

where $\langle \tau_0\rangle$ is the average lifetime in the absence of energy transfer and $k_{ET}$ is the apparent rate constant of energy transfer (in principle, $k_{ET}$ may also contain contributions caused by conformational changes resulting from the replacement W126F). Using the data of Table 6.2, we calculated $k_{ET} = 0.07\,\mathrm{ns}^{-1}$ for the reduced F26L variant. Applying the same consideration to the reduced WT and the W126F variant gives a value for $k_{ET}$ of $0.03\,ns^{-1}$ (Table 6.4).

Because the difference between the fluorescence of oxidized WT and the oxidized W126 variant should only be due to the energy transfer from W76 to W126, we calculated an energy transfer rate constant for oxidized WT of $k_{ET} = 0.13\,ns^{-1}$. The nonradiative energy transfer is thus more efficient in the oxidized state than in the reduced state. Possibly due to a different orientation of W76 relative to W126. However we have compared the two X-ray structures and calculated the root mean square positional difference (RMSPD) of the two tryptophans in the two structures and did not find any substantial difference (RMSPD W76: 0.12, RMSPD W126: 0.18). The differences between the lifetimes of the reduced variant W126F and the oxidized variant F26L can only be due to nonradiative energy transfer ($k_{ET}$) from W76 to W126 and the quenching by the disulphide bridge ($k_Q$). The sum of $k_{ET}$ and $k_Q$ for the oxidized variant F26L can thus be calculated and is $0.01\,ns^{-1}$. The quenching constant of the disulphide bridge is thus between 0 and $0.01\,ns^{-1}$ and therefore strongly reduced in the absence of F26 (Table 6.4). The overall conclusion is that energy transfer from W76 to W126 appears in both redox states, but is more pronounced in the oxidized state. The disulphide bridge is able to create a dynamic quenching of W76, and it largely needs F26 for this effect. The overall situation of dynamic quenching and energy transfer processes in oxidized and reduced DsbA WT can therefore be represented by Scheme 6.1. It should be noted that in this scheme the change in the balance of microconformations of W76, due to oxidation, is not represented, while the effect of Q74 and N127 on W126 is uniquely due to this effect. Since $k_{intr} = 0.27\,ns^{-1}$ can be considered as a lower limit, the rate constants for $k_Q$ and $k_{ET}$ are upper limits. Only in the oxidized variant W126F the absence of a static quenching component of the disulphide-dependent quenching of W76 was observed. It is not clear why there is static quenching in the other variants. From the known 2.0 Å X-ray structure of oxidized

Table 6.4. Apparent Rate Constants of Energy Transfer ($k_{ET}$) and Dynamic Quenching ($k_Q$)

| | $k_{ET}$ | $k_Q$ |
|---|---|---|
| WT ox | 0.13 | 0.13 |
| WT red | 0.03 | / |
| F26L ox | 0.01 | 0 |
| F26L red | 0.07 | / |
| W126F ox | / | 0.13 |
| W126F red | / | / |

DsbA, we calculated a value of 0.122 for the efficiency of energy transfer (E) from W76 to W126 in the oxidized WT ($J_{AD}$ = 1.64 × $10^{-16}$ $cm^6 mmol^{-1}$, $F_D = 4.49 \times 10^{-2}$, $n = 1.5$, $\kappa^2 = 0.709$, $R_0 = 9.84$, $R = 13.66$). The efficiency of energy transfer was also calculated from experimental rate constants and yields values of 0.20 and 0.05 in the reduced and oxidized variant F26L and 0.12 and 0.24 in the reduced and oxidized WT, respectively. This value of the oxidized state is about 2-fold higher than theoretically expected. This can principally result from an underestimation of $k_{ET}$ due to small conformational changes caused by the mutations, from an overestimation of the average distance between W76 and W126 in the solution structure of DsbA, and from the error in the calculation of $J_{AD}$.

## 6.3. Fluorescence Properties of W126

The fluorescence properties of W126 were not only investigated by replacing W76 by a phenylalanine, but also by replacing the possible quenchers Q74 and N127 by alanine, yielding the following set of variants: W76F, W76F/Q74A, W76F/N127A and W76F/Q74A/N127A29.

### 6.3.1. Quenching Analysis

Compared to Trp in solution which has a quantum yield of 0.14 the fluorescence of W126 is highly quenched in both the oxidized ($Q = 0.013$) and reduced state ($Q = 0.012$) of DsbA (Table 6.5). This seems to be largely due to an increase of dynamic quenching because the apparent radiative rate constant is the same as the radiative rate constant of tryptophan (0.053 $ns^{-}$),[30] whereas the nonradiative rate constant is 3.8 $ns^{-1}$ compared to 0.33 $ns^{-1}$ for

Table 6.5. Molar Absorption Coefficients ($\varepsilon_{295}$), Quantum Yields (Q), Average Lifetimes ($\langle\tau\rangle_\alpha$) and Radiative Rate Constants ($\langle k_r\rangle$) of W126 in the Different DsbA Variants

| | $\varepsilon_{295}$ ($M^{-1} cm^{-1}$) | Q | $\langle\tau\rangle_\alpha$ (ns) | $\langle k_r\rangle$ ($ns^{-1}$) |
|---|---|---|---|---|
| $W76F_{ox}$ | 2656 ± 496 | 0.013 ± 0.003 | 0.26 | 0.051 |
| $W76F_{red}$ | 2999 ± 420 | 0.012 ± 0.003 | 0.27 | 0.044 |
| $W76F/N127A_{ox}$ | 2572 ± 247 | 0.015 ± 0.002 | 0.29 | 0.052 |
| $W76F/Q74A_{ox}$ | 2442 ± 166 | 0.030 ± 0.002 | 0.64 | 0.046 |
| $W76F/N127A/Q74A_{ox}$ | 2877 ± 143 | 0.036 ± 0.005 | 0.86 | 0.042 |

Table 6.6. Lifetimes ($\tau$) and Wavelength Independent Amplitude Fractions ($\alpha$) and $\chi^2_R$ as Obtained by Global Analysis and Decay Associated Spectra[29] of the Fluorescence Decay of DsbA and its Variants in the Oxidized and Reduced States

| | $\tau_1$ (ns) ($\alpha_1$) | $\tau_2$ (ns) ($\alpha_2$) | $\tau_3$ (ns) ($\alpha_3$) | $\chi_R^{2}$* |
|---|---|---|---|---|
| $W76F_{ox}$ | 0.14 ± 0.01 | 1.81 ± 0.03 | 3.94 ± 0.01 | 3.9 |
| | (0.94 ± 0.04) | (0.050 ± 0.001) | (0.01 ± 0.04) | |
| $W76F_{red}$ | 0.14 ± 0.03 | 1.73 ± 0.1 | 3.96 ± 0.1 | 2.8 |
| | (0.93 ± 0.02) | (0.06 ± 0.01) | (0.01 ± 0.02) | |
| $W76F/N127A_{ox}$ | 0.12 ± 0.01 | 1.33 ± 0.2 | 3.16 ± 0.1 | 3.8 |
| | (0.92 ± 0.02) | (0.037 ± 0.005) | (0.04 ± 0.02) | |
| $W76F/Q74A_{ox}$ | 0.14 ± 0.01 | 0.83 ± 0.1 | 3.07 ± 0.06 | 3.1 |
| | (0.77 ± 0.02) | (0.07 ± 0.008) | (0.15 ± 0.02) | |
| $W76F/N127A/Q74A_{ox}$ | 0.13 ± 0.01 | 1.03 ± 0.1 | 3.57 ± 0.06 | 2.4 |
| | (0.75 ± 0.02) | (0.05 ± 0.004) | (0.20 ± 0.02) | |

*The high $\chi^2_R$ is due to the very low intensity cfr. the quantum yields.

Trp in solution. We therefore looked for dynamic quenchers in the neighbourhood of W126. The only two candidates within collisional distances were the amide groups of Q74 and N127. Replacing Q74 and N127 by alanine indeed reduced the nonradiative rate constant to lower values.

The remaining questions are: how does N127 and Q74 quench the fluorescence of W126 and why is the remaining nonradiative rate constant still quite high ($1.12\,ns^{-1}$)? Inspection of the lifetime data (Table 6.6) reveals that the lifetimes themselves hardly change or even decrease upon replacement of Q74 or N127. A detailed quenching analysis (Table 6.7) can reveal the origin of the increase in quantum yield. Upon removal of the amide of Q74 or N127 there is no or only a small decrease in quantum yield due to static quenching ($f_{kr}$) and also a decrease in the quantum yield due to

Table 6.7. Relative Quenching Analysis: Static Quenching ($k_r/k_{r0}$), Dynamic Quenching ($f_{DQ}$) and Decrease in the Fluorescence Intensity of W126 by the Change of Microconformations ($f_{PR}$) with the DsbA Variant W76F as Reference State

| | $Q/Q_0$ | $f_{kr}$ | $f_{PR}$ | $f_{DQ}$ |
|---|---|---|---|---|
| W76F → W76F/ N127A | 1.15 | 1.05 | 1.35 | 0.81 |
| W76F → W76F/ Q74A | 2.23 | 0.93 | 3.16 | 0.76 |
| W76F → W76F/N127A/Q74A | 2.73 | 0.82 | 3.76 | 0.88 |

dynamic quenching. The only reason why the total quantum yield increases upon removal of the amide of Q74 or N127 is due to the factor $f_{PR}$ which represents a fluorescence increase due to population reshuffling. This indicates that amide groups are no direct quenchers of tryptophan fluorescence. The cause of the strong quenching of W126 in WT is due to the high population of the shortest lifetime.

Replacement of Q74 or N127 allow for micro conformations with higher lifetimes to become more populated and thus increases the fluorescence. A similar phenomenon, where a thermally induced increase in fluorescence intensity of Trp-X peptides is due to the higher population of the longer lifetime has been reported before.[31]

### 6.3.2. Molecular Mechanics

The micro conformation of W126 with the shortest lifetime is 95% populated. Replacement of N127 increases the population of the longest lifetime from 1 to 4%, replacement of Q74 increases the population of the longest lifetime to 16% and replacements of both to 20%. To investigate if it is possible for Trp to change its conformation an energy map was calculated. The energy map calculation reveals that there are two energetically possible conformations of W126 in DsbA in both the wild type and the W76F/Q74A variant (Figure 6.2). Calculated energy map of W126 ($\chi_1$ and $\chi_2$) in DsbA wild type, W76F and W76F/Q74A variant. Energy is expressed in kcal/mol and is relative to the lowest energy. It is interesting to note that in the X-ray structure of reduced DsbA one energy minimum is populated (anti-perpendicular)[10] while in the NMR structure of reduced DsbA the other energy minimum is populated(perpendicular).[11] The middle lifetime has to result from other conformations, not highly populated (compare with ±5% in the fluorescence measurements) in the experimentally determined structures nor in the calculations.

### 6.3.3. Linking the Conformations with the Lifetimes

N-bromosuccinimide (NBS) reacts irreversibly with tryptophan generating a totally nonfluorescent oxindole product.[32] NBS reacts with the pyrrole ring of tryptophan.[33] Thus for tryptophan in proteins NBS will react preferentially with those tryptophans which have a solvent exposed pyrrole ring. Analysis of the NMR structure in the reduced DsbA reveals that in the anti conformation the pyrrole ring of W126 is the most exposed. This structure in the vicinity of W126 is the same in both the reduced as the oxidized state.[29]

Lifetime determination of DsbA W76F/N127A/Q74A that had reacted with increasing amounts of NBS reveals that the amplitude fraction of the longest lifetime shows the strongest decrease. Thus, the reaction with NBS identifies the longest lifetime of W126 with the most exposed and therefore with the anti conformation ($\chi_1 = 139°$ and $\chi_2 = -103°$). When the steady state fluorescence of the DsbA W76F/N127A/Q74A variant is followed upon reaction with a tenfold excess of NBS an initial fluorescence decrease is followed by a slow fluorescence increase (Figure 6.3).

Lifetime measurements in the course of the reaction show again that it is the long lifetime component that recovers. Our interpretation is that NBS reacts preferentially with the exposed anti conformation, and that reshuffling from the other microstates is responsible for the fluorescence recovery. (It should be noted that NBS reagents slowly hydrolyzes.) A molecular dynamics simulation of the reduced variant W76F/Q74A reveals that the carbonyl carbon of the backbone of W126 is closer in the perpendicular conformation ($\chi_1 = 169°$ and $\chi_2 = 77°$). Because carbonyl quenches the fluorescence of tryptophan[34] the lifetime of this conformation could be lower.[37] Thus this conformation is linked to the smallest and/or middle lifetime.

## 6.4. Overall Scheme of the Quenching in DsbA

The overall scheme of rate constants (Scheme 6.1) and energy transfer (Table 6.4) gives a picture of the fluorescence decay pathway of the two tryptophans in DsbA. Tryptophan 76 is quenched by both energy transfer to W126 and by dynamic quenching by the disulfide, mediated by F26. There is an additional fluorescence change of W76 upon reduction of the oxidized DsbA, most likely due to a conformational change of W76 (not shown in scheme 6.1). W126 has only a weak fluorescence intensity due to the high population of the smallest lifetime and population of other conformations appears to be hindered by Q74 and N127. Removing Q74 or N127 increases the fluorescence of W126, giving rise to a virtual quenching constant shown in Scheme 6.1.

## 6.5. Conclusion

In conclusion, we have shown that only one tryptophan, W76, is responsible for the fluorescence increase of DsbA upon reduction of the active-site disulphide. In oxidized DsbA, the fluorescence of W76 is diminished by an intramolecular, dynamic quenching mechanism, involving contacts with F26

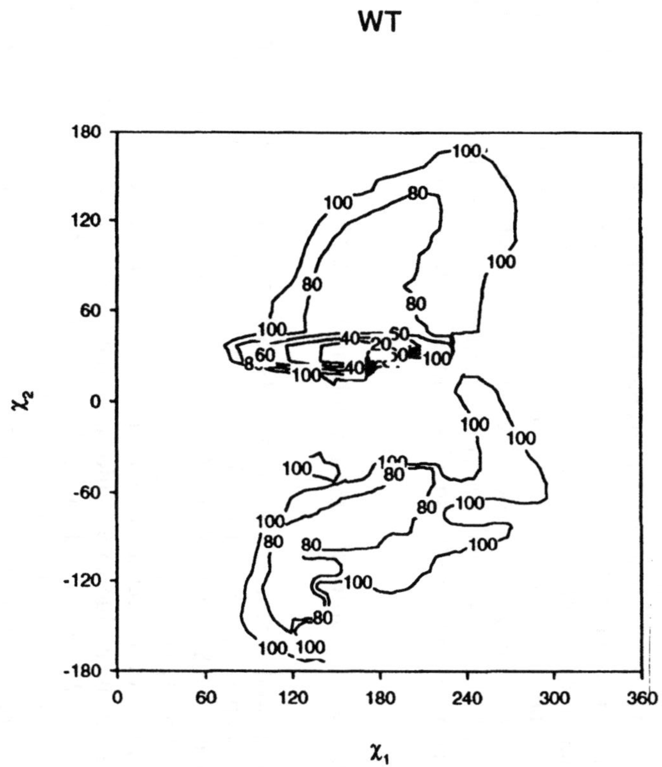


Figure 6.2. Calculated energy map of W126 ($\chi_1$ and $\chi_2$) in DsbA wild type, W76F and W76F/Q74A variant.

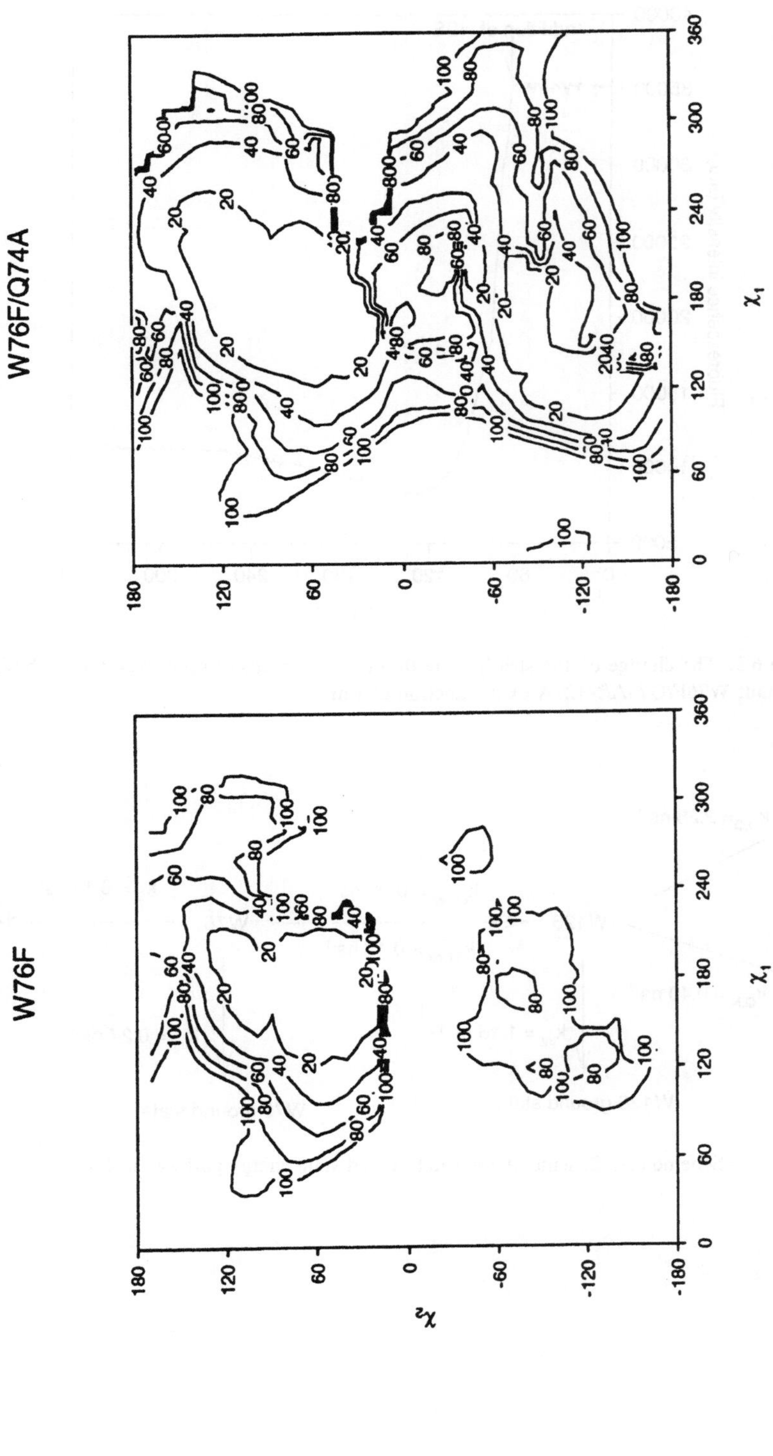


Figure 6.2. *Continued*

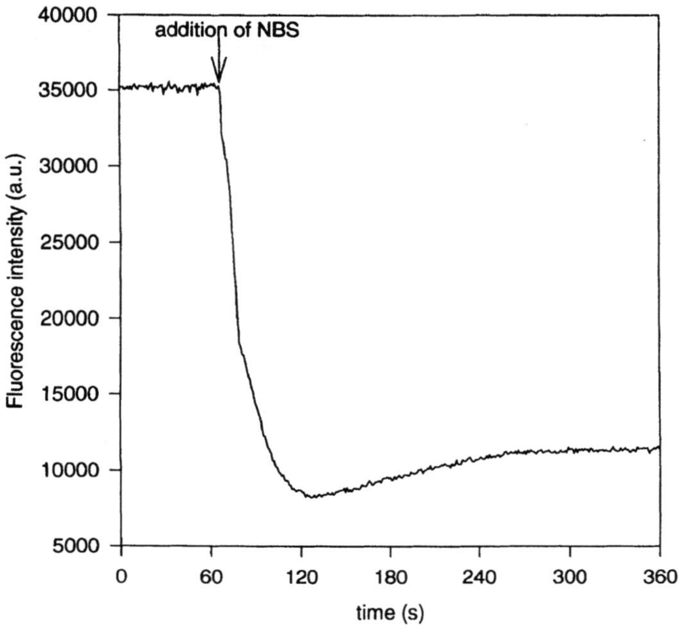


Figure 6.3. The change of the steady state fluorescence intensity upon reaction of NBS with thevariant W76F/Q74A/N127A ox as function of time.

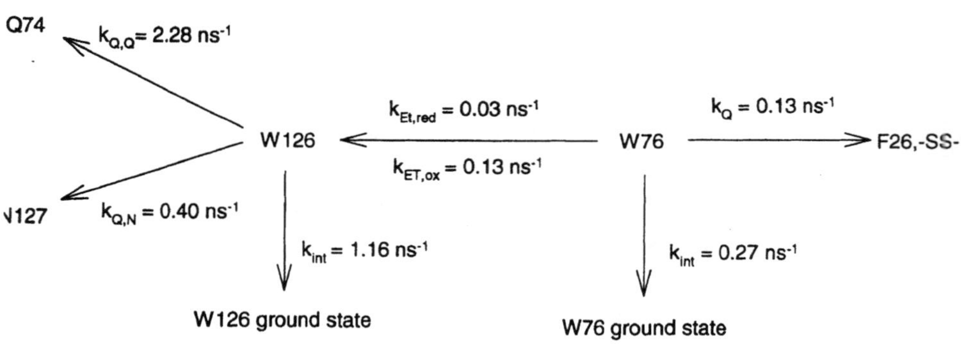


Scheme 6.1. Scheme of the total excited state energy pathway in DsbA.

and the disulphide, and by energy transfer to W126. In the reduced WT, only the energy transfer to W126 remains as the major quenching mechanism. The increase in fluorescence intensity of W126 upon removal of the neighboring amides of N127 and Q76 is not due to the removal of collisional quenchers but due to the fact that more space becomes available around W126 making it possible for the tryptophan to populate conformations which are less quenched. The high $k_{nr}$ in the triple mutant is due to the fact that the conformation with the lowest lifetime is still highly populated (75%). The high $k_{nr}$ is due to the proximity of the carbonyl carbon of the backbone of W126. The amide groups do not quench tryptophan fluorescence directly. Our results indicate that the multiple exponential fluorescence decay is observed for DsbA caused by multiple micro conformations of tryptophan in the protein matrix[35] that slowly interchange from one conformation to the other or due to different conformations of DsbA itself with different conformations of the tryptophan. Our NBS experiments give an idea of the timescale on which large amino acids like tryptophan which are partially buried in the protein matrix change conformation. The timescale of the process is in the seconds range in this protein, indicating that microstate reshuffling could be linked to major reorganization within the protein. This would also explain why so many conformational changes in proteins are accompanied by fluorescence changes.

## References


1. H. F. Gilbert. Molecular and cellular aspects of thiol-disulfide exchange. Adv. Enzymol. Relat. Areas Mol. Biol. 63, 69–172 (1990).
2. J. C. A. Bardwell and J. Beckwith. The bonds that tie: catalyzed disulfide bond formation. Cell 74, 769–771 (1993).
3. H. Loferer and H. Hennecke. Protein disulphide oxidoreductase in bacteria. Trends Biochem. Sci. 19, 169–171 (1994).
4. A. Holmgren and B. M. Sjöberg. Immunochemistry of thioredoxin. I. Preparation and cross-reactivity of antibodies against thioredoxin from Escherichia coli and bacteriophage T4. J. Biol. Chem. 247(13), 4160–4164 (1972).
5. M. Wunderlich and R. Glockshuber. Redox properties of protein disulfide isomerase (DsbA) from Escherichia coli. Protein Sci. 2(5), 717–726 (1993).
6. H. Loferer, M. Wunderlich, H. Hennecke and R. Glockshuber. A bacterial thioredoxin-like proteinthat is exposed to the periplasm has redox properties comparable with those of cytoplasmic thioredoxins. J. Biol. Chem. 270 (44), 26178–26183 (1995).
7. J. C. A. Bardwell, K. McGovern and J. Beckwith. Identification of a protein required for disulfidebond formation in vivo. Cell 67, 581–589 (1991).
8. S. Kamitani, Y. Akiyama and K. Ito. Identification of an Escherichia coli gene required for the formation of correctly folded alkaline phosphatase, a periplasmic enzyme. EMBO J. 11, 57–62 (1992).

9. L. W. Guddat, J. C. Bardwell, T. Zander and J. L. Martin. The uncharged surface features surrounding the active site of Escherichia coli DsbA are conserved and are implicated in peptide binding. Protein Sci. 6, 1148–1156 (1997).
10. L. W. Guddat, J. Bardwell and J. L. Martin. Crystal structures of reduced and oxidized DsbA: investigation of domain motion and thiolate stabilization. Structure 6, 757–767 (1998).
11. H. J. Schirra, C. Renner, M. Czisch, M. Huber-Wunderlich, T. A. Holak and R. Glockshuber. Structure of reduced DsbA from Escherichia coli in solution. Biochemistry 37, 6263–6276 (1998.).
12. J. L. Martin. Thioredoxin—a fold for all reasons. Structure 3, 245–250 (1995).
13. A. Zapun, J. C. Bardwell and T. E. Creighton. The reactive and destabilizing disulfide bond of DsbA, a protein required for protein disulfide bond formation in vivo. Biochemistry 32(19), 5083–5093 (1993).
14. M. Wunderlich, A. Otto, R. Seckler and R. Glockshuber. Bacterial protein disulfide isomerase: efficient catalysis of oxidative protein folding at acidic pH. Biochemistry 32(45), 12251–12256 (1993).
15. U. Grauschopf, J. R. Winther, P. Korber, T. Zander, P. Dallinger and J. C. A. Bardwell. Why is DsbA such an oxidizing disulfide catalyst? Cell 83, 947–955 (1995).
16. Y. Chen and M. D. Barkley. Toward understanding tryptophan fluorescence in proteins. Biochemistry 37, 9976–9982 (1998).
17. N. Rouviere, M. Vincent, C. T. Craescu and J. Gallay. Immunosupressor binding to the immunophilin FKBP59 affects the local structural dynamics of a surface beta-strand: time resolved fluorescence study. Biochemistry 36, 7339–7352 (1998).
18. F. Merola, R. Rigler, A. Holmgren and J.-C. Brochon. Picosecond Tryptophan fluorescence of thioredoxin: evidence for discrete species in slow exchange. Biochemistry 28, 3383–3398 (1989).
19. J. Hennecke, A. Sillen, M. Huber-Wunderlich, Y. Engelborghs and R. Glockshuber. Quenching of tryptophan fluorescence by the active-site disulfide bridge in the DsbA protein from Escherichia coli. Biochemistry 36, 6391–6400 (1997).
20. S. K. Katti, D. M. LeMaster and H. Eklund. Crystal structure of thioredoxin from Escherichia coli at 1.68 D resolution. J. Mol. Biol. 212(1), 167–184 (1990).
21. M. F. Jeng, A. P. Campbell, T. Begley, A. Holmgren, D. A. Case, P. E. Wright and H. J. Dyson. High-resolution solution structures of oxidized and reduced Escherichia coli thioredoxin. Structure 2(9), 853–868 (1994).
22. G. Krause and A. Holmgren. Substitution of the conserved tryptophan 31 in Escherichia coli thioredoxin by site-directed mutagenesis and structure-function analysis. J. Biol. Chem. 266(7), 4056–4066 (1991).
23. I. Slaby, V. Cerna, M. F. Jeng, H. J. Dyson and A. Holmgren. Replacement of Trp28 in Escherichia coli thioredoxin by site-directed mutagenesis affects thermodynamic stability but not function. J. Biol. Chem. 271(6), 3091–3096 (1996).
24. R. Loewenthal, J. Sancho and A. R. Fersht. Fluorescence spectrum of barnase: contribution of three tryptophan residues and a histidine-related pH dependence. Biochemistry 30(27), 6775–6779 (1991).
25. K. Willaert, R. Loewenthal, J. Sancho, M. Froeyen, A. R. Fersht and Y. Engelborghs. Determination of the excited-state lifetimes of the tryptophan residues in barnase, via multifrequency phase fluorometry of tryptophan mutants. Biochemistry 31(3), 711–716 (1992).
26. R. W. Cowgill. Fluorescence and protein structure XI. Fluorescence quenching by disulfide and sulfhydryl groups. Biochim. Biophys. Acta 140, 37–44 (1967).
27. A. G. Szabo and D. M. Rayner. Fluorescence decay of tryptophan conformers in aqueous solutions. J. Am. Chem. Soc. 102, 554–563 (1980).

28. A. Sillen and Y. Engelborghs. The correct use of "Aaverage" fluorescence parameters. Photochem. Photobiol. 67(5), 475–486 (1998).
29. A. Sillen, J. Hennecke, D. Roethlisberger, R. Glockshuber and Yves Engelborghs. Fluorescence quenching in the DsbA protein from Escherichia coli. The complete picture of the excited state energy pathway and evidence for the reshuffling dynamics of the microstates of tryptophan. Proteins: Struc. Func. Genet. 37, 253–263 (1999).
30. M. R. Eftink, Y. Jia, D. Hu and C. A. Ghiron. Fluorescence studies with tryptophan analogues: excited state interactions involving the side chain amino group. J. Phys. Chem. 99, 5713–5723 (1995).
31. L. Brancaleon, G. Gasparini, M. Manfredi and A. Mazzini. A model for the explanation of the thermally induced increase of the overall fluorescence in tryptophan-X peptides. Archiv. Biochem. Biophys. 348, 125–133 (1997).
32. T. Imoto, L. S. Forster, J. A. Ruplay and F. Tanaka. Fluorescence of lysozyme: Emission from tryptophan residues 62 and 108 and energy migration. Proc. Natl. Acad. Sci. USA 69, 1151–1155 (1972).
33. N. M. Green and B. Witkop. Oxidation studies of indoles and the tertiary structure of proteins. Trans. N.Y. Acad. Sci. 26, 659–669 (1964).
34. Y. Chen, B. Liu, H.-T. Yu and M. D. Barkley. The peptide bond quenches indole fluorescence. J. Am. Chem. Soc. 118, 9271–9278 (1996).
35. T. E. S. Dahms, K. J. Willis and A. G. Szabo. Conformational heterogeneity of tryptophan in a protein crystal. J. Am. Chem. Soc. 117, 2321–2326 (1995).
36. Kraulis PJ. MOLSCRIPT: A program to produce both detailed and schematic plots of protein structures. J. App. Crystalogr. 24, 946–950 (1991).
37. A. Sillen, J. F. Díaz and Y. Engelborghs. A step toward the prediction of the fluorescence lifetimes of tryptophan residues in proteins based on structural and spectral data. Protein Sci. 9, 158–169 (2000).

# 7

# The Conformational Flexibility of Domain III of Annexin V is Modulated by Calcium, pH and Binding to Membrane/Water Interfaces

Jacques Gallay, Jana Sopková, and Michel Vincent

## 7.1. Introduction

Annexin V belongs to a family of water-soluble proteins, which bind reversibly to negatively charged phospholipid model membranes and to specific cellular membranes.[1-3] This binding is calcium-dependent and is reversible by EDTA at neutral pH. Annexins are widely distributed in different species, tissues and cell types. They are abundant in most eukaryotic cells, where they represent up to 1% of the total cell proteins. They are likely involved in important physiological functions, related most probably to their ability to bind to membranes, although the particular physiological roles of each member of the family still remains precisely unknown. Some annexins appear to be involved in various types of membrane fusion events occurring in endo- and exocytosis; others exhibit anti-inflammatory and anticoagulant properties *in vitro*.[2] Some of these proteins display ion channel activity *in vitro*.[3]

Initially solved for annexin V[4-11] and later for annexins I, II, III, IV, VI and XII,[12-18] the crystal structures of many of these proteins show that all these proteins are constituted by a conserved core of about 300 amino acids in length, organized in a cyclic array with four-fold repeats of 70 residues, each constituting a structural domain, with the exception of annexin VI which contains two conserved cores. The core exhibits a compact bent disk

---

*Jacques Gallay, Jana Sopková, and Michel Vincent* • Laboratoire pour l'Utilisation du Rayonnement Electromagnétique, Université Paris-Sud, Orsay cedex, France.

*Topics in Fluorescence Spectroscopy, Volume 6: Protein Fluorescence*, edited by Joseph R. Lakowicz. Kluwer Academic / Plenum Publishers, New York, 2000

shape with convex and concave faces. Each domain comprises five $\alpha$-helices (named from A to E), wrapped into a right-handed super-helix and a principal calcium binding site situated on the convex face of the molecule. This led to surmise that this side is oriented towards the membrane surface, the calcium ions making bridges between the negatively charged head groups of the phospholipid molecules and the protein. This hypothesis is compatible with the results of two-dimensional electron microscopy studies.[19–21] The N-terminal segment is more variable and contains specific sites of phosphorylation and of interaction with other proteins.[2]

The knowledge of these structures has allowed a better understanding of the molecular basis of the mechanism of interaction of annexins with calcium ions and membranes. The calcium ion is bound to carbonyl oxygens of the loops connecting helices A and B, and to a carboxyl of the negatively charged amino acid side-chain (Glu or Asp) about 40 residues downstream, in the loop connecting helices D and E in the same domain. The calcium-binding configuration is different from the classical E-F hand type[22] but resembles that found in phospholipase $A_2$.[23] Nevertheless, the annexin calcium-binding sites are highly exposed on the surface of the molecule, while the single calcium site of phospholipase $A_2$ lies within the enzymatic site cavity. This suggests different modes of interaction of these two proteins with phospholipids.

In annexin V, a particular situation prevails however. The occurence of the calcium-binding site in domain III requires a large conformational change to take place. This change was observed by X-ray diffraction studies,[8,9,24] showing that the IIIA-IIIB loop is brought from a buried position onto the surface of the protein. At the same time, the unique tryptophan residue (Trp187) present in the IIIA-IIIB loop becomes exposed to the solvent at the protein surface. This conformational change was detected in solution by a large red shift of the steady-state fluorescence emission spectrum at high calcium concentrations, which questions the specificity of the effect of the divalent ion.[25]

A model of annexin V-membrane complexes has been proposed from X-ray diffraction[11] and steady-state fluorescence studies.[26–30] In the crystal structure of complexes of annexin V with glycerophosphoserine, used as an analogue of the negatively charged phospholipid polar head group, the Trp187 residue has been found to be situated in close contact with the glycerol moiety.[11] This observation was extrapolated to the real membrane bilayer. In this model, the indole ring is expected to be inserted into the first carbon region of the phospholipid and to participate by hydrophobic interactions to the stabilization of the annexin/membrane complex. This model contains predictive features which can be tested, regarding in particular the mobility of the tryptophan residue, of its environment (protein and acyl chains), the position of the indole ring inside the lipid bilayer and the hydrophobicity of its micro-environment.

In order to check these predictive features, we evaluated the effect of calcium binding and of the protein interaction with model membranes on the conformation and dynamics of domain III of the protein. For this purpose, time-resolved fluorescence intensity and anisotropy decay measurements of the single tryptophan residue Trp187 were performed. These techniques allow to quantify specifically the changes of the local dynamics around the Trp187 residue induced in different experimental conditions. Two membrane systems were used: small unilamellar vesicles of phosphatidylcholine/phosphatidylserine (SUV) at different lipid/protein molar ratios (L/P) and reverse micelles of surfactant in organic solvent.[31,32] This last system provides an experimental model of membrane/water interface optically transparent, in which the proton activity is high and the availability of water molecules for hydration is limited. We also studied the effect of pH on the Trp187 fluorescence parameters in order to define a plausible mechanism of the calcium-induced conformational change of domain III.

## 7.2. Experimental Procedures

### 7.2.1. Protein Preparation and Chemicals

Phospholipids (1-palmitoyl-2-oleoyl-*sn*-phosphocholine, POPC, and 1-palmitoyl-2-oleoyl-*sn*-phosphoserine, POPS) were obtained from Serdary Research. Sodium bis(2-ethylhexyl) sulfosuccinate (Aerosol OT, AOT) was purchased from Sigma and used as supplied. Recombinant human annexin V was prepared as described.[33] In this procedure, all calcium is removed during the purification by EDTA and the protein is stored in the absence of calcium. For measurements of absorbance, circular dichroism and fluorescence, the protein solutions were prepared in 50 mM Tris-HCl pH 7.5, 0.15 M NaCl. All chemicals were of analytical grade purity, obtained from Merck, France.

### 7.2.2. Preparation of Phospholipidic Vesicles and Reverse Micelles

The phospholipid suspensions were prepared by the sonication method. The chloroformic solution containing POPC and/or POPS was evaporated to dryness in a glass tube under a stream of nitrogen followed by primary vacuum during several hours. Hydration of the sample was achieved with buffer and after vortexing, the multilamellar vesicles formed were sonicated at room temperature with the micro-tip of a Branson-B12 sonicator during five minutes with half-duty cycles. The POPS/POPC molar ratio of the vesicles was varied from 10% to 25%. Reverse micelles were prepared as previously described[34] with 0.1 M AOT in isooctane and the desired water/surfactant molar ratio ($w_0$) from 2.8 to 50. Solubilization of the protein in reverse micelles (0.6 mg/ml for fluorescence, 0.15 mg/ml for far-UV CD) was achieved by sonication in a Branson-type bath sonicator for few minutes.

### 7.2.3. Steady-State Fluorescence Measurements

Tryptophan fluorescence emission, excitation and excitation anisotropy spectra were recorded on a SLM 8000 spectrofluorometer, using 5 × 5 mm (for the samples containing lipid vesicles) or 10 × 10 mm (for the other samples) optical path cuvettes. Blanks were always subtracted in the same experimental conditions. To remove polarization artifacts, the fluorescence emission spectra were reconstructed from the four polarized spectra as described previously.[35]

### 7.2.4. Time-Resolved Fluorescence Measurements

Fluorescence intensity decays were obtained by the time-correlated single photon counting technique from the polarized components $I_{vv}(t)$ and $I_{vh}(t)$ on the experimental set-up installed on the $SB_1$ window of the synchrotron radiation machine Super-ACO (Anneau de Collision d'Orsay), which has been described elsewhere.[35,36] The storage ring provides a light pulse with a full width at half maximum (FWHM) of ~500 ps at a frequency of 8.33 MHz for a double bunch mode. A Hamamatsu microchannel plate R1564U-06 was utilized to detect the fluorescence photons. Data for $I_{vv}(t)$ and $I_{vh}(t)$ were stored in separated 2 K memories of a plug-in multichannel analyzer card (Canberra). The automatic sampling of the data was driven by the microcomputer. The instrumental response function was automatically collected each 5 minutes by measuring the scattering of a glycogen solution at the emission wavelength during 30 s, in alternation with the parallel and perpendicular components of the polarized fluorescence decay, which were cumulated during 90 s. The time resolution was usually in the range of 10–20 ps per channel. The light scattering by the lipid vesicles was strongly reduced by interposing a 1 M $CuSO_4$ filter (1-cm optical path) on the emission side. Blanks were substracted.

### 7.2.5. Analysis of the Time-Resolved Fluorescence Data

Analyses of fluorescence intensity and anisotropy decay as sums of exponentials were performed by the maximum entropy method.[37–39] The programs use the commercially available library of subroutines MEMSYS 5 (MEDC Ltd., U.K.). Details of the principles and application of the method to fluorescence decays have been previously published.[40–46] They will be summarized in the following.

#### 7.2.5.1. *Fluorescence Polarized Fluorescence Intensity Decays*

In the general case where a chromophore is emitting with a lifetime $\tau$ and rotates with a rotational correlation time $\theta$, the expression of each impulse polarized fluorescence intensity decay is:

$$I_{vv}(t)=\frac{1}{3}\int_0^{\infty}\int_0^{\infty}\int_{-0.2}^{0.4}\gamma(\tau,\theta,\mathrm{A})\,\mathrm{e}^{-t/\tau}\left(1+2Ae^{-t/\theta}\right)d\tau d\theta dA \tag{7.1}$$

and

$$I_{vh}(t)=\frac{1}{3}\int_0^{\infty}\int_0^{\infty}\int_{-0.2}^{0.4}\gamma(\tau,\theta,\mathrm{A})\,\mathrm{e}^{-t/\tau}\left(1-Ae^{-t/\theta}\right)d\tau d\theta dA \tag{7.2}$$

where $\gamma(\tau, \theta, A)$ is the chromophore population with lifetime $\tau$, rotational correlation time $\theta$ and intrinsic anisotropy $A$. If a single intrinsic anisotropy value $A$ is expected, like for the case of a single chromophore, the above expressions can be simplified to:

$$I_{vv}(t)=\frac{1}{3}\int_0^{\infty}\int_0^{\infty}\Gamma(\tau,\theta)\,\mathrm{e}^{-t/\tau}\left(1+2Ae^{-t/\theta}\right)d\tau d\theta \tag{7.3}$$

and

$$I_{vh}(t)=\frac{1}{3}\int_0^{\infty}\int_0^{\infty}\Gamma(\tau,\theta)\,\mathrm{e}^{-t/\tau}\left(1-Ae^{-t/\theta}\right)d\tau d\theta \tag{7.4}$$

To obtain the target distribution $\Gamma(\tau, \theta)$, the entropy function $S$:[47,48]

$$S=\int_0^{\infty}\int_0^{\infty}\left[\Gamma(\tau,\theta)-m(\tau,\theta)-\Gamma(\tau,\theta)\log\frac{\Gamma(\tau,\theta)}{m(\tau,\theta)}\right]d\tau d\theta \tag{7.5}$$

is maximized. In this expression, $m(\tau, \theta)$ is the starting distribution chosen as a flat surface over the explored $(\tau, \theta)$ domain, which corresponds to the lowest *a priori* knowledge about the final distribution. A global analysis of $I_{vv}(t)$ and $I_{vh}(t)$ is performed which is constrained by:

$$\sum_{k=1}^{M} \frac{\left(I_{vv,k}^{\ calc} - I_{vv,k}^{\ obs}\right)^2}{\sigma^2_{\ vv,k}} + \frac{\left(I_{vh,k}^{\ calc} - I_{vh,k}^{\ obs}\right)^2}{\sigma_2^{\ vh,k}} \leq 2M \tag{7.6}$$

where $I_{vv,k}^{\ calc}$ and $I_{vh,k}^{\ obs}$ are the $k^{th}$ calculated and observed intensities. $\sigma^2_{\ vv,k}$ and $\sigma^2_{\ vh,k}$ are the variances of the $k^{th}$ point for $I_{vv}(t)$ and $I_{vh}(t)$ respectively.[49] $M$ is the number of (independent) observations of the fluorescence intensity at times $t$. In principle, this analysis allows to describe the lifetime distribution and the association between one particular excited state lifetime and specific rotational correlation time(s). There is nevertheless an inherent limit to this method, since as shown from formula 7.3 and 7.4, the parallel and the perpendicular components of the polarized decay involve in their expressions the harmonic mean $\kappa_i$ between $\tau_i$ and $\theta_i$:

$$1/\kappa_i = 1/\tau_i + 1/\theta_i \tag{7.7}$$

where $\tau_i$ and $\theta_i$ can be exchanged without any modification in the $\kappa_i$ value, leading to construction of iso-kappa curves.[39] Such curves were constructed and represented as dotted lines in the different figures of the paper. The improvements of the computer power and calculation rate however allow now to reduce this bias for most of its part. Calculations were performed on a DEC alpha computer Vax 7620. The program including the MEMSYS 5 subroutines was written in double precision FORTRAN 77. CPU time of ~2 hours ($10^3$ iterations) was required to achieve the global analysis of $I_{vv}(t)$ and $I_{vh}(t)$ with 40 values respectively for $\tau$ and $\theta$.

### *7.2.5.2. Excited State Lifetime Distribution*

In practice, an analysis of the fluorescence intensity decay is first performed. For this purpose, the intensity is classically reconstructed from the polarized fluorescence decays by adding the parallel and twice the perpendicular components:

$$T(t) = I_{vv}(t) + 2\beta_{corr} I_{vh}(t) = \int_0^{\infty} \alpha(\tau)\exp(-t/\tau)d\tau \tag{7.8}$$

where $\beta_{corr}$ is the correction factor[49] taking into account the difference of transmission of the polarized light components by the optics and $\alpha(\tau)$ is the lifetime distribution. The recovered distribution $\alpha(\tau)$ which maximizes the entropy function S:

$$S = \int_0^\infty \left[ \alpha(\tau) - m(\tau) - \alpha(\tau) \log \frac{\alpha(\tau)}{m(\tau)} \right] d\tau \qquad (7.9)$$

is chosen. In this expression, $m(\tau)$ is the starting model for which a flat map over the explored ($\tau$) domain is chosen since no *a priori* knowledge about the final distribution is available. The analysis is bound by the constraint:

$$\sum_{k=1}^{M} \frac{\left(T_k^{\,calc} - T_k^{\,obs}\right)^2}{\sigma^2{}_k} \leq M \qquad (7.10)$$

where $T_k^{calc}$ and $T_k^{obs}$ are the k$^{th}$ calculated and observed intensities. $\sigma^2_k$ is the variance of the k$^{th}$ point.[49] $M$ is the number of (independent) observations of the fluorescence intensity at times $t$. The center $\tau_j$ of a single class j of lifetimes over the $\alpha(\tau_i)$ distribution is defined as:

$$\tau_j = \frac{\sum_i \alpha_i(\tau_i)\tau_i}{\sum_i \alpha_i(\tau_i)} \qquad (7.11)$$

the summation being performed on the significant values of the $\alpha(\tau_i)$ for the j class.

### 7.2.5.3. *Rotational Correlation Time Distribution*

If all the emitting species are assumed to display the same intrinsic anisotropy and rotational dynamics, equations 7.1 and 7.2 can be rewritten as:

$$I_{vv}(t) = \frac{1}{3}\int_0^\infty \alpha(\tau)\, e^{-t/\tau} d\tau \left[1 + 2\int_0^\infty \beta(\theta)\, e^{-t/\theta} d\theta\right] \qquad (7.12)$$

and

$$I_{vh}(t) = \frac{1}{3}\int_0^\infty \alpha(\tau)\, e^{-t/\tau} d\tau \left[1 - \int_0^\infty \beta(\theta)\, e^{-t/\theta} d\theta\right] \qquad (7.13)$$

with

$$A = \int_0^{\infty} \beta(\theta)d\theta \tag{7.14}$$

where $\beta(\theta)$ is the rotational correlation time distribution and the other symbols have the same meaning as in equation 7.1. The $\alpha(\tau)$ profile is given from a first analysis of $T(t)$ by MEM and is held constant in a subsequent and global analysis of $I_{vv}(t)$ and $I_{vh}(t)$ which provides the distribution $\beta(\theta)$ of correlation times.[39,50,51] 100 rotational correlation time values, equally spaced in logarithmic scale and ranging from 0.01 to 50 ns were used for the analysis of $\beta(\theta)$. The barycenters of the correlation time distribution are calculated as:

$$\theta_j = \frac{\sum_i \beta_i(\theta_i)\theta_i}{\sum_i \beta_i(\theta_i)} \tag{7.15}$$

$\beta_i$ is the contribution of the rotational correlation time i to the class j.

*7.2.5.4. Wobbling-in-Cone Angle Calculation*

Following the Karplus formalism,[52] if the indole ring is subjected to a fast rotational motion which decays exponentially with a relaxation time $\theta$ and reaches a plateau value $P_\infty$, we have:

$$A(t) = A\left[(1-P_\infty)\exp\left(-t/\theta\right)+P_\infty\right]\exp\left(-t/\theta_m\right) \tag{7.16}$$

where $\theta_m$ is the Brownian rotational correlation time of the protein (taken as a sphere) and $A$ the intrinsic anisotropy. If the fast rotational motion corresponds to a correlation function that separates into two time scales ($\theta_1$ and $\theta_2$), the expression of the anisotropy can be written as:

$$A(t) = A\left[(1-P_{\infty,1})\exp\left(-1/\theta_1\right)+P_{\infty,1}\right]\left[(1-P_{\infty,2})\exp\left(-1/\theta_2\right)+P_{\infty,2}\right]\exp\left(-1/\theta_m\right) \tag{7.17}$$

with $\theta_1 << \theta_2 << \theta_m$. $P_{\infty,i}$ are the plateau values of the correlation function describing these internal motions. The above expression of the anisotropy decay can be approximated by:

$$A(t) = \beta_1 \exp(-t/\theta_1) + \beta_2 \exp(-t/\theta_2) + \beta_3 \exp(-t/\theta_m) \qquad (7.18)$$

where $\beta_1 = (1 - P_{\infty,1}) \times A$, $\beta_2 = P_{\infty,1}(1 - P_{\infty,2}) \times A$ and $\beta_3 = P_{\infty,1} \times P_{\infty,2} \times A$

In principle, if the time resolution of the experiment is such that it allows the description of all the rotational motions, the time-zero value of the anisotropy ($A_{t=0} = \Sigma\beta_j$) must be equal to the $A$ value measured in the total absence of molecular motion. If short rotations (<50 ps) cannot be resolved, the extrapolated $A_{t=0}$ value can be smaller than the A value. The order parameter ($S_1$) associated to the subnanosecond motion of the indole ring and the cone semi-angle ($\omega_{max}$) of this motion[53,54] can be calculated from:

$$\sum_{i=n-1}^{i=n} \beta_i / A = P_{\infty,1} = S_1^2 = \left[\tfrac{1}{2}\cos\omega_{max}(1+\cos\omega_{max})\right]^2 \qquad (7.19)$$

Such a modeling of the Trp motion inside a protein is only for the sake of comparison, since the geometry of the motion remains unknown. Moreover, strictly speaking, this model is limited to single lifetime decays. If multiexponential intensity decays are expected, the two-dimensional analysis should be tested to detect the possible associations between lifetime and rotational dynamics. If one excited-state lifetime class is associated with both fast and slow rotations, a similar calculation of the cone semi-angle of wobbling can be performed, taking into account the respective $\Gamma(\tau, \theta)$ coefficients.

### 7.2.6. Absorbance and Circular Dichroism Measurements

UV-difference absorption spectra were measured with a Specord M40 spectrophotometer (Carl Zeiss, Jena). The same concentration of annexin V (~1 mg/ml) in 50 mM Tris-HCl, pH 7.45 was placed in both the sample and the reference beams. $CaCl_2$ was added to the sample and an equivalent volume of buffer A to the reference solution. Spectra were recorded in the 250–340 nm wavelength range, using quartz cuvettes of 1-cm path length. CD spectra were recorded either with a dichrograph Mark V, Jobin Yvon (Longjumeau, France) or with a J-710 spectropolarimeter (Jasco, Japan). The far UV CD spectra were measured between 200–270 nm with a concentration of annexin V of 0.37 mg/ml, in a 0.1 cm optical path cuvette. The near UV CD spectra were measured between 250–310 nm with a protein concentration of 1.8 mg/ml, in a 1 cm optical path cuvette. The bandwidth was 2 nm and the spectra were averaged over 10 scans of 100 nm per minute with an integration time of 0.5 s.

Series of absorption, difference absorption, steady-state fluorescence and circular dichroism (CD) spectra, as obtained in Ca-titration experiments, were analyzed by the principal component variant of factor analysis.[55] As the result of this treatment, spectra of each series are expressed as a linear combination of common orthogonal basic spectra. The coefficients in these linear combinations quantify the contributions of the respective basic spectra to any individual measured spectrum. By relating the coefficients to experimental variables (calcium concentration, temperature) the commonly used one-wavelength dependencies are replaced by the quantities depending on the correlated intensities in all measured wavelengths. The number of independent spectral components was determined using the indicator function[56] and provides information about the complexity of the molecular process underlying the observed spectral changes.

## 7.3. Results

### 7.3.1. Effect of Calcium on the Structure and Dynamics of Domain III of Annexin V

#### *7.3.1.1. UV-Difference Absorption Spectra*

The addition of calcium to annexin induces differences in the aromatic chromophore region (250–330 nm) with three negative maximum at 293, 285 and 275 nm. The increase in calcium concentration does not change in any observable extent either the positions of the difference absorption peaks or their relative intensities. This is shown by the results of the principal component decomposition of the 11 difference absorption spectra, which gave only one significant component. The coefficients of this component reflect the Ca-dependent increase of the difference as shown in Figure 7.1. The plot of the dependence of the band magnitude at 293 nm on calcium addition is shown in Figure 7.1 (inset).

#### *7.3.1.2. Circular Dichroism*

The overall band shape of the CD spectrum in the n-$\pi^*$ and $\pi$-$\pi^*$ transition region of the amide chromophores corresponds to 70% of $\alpha$-helical secondary structure as expected from the crystal structure of annexin V (Figure 7.2A). Upon the first addition of calcium (0.09 mM Calcium into a 0.34 mg/ml (0.01 mM) solution of annexin V) we observe a minor increase of negative CD intensity at 221.5 nm (by ~5% of the original value) (Figure

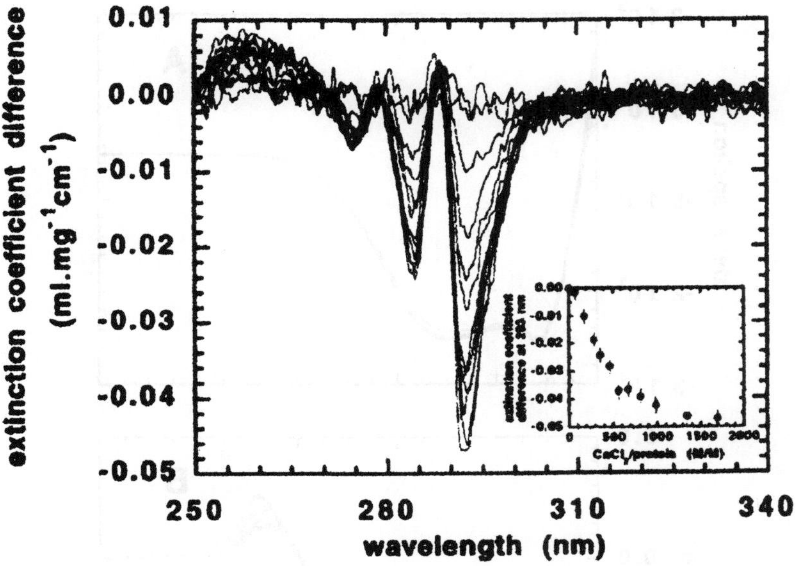


Figure 7.1. *UV-absorption difference spectra of annexin V* as a function of calcium concentration (from 0 to 0.046 M). Inset: titration at 293 nm resulting from the difference spectra (obtained from ref. 25).

7.2A). Upon further addition of calcium, the band shape of the amide UV CD spectrum remains highly conserved; there is no observable band shape change even in the difference CD spectra in this region.

The long-wavelength region of the calcium-free annexin V CD spectrum exhibits low-intensity bands of aromatic chromophores: a positive peak at 292 nm and negative ones at 286, 277, 268 and 262 nm. By contrast to the amide region, the calcium titration induces monotonic changes of CD band shape in this region (Figure 7.2B). By principal component analysis, the series of 11 spectra in this region was found to be composed of two basic spectra, the second one being identical to the difference CD as calculated from the original CD curves (Figure 7.2C). The band shape of this difference basic spectrum describes the correlated intensity decrease of positive and negative maxima at 292 and 286 nm respectively (at 58 mM Calcium it is down to 50% of the original intensities) accompanied by a small blue shift (1–2 nm, Figure 7.2B). No significant changes were observed for the Calcium titration in the negative CD bands at 262, 268 and 277 nm. The plot of the dependence of the band magnitude at 292 nm on calcium addition is shown in Figure 7.2C (inset). The band at 292 nm corresponds to tryptophanyl 0-0 $^1L_b$ band and that at 286 nm is probably the tryptophanyl 0-1 $^1L_b$ band.[56]

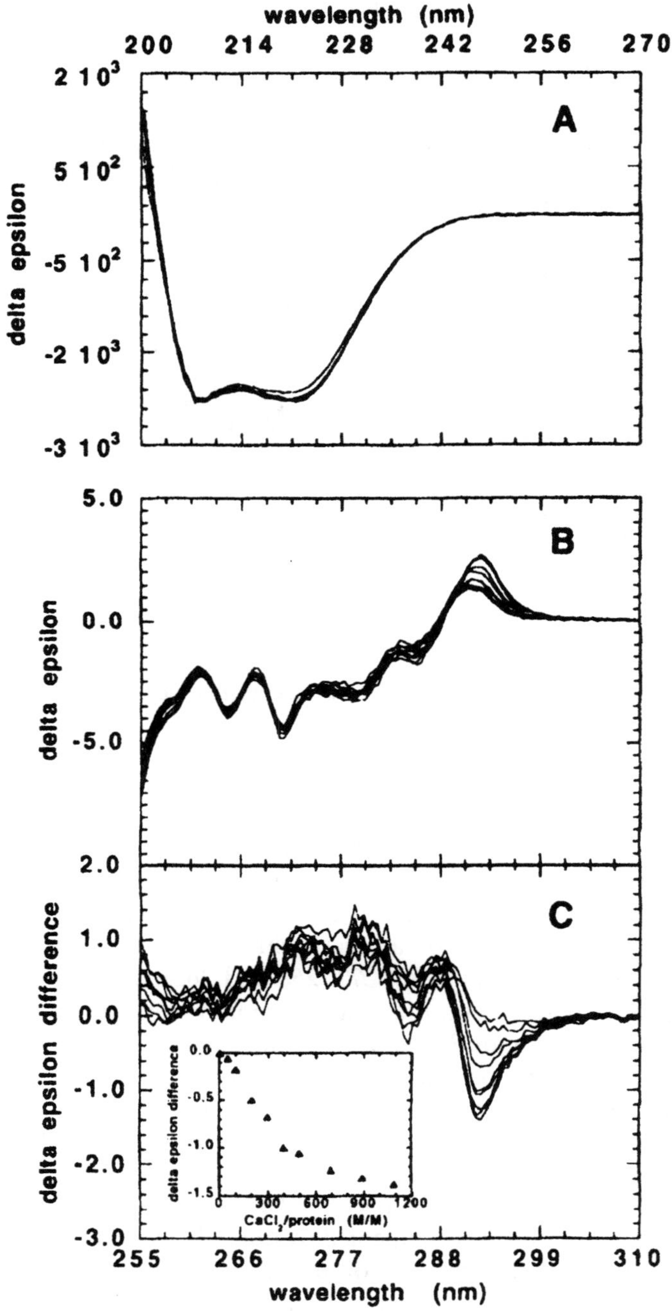


Figure 7.2. *Circular dichroism of annexin V.* A) Far-UV CD spectra of the peptide groups of annexin V as a function of calcium concentration (from 0 to 0.375 M). B) Near-UV CD spectra of the aromatic residues of annexin V as a function of calcium concentration (from 0 to 0.062 M). C) Near-UV CD difference spectra as a function of calcium concentration. Inset: titration of the near-UV CD difference spectra at 292 nm resulting from the difference spectra (from ref. 25).

*7.3.1.3. Steady-State Fluorescence of Trp187*

The steady-state fluorescence spectrum of Trp187 in annexin V at neutral pH is maximum at 326 nm (Figure 7.3). This indicates that the aromatic residue is located in a position weakly accessible to the solvent, in agreement with the crystal structure of the protein without the calcium ion in domain III.[4,10]

Detailed examination of the three-dimensional structure of domain III in the absence of bound calcium (structure A, resolution of 2 Å)[10] shows that

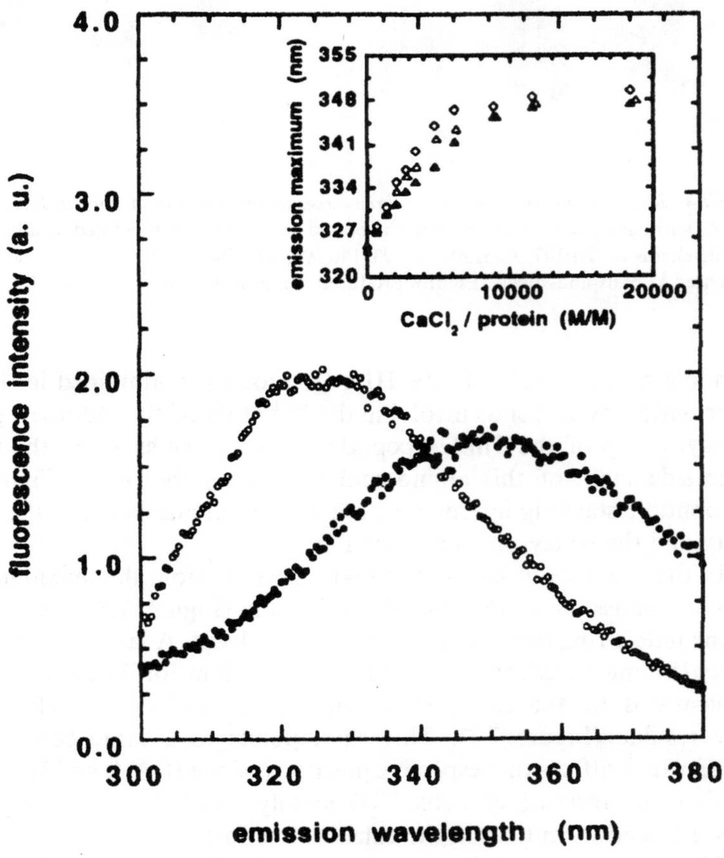


Figure 7.3. *Fluorescence emission spectrum of Trp187 in annexin V.* Full line: calcium-free protein, doted line: calcium-bound protein. Inset: variation of the fluorescence emission maximum as a function of the total calcium/protein mole ratio; Temperature: (△) 10 °C, (▲) 20 °C, (◇ 30 °C. Excitation wavelength: 295 nm. Protein concentration: 0.4 mg/ml. (from ref. 25).

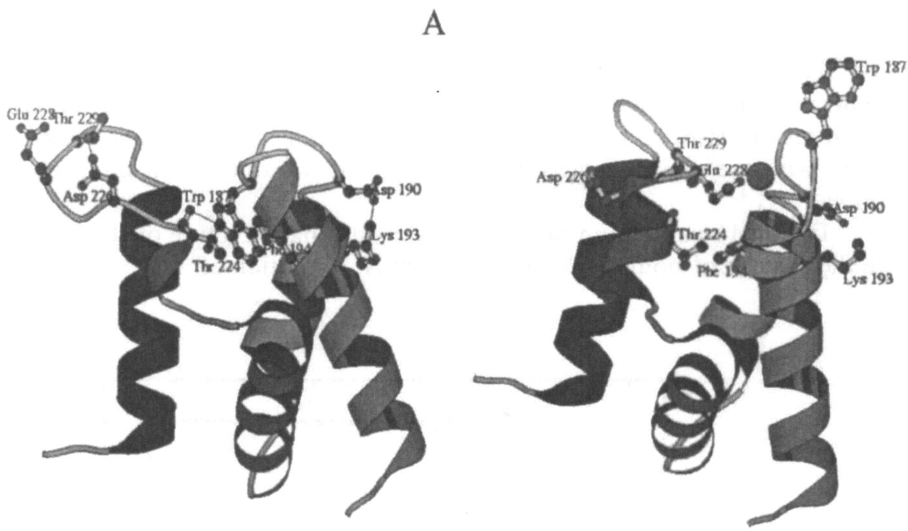


Figure 7.4. *Ribbon representation of the three-dimensional structure of domain III in annexin V.* Form-A: without calcium; form-B: with the bound calcium ion (represented as a ball). Amino-acid side-chains of Trp187, Asp190, Lys193, Phe194, Thr224, Asp226, Glu228 and Thr229 are represented as balls-and-sticks. (Figures produced using Molscript)[57] (from ref. 58).

the indole ring of Trp187 in the IIIA-IIIB loop is maintained in the buried conformation by H-bonds involving the $N_\varepsilon^1$ atom of the indole ring and the carbonyl group of the Thr224 peptide bond on one side and the $O_\gamma$ group of the side-chain of this amino-acid residue on the other (Figure 7.4A). Hydrophobic stacking interactions are also occurring between the aromatic moiety and the benzyl side-chain of Phe194.

In the presence of calcium at high concentration, the maximum of fluorescence emission is red-shifted to 350 nm (Figure 7.3), demonstrating that the indole ring becomes exposed to the solvent. A large perturbations of the Trp187 microenvironment induced by calcium ion binding is therefore demonstrated by the steady-state fluorescence emission spectrum of the W187 residue (Figure 7.3). This corresponds to a large conformational change, which affects the respective positions of the IIIA-B and IIIC-D loops and also the structure of helix IIID mainly, which becomes longer in the calcium-bound form than in the calcium-free form.

The conformational change can be described according to the crystalline structure of the calcium-bound form (P1 structure at 1.9 Å resolution),[9] by a concerted motion of the two loops mentioned above. It brings Glu228 from a surface position in IIID-E loop in the calcium-free form, to a more internal position, in interaction with the bound ion, occupying one valence of the divalent ion (Figure 7.4B).

In this conformation, the indole ring appears no more in contact with protein moieties. In particular, the H-bonds mentioned above with Thr224 are disrupted as well as the possible stacking interactions with Phe residues. In this conformation, the W187 indole ring should be extremely mobile and the IIIA-B loop should be extremely flexible. This should modify also considerably the fluorescence intensity and the anisotropy decays as well.

### *7.3.1.4. Time-Resolved Fluorescence Intensity Decay of Trp187*

The time-resolved fluorescence decay of Trp187 is strongly modified upon calcium binding. In the absence of calcium at neutral pH, three excited state lifetime populations are detected by MEM analysis of the fluorescence intensity decay of Trp187 (Figure 7.5A).

A major lifetime population of 0.9–1 ns represents 72% of the total excited state populations while a shorter one corresponds to 20–24% and a long minor one to only 4–8% (Table 7.1). Decay associated spectra show that the two major excited states display similar emission spectra with a maximum around 320–325 nm (Figure 7.5). The most likely interpretation of the existence of this emission heterogeneity in this case, is a ground-state conformer

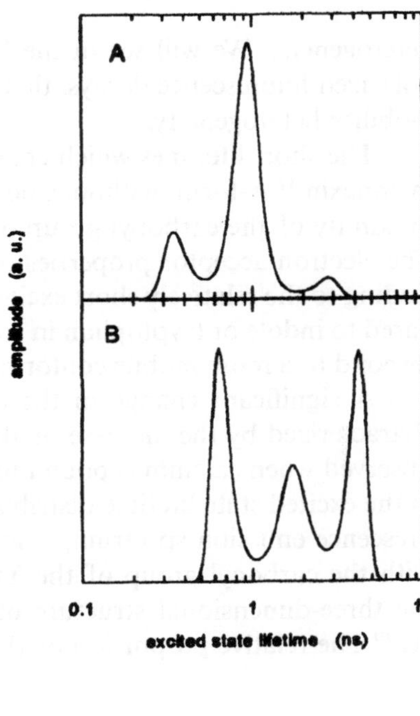


Figure 7.5. MEM reconstituted excited state lifetime distribution of Trp187 in annexin V in the absence of calcium (A) and in the presence of calcium 0.1 M. Excitation wavelength: 295 nm (bandwidth 5 nm). Emission bandwidth: 10 nm. Temperature: 20 °C.

Table 7.1. Fluorescence intensity and anisotropy decay parameters for Trp187 in annexin V in buffer pH 7, in the presence of $CaCl_2$ 10 mM without membranes and as a function of the lipid/protein molar ratio (L/P). Excitation wavelength: 295 nm (bandwidth 4 nm), emission wavelength: 340 nm (bandwidth 8 nm) except for the protein at neutral pH in the absence of calcium (325 nm) (data from ref. 61). $C_j$ is the normalized contribution of the lifetime class $j$

| L/P | $CaCl_2$ M | $\tau_{1(ns)}$ $C_1$ | $\tau_{2(ns)}$ $C_2$ | $\tau_{3(ns)}$ $C_3$ | $\tau_{4(ns)}$ $C_4$ | $\langle\tau\rangle$ (ns) | $\theta_{1(ns)}$ $\beta_1$ | $\theta_{2(ns)}$ $\beta_2$ | $\theta_{3(ns)}$ $\beta_3$ | $A_{t=0}$ | $S$ | $\omega_{max}$ (°) |
|---|---|---|---|---|---|---|---|---|---|---|---|---|
| 0 | 0 | 0.32 | 0.95 | 2.67 | — | 0.94 | — | — | 14.9 | 0.174 | 0.93 | 17 |
| | | *0.20* | *0.72* | *0.07* | | | | | *0.174* | | | |
| | 0.01 | 0.46 | 1.18 | 3.57 | — | 1.23 | — | 3.4 | 29.8 | 0.160 | 0.89 | 46 |
| | | *0.45* | *0.40* | *0.15* | | | | *0.089* | *0.071* | | | |
| 0 | 0.1 | 0.74 | 1.91 | 4.09 | — | 2.17 | 0.08 | 1.20 | 13.1 | 0.244 | 0.75 | 35 |
| | | *0.39* | *0.28* | *0.33* | | | *0.106* | *0.073* | *0.065* | | | |
| 27 | 0.01 | 0.46 | 1.18 | 2.93 | 5.68 | 1.84 | — | 2.9 | ∞ | 0.153 | 0.75 | 35 |
| | | *0.35* | *0.37* | *0.13* | *0.15* | | | *0.043* | *0.112* | | | |
| 54 | id | 0.55 | 1.68 | — | 5.39 | 2.42 | 1.7 | 7.5 | ∞ | 0.152 | 0.69 | 39 |
| | | *0.40* | *0.29* | | *0.31* | *0.035* | *0.022* | *0.095* | | | | |
| 78 | id | 0.52 | 1.71 | — | 5.50 | 2.92 | 0.5 | 4.5 | ∞ | 0.164 | 0.74 | 36 |
| | | *0.32* | *0.27* | | *0.40* | | *0.027* | *0.028* | *0.109* | | | |
| 342 | id. | 0.45 | 1.76 | — | 5.68 | 3.50 | 3.2 | 18.0 | ∞ | 0.130 | 0.64 | 43 |
| | | *0.22* | *0.26* | | *0.52* | | *0.012* | *0.037* | *0.081* | | | |
| 730 | id | 0.72 | — | 2.70 | 6.08 | 3.68 | 5.7 | — | ∞ | 0.100 | 0.59 | 46 |
| | | *0.26* | | *0.30* | *0.44* | | *0.030* | | *0.070* | | | |

heterogeneity. We will see in the following paragraph, after analysis of the polarized fluorescence decays, that this lifetime heterogeneity is coupled to a mobility heterogeneity.

The short lifetimes which characterize the fluorescence decay of Trp187 in annexin V A-form (without calcium in domain III) is likely due to the close proximity of the carbonyl group of the Thr224 peptide bond (Figure 7.4A). The electron acceptor properties of this group, explain the quenching effect leading to the relatively short excited state lifetime for this conformer as compared to indole or tryptophan in solution.[59,60] The shortest lifetime could correspond to a more mobile conformation and the longest to a less mobile one.

A significant change in the excited state lifetime distribution, mainly characterized by the increase of the long lifetime contribution and value, is observed when calcium is bound to domain III (Figure 7.5B). Such a change in the excited state lifetime distribution, coupled with the red shift of the fluorescence emission spectrum, is in line with the breakage of the interaction with the carbonyl group of the Thr224 peptide bond that was observed in the three-dimensional structure of the protein without calcium in domain III.[10] The relative proportion of the long excited state lifetime increases by a factor 3.6 as a function of the emission wavelength (from 0.18 at 310 nm to 0.66 at 400 nm), at the expense of the shortest one, which indicates that this conformer corresponds to the exposed conformation. However, three lifetime populations are present, which may indicate the existence of several conformers in slow exchange with respect to the nanosecond time-scale. Analysis of the coupling between lifetimes and rotational correlation times can bring support to this hypothesis.

### *7.3.1.5. Fluorescence Anisotropy of Trp187*

The steady-state fluorescence anisotropy excitation spectrum of calcium-free annexin V, measured at the maximum emission wavelength (325 nm), shows high anisotropy values whatever the excitation wavelength (Figure 7.7, spectrum 4). This is a characteristic feature of a quasi-immobilized Trp with a short mean excited state lifetime (the ratio of the mean lifetime versus the mean correlation time is small). We observe the characteristic minimum of the Trp anisotropy excitation spectrum near 290 nm and a steep increase at excitation wavelengths ranging from 290 nm to 300 nm. The anisotropy value at 305 nm is between 0.25 and 0.30, close to the maximum value measured in vitrified medium in the absence of motion both for NATA.[62] Addition of calcium in the millimolar range of concentration leads to only a small change of the fluorescence excitation spectrum and a slight decrease of the anisotropy (Figure 7.6, spectrum 5).

Time-resolved fluorescence anisotropy studies were performed. The fluorescence polarized decay data of Trp187 in annexin V at neutral pH, analyzed by the one-dimensional model of the anisotropy (which correlates all the lifetimes with all the rotational correlation times), did not detect any fast rotational motion of the indole ring at pH 7 (Figure 7.8A). Only the Brownian rotational correlation time of the molecule is observed (Table 7.1). A

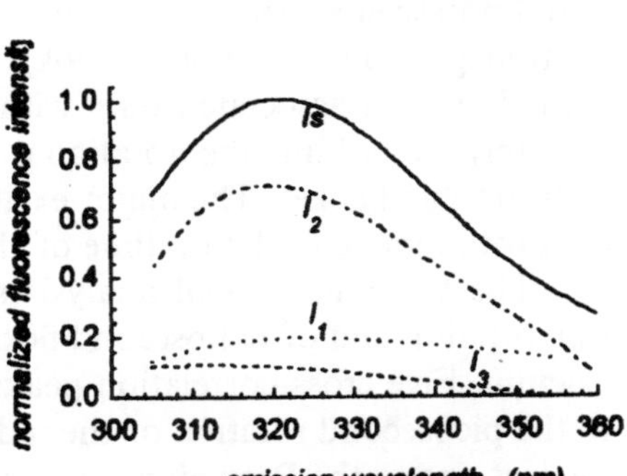


Figure 7.6. Decay associated spectra of Trp187 of annexin V in the absence of calcium ion. *Is*: steady-state fluorescence emission spectrum; $I_1$: DAS of the 0.3 ns component; $I_2$: DAS of the 1 ns component; $I_3$: DAS of the long lifetime component.

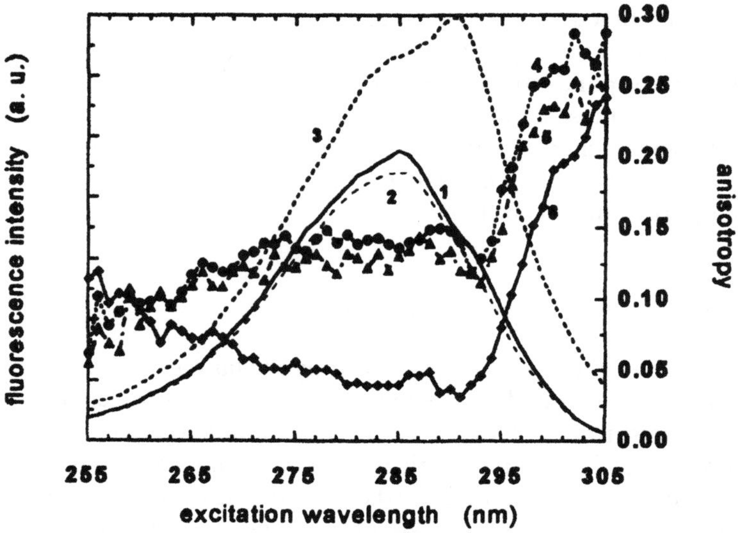


Figure 7.7. *Steady-state fluorescence excitation spectra* of (1) annexin V in neutral buffer pH 7, (2) annexin V in the presence of 5.5 mM $CaCl_2$ and (3) in the presence of 5.5 mM $CaCl_2$ and with a L/P ratio of 500. *Steady-state anisotropy excitation spectra* of (4) annexin V in neutral buffer, (5) annexin V in the presence of 10 mM $CaCl_2$ and (6) in the presence of 5.5 mM $CaCl_2$ and with a L/P ratio of 500. Protein concentration: 4.5 μM. Emission wavelength: 325 nm for calcium-free and calcium-bound annexin V, 340 nm for membrane-bound annexin V. Temperature: 20 °C (from ref. 61).

wobbling-in-cone semi-angle of 17° can be calculated.[54] This feature can be due either to the absence of any rotational motion of the indole ring or to a biased analysis owing to a possible coupling between short lifetime and fast rotations.

The existence of such specific coupling between lifetimes and correlation times can however be detected by a two-dimensional analysis of the polarized fluorescence decays. The results of the analysis are represented as contour plots $\Gamma(\tau, \theta)$ (Figure 7.9A). The results show that the shortest-lived excited state is associated only with a fast rotational motion (200–300 ps), probably describing the rotational motion of the indole ring within its hydrophobic pocket. The major excited state of 0.9 ns is associated with the long rotational correlation time of the protein.

The two-dimensional analysis shows also the presence of fast indole ring rotation and of nanosecond flexibilities of domain III in the presence of calcium. Five cross-correlation peaks dominate the picture: two correspond to the picosecond rotation of the indole, two to a nanosecond flexibility and the last one to the Brownian rotational motion of the protein (Figure 7.9B).

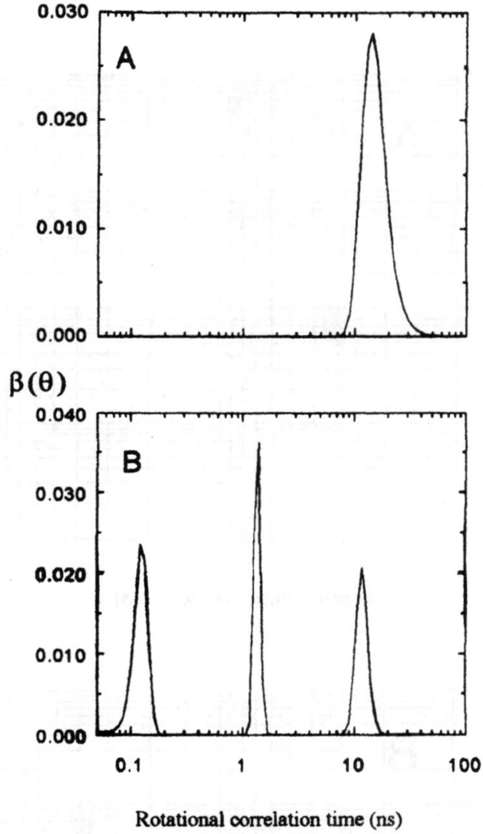


Figure 7.8. *MEM reconstituted rotational correlation time distribution of Trp187 in annexin V.* Upper panel: in the absence of calcium. Lower panel: in the presence of calcium 100 mM. Excitation wavelength: 295 nm (bandwidth 5 nm), emission wavelength: 335 nm (bandwidth: 10 nm). Temperature: 20 °C.

The shortest excited state is associated only with the fast subnanosecond rotation whereas the intermediate lifetime is coupled also with a slower nanosecond rotational motion (Figure 7.9B). The long lifetime is associated both to the Brownian rotational motion of the molecule and to the nanosecond local flexibility. This pattern suggests a much larger flexibility of the IIIA-B loop in this opened conformation than in the closed one.

Calcium binding to domain III of annexin V increases the flexibility of domain III in the region of Trp187. Rotational motions of the indole ring in the pico/nanosecond time range can be detected as shown by the one-dimensional model (Figure 7.8B). The wobbling-in-cone semi-angle angle of

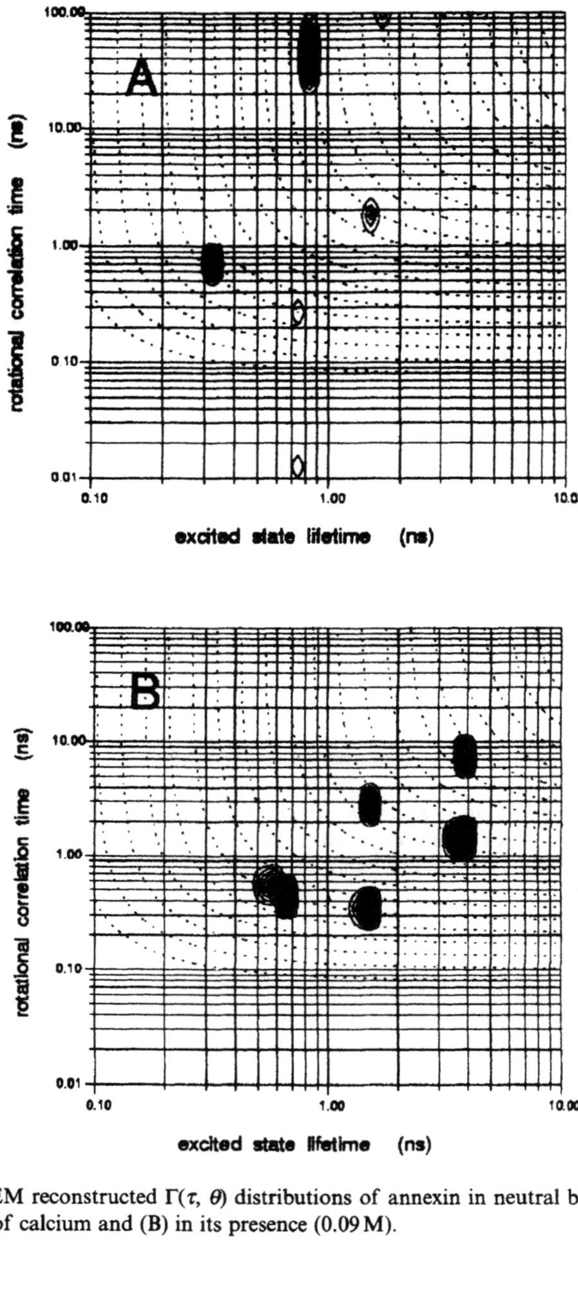


Figure 7.9. MEM reconstructed $\Gamma(\tau, \theta)$ distributions of annexin in neutral buffer pH 7.5. (A) in the absence of calcium and (B) in its presence (0.09 M).

rotation of the indole ring is increased as compared to that in the absence of calcium (Table 7.1).

### 7.3.2. Effect of pH on the Conformation and Dynamics of Domain III of Annexin V

#### *7.3.2.1. Steady-State Fluorescence Emission Spectrum of Trp187*

The need for a high calcium concentration to shape the calcium binding site in domain III of annexin V casts some doubt on the specificity of the binding. Part of the effect can have its origin in the screening of specific electrostatic interactions occurring between charged amino acid side-chains in this domain.

This is strongly suggested by the simulation of the conformational change pathway (*Sopkova et al., in preparation*). If this is true, a pH effect should be observed. pH titration from the neutral region down to pH 3.5 in the absence of calcium shows indeed a large progressive red shift of the emission maximum of 12 nm, from 326 to 338 nm (Figure 7.10).

This spectral shift is however smaller than that observed at the highest concentration of calcium. The shift starts to occur at pH 6 and ends at pH ~4.5 with a mid-point situated around pH 5. This pH range indicates that carboxylic side chains are probably involved. The width of the pH range in which the spectral change takes place suggests that more than one titratable acidic group may be involved in the process. Two residues with pK around

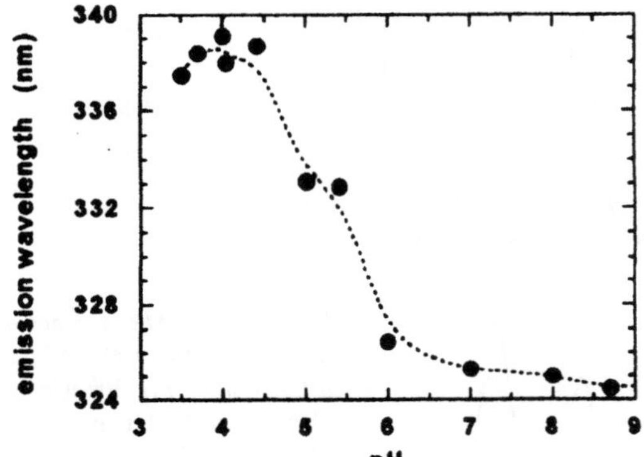


Figure 7.10. *Variation of the maximum emission wavelength of Trp187 in annexin V* as a function of pH. Excitation wavelength: 295 nm. Temperature: 20 °C (from ref. 58).

4.6 and 5.6 can be implicated (Figure 7.10). It appears therefore that a conformational change similar to that provoked by calcium at high concentration is induced in mild acidic pH conditions.

### 7.3.2.2. *Excited State Lifetime Heterogeneity of Trp187 at Different pH*

Decreasing the pH leads to changes in the relative proportions of the excited state lifetime peaks (Figure 7.11). The values of the two longest excited state lifetime increase also. At pH 3.8, the longest lifetime dominates the fluorescence emission and corresponds to 72% of the fluorescence intensity whereas at pH 7, it represents only 16% of the fluorescence intensity. On

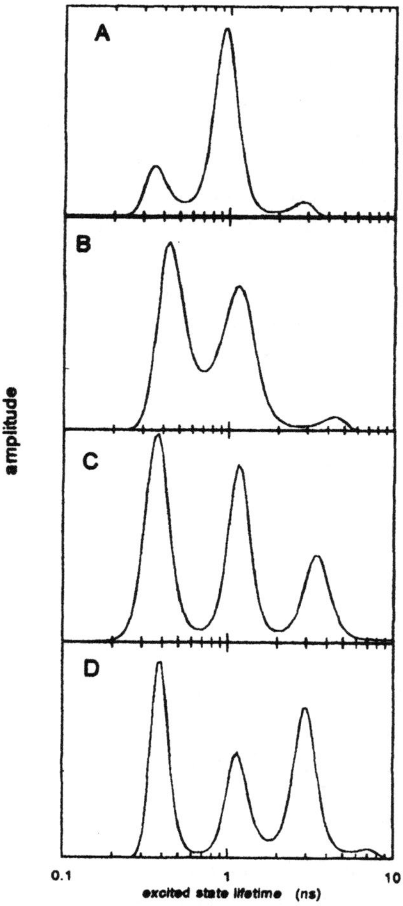


Figure 7.11. *MEM reconstructed excited state lifetime spectra of Trp187 fluorescence emission as a function of pH.* Protein concentration: 10 μM. Cacodylate buffer at pH 7.5 and acetate buffer pH 6, 5 and 4. Excitation wavelength: 295 nm (bandwidth 4 nm), emission wavelength: 335 nm (bandwidth 8 nm). A) pH 7.5; B) pH 6; C) pH 5 and D) pH 4 (from ref. 58).

the contrary, the proportion of the major lifetime of 0.9–1 ns at pH 7 decreases strongly. It is responsible for 75% of the relative fluorescence intensity at pH 7 and for only 29% at pH 3.8.

The lifetime profile at pH 4 (Figure 7.11D) looks very similar to the one observed in the presence of high calcium concentrations (Figure 7.5B). This strongly suggests that the local interactions of the indole ring with the intimate surrounding groups are the same as that of the "opened" form of the domain III of the protein. The decrease of pH down to 4, which leads to protonation of the acidic amino-acid residues, breaks the salt-bridges involving in particular Asp226 in the IIID–E loop. This last residue seems to be of crucial importance in the path of the conformational change. It undergoes a H-bond interaction with Trp187 in the transient structure of the "saddle-point" (Sopkova *et al.*, in preparation). In the "closed " conformation, it is H-bonded to Thr229. Asp190, in the IIIA-B loop, is also H-bonded to Lys193. Both acidic amino-acid residues stabilize therefore the "closed" conformation (form A) (Figure 7.4A), whereas these interactions disappear in the "opened" conformation of the protein (form B) (Figure 7.4B). This changes the local structure in a similar way as high calcium concentrations, leading to a more opened conformation in which Trp187 is not anymore in H-bond interaction with amide carbonyl groups.

#### *7.3.2.3. Time-Resolved Fluorescence Anisotropy Study as a Function of pH*

At pH 4, the analysis by the one-dimensional model of the anisotropy shows the existence of a hindered rotational motion in the nanosecond time range which was not visible at pH 7. The local flexibility is higher therefore at this pH. Moreover, a high plateau value of the anisotropy at long times is seen but no evidence for the correlation time of the monomeric form (Table 7.2). This last observation indicates that the protein probably oligomerizes at this pH.

Table 7.2. Fluorescence Anisotropy Decay Parameters of Trp187 in Annexin V as a Function of pH. Excitation Wavelength: 295 nm (Bandwidth 4 nm), Emission Wavelength: 325 nm at pH 7 and 335 nm at pH 3.8 (Bandwidth: 8 nm) (Data from ref. 58)

| pH | $A_s$ | $\theta_1$ (ns) | $\theta_2$ (ns) | $\beta_1$ (ns) | $\beta_2$ (ns) |
|---|---|---|---|---|---|
| 7 | 0.221 ± 0.003 | 14.7 ± 0.4 | — | 0.224 ± 0.008 | — |
| 6 | 0.214 | 15.3 | — | 0.221 | — |
| 5 | 0.198 | 13.2 | $\infty$ | 0.133 | 0.057 |
| 4 | 0.170 | 5.1 | $\infty$ | 0.069 | 0.112 |

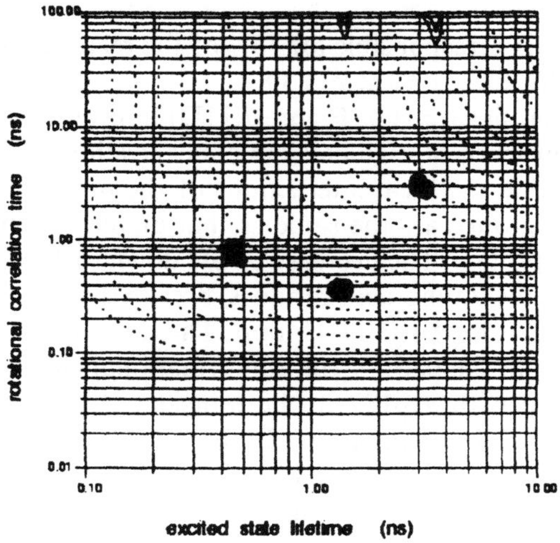


Figure 7.12. MEM reconstructed $\Gamma(\tau, \theta)$ distributions of annexin at pH 4. Protein concentration: $10\,\mu$M (from ref. 58).

The two-dimensional analyses reveal specific connectivities between the short lifetime and the fastest rotational motion (Figure 7.12).

The short lifetime corresponds to a conformer in which the indole ring freely rotates in the nanosecond time range. The two long lifetimes are associated with the infinite correlation time values which were detected in the one-dimensional analysis. These two excited state lifetime populations correspond therefore to conformers in which the indole ring motion is partially hindered. The wobbling angle for the sub-nanosecond/nanosecond motions displays a value of 41–44°. These results show that at pH 4, the local structure of the IIIA-B loop is more flexible than at pH 7, but similar to that of the calcium-bound form.

#### *7.3.2.4. Accessibility of Trp187 to Acrylamide, a Water-Soluble Fluorescence Quencher*

In order to estimate the solvent-accessible surface of indole, quenching experiments were performed by measuring the fluorescence decay as a function of acrylamide concentration for the protein at pH 7.5 in the absence of calcium, in the presence of 0.02 M calcium and at pH 4.

The Stern-Volmer plots of the ratios of the mean excited state lifetime values without quencher and in the presence of increasing concentration of acrylamide are linear in the three cases. The Stern-Volmer and bimolecular

Table 7.3. Acrylamide Quenching Constants Obtained by the Stern-Volmer Representation of the Mean Excited State Lifetime Ratio $\langle\tau_0\rangle/\langle\tau\rangle$ as a Function of Acrylamide Concentration, for Annexin V in Acetate Buffer pH 4 in the Absence of Calcium, in Tris-HCl Buffer pH 7.5 (in the Absence or in the Presence of 0.01 M Calcium) and Bound to Phospholipid Membranes (POPC/POPS 80/20, L/P = 150). Protein Concentration: 10 $\mu$M

| sample | $K_{sv}$ ($M^{-1}$) | $\langle\tau\rangle$ (ns) | $k_q$ ($M^{-1}s^{-1}$) |
|---|---|---|---|
| pH 4 | 2.44 | 1.74 | 1.40 $10^9$ |
| pH 7.5 | 0.39 | 0.97 | 4.02 $10^8$ |
| pH 7.5 + $CaCl_2$ | 3.51 | 2.09 | 1.68 $10^9$ |
| pH 7.5 L/P = 150 | 1.03 | 3.68 | 2.79 $10^8$ |

quenching constant values are listed in Table 7.3. The bimolecular quenching constant value for Trp187 in annexin V at neutral pH in the absence of calcium is low as compared to that for solvent accessible Trp residues like N-acetyltryptophan amide in water (6–7 $10^9\,M^{-1}s^{-1}$),[63] small water-soluble peptides like melittin, ACTH or glucagon, but it is comparable to that of proteins with buried Trp residues like RNAse T1 and the B sub-unit of cholera toxin.[63] The Trp187 of annexin V at neutral pH is therefore weakly accessible to acrylamide. According to Johnson and Yguerabide,[64] a surface area accessible to the quencher of ~5% can be estimated. At pH 4, in contrast, the bimolecular quenching constant value increases strongly by a factor of ~4, which indicates an accessibility of about 60%. In the presence of high calcium concentration at neutral pH, the accessibility of the Trp187 is also high, of the order of 80%. These results are in complete agreement with the red shift of the maximum of fluorescence emission observed by decreasing the pH. The Trp187 residue becomes more solvent-exposed in the conformation of domain III in mild acidic conditions.

#### *7.3.2.5. Secondary Structure of Annexin V as a Function of pH: Circular Dichroism Study*

The overall secondary structure of the protein is not significantly modified upon decreasing the pH from 7 down to 4 (Figure 7.13).

The dichroic bands characteristic of an $\alpha$-helical structure are even reinforced at acidic pH as compared to neutral pH. On the other hand, the

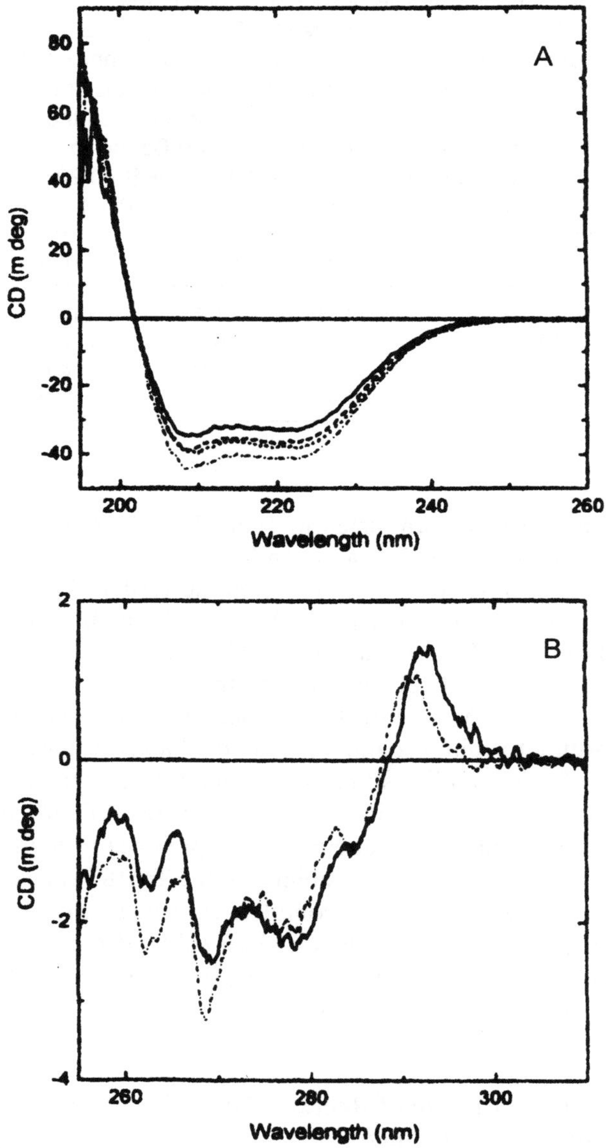


Figure 7.13. *Circular dichroism spectra of annexin V as a function of pH.* A) far-UV spectra: from top to bottom pH 7, pH 6, pH 5 and pH 4; B) near-UV spectra: plain line pH 7, dotted line pH 4 (from ref. 58).

dichroic bands in the near-UV region, which is related to the environment of the aromatic aminoacids and more particularly to the Trp residue, follow a similar variation as a function of pH as with increasing calcium concentration (Figure 7.2C).

### 7.3.3. The Interaction of Annexin V with Small Unilamellar Vesicles

#### *7.3.3.1. Polarity Change Around Trp187 Induced by the Interaction with Membranes: Steady-state Fluorescence Spectra of Trp187*

In the presence of a calcium concentration of 0.5 mM, which does not have any effect on the fluorescence spectrum of annexin V in the absence of phospholipids, the addition of SUV (POPC/POPS 80/20 M/M) induces a saturatable red-shift of 10 nm (from 325 nm to about 335 nm) for a concentration of phospholipids of 8.5 $10^{-4}$ M (L/P = 85) above which the emission maximum remains at a constant value (Figure 7.14A). Further $CaCl_2$ addi-

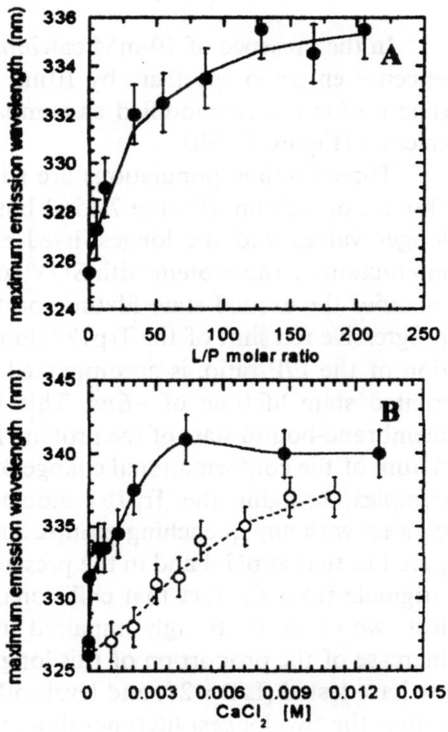


Figure 7.14. A) Variation of the maximum emission wavelength of Trp187 of annexin V as a function of the phospholipid concentration (POPC/POPS 80/20). Protein concentration: 10 $\mu$M, $CaCl_2$ concentration: 0.5 mM. Excitation wavelength: 295 nm (bandwidth: 2 nm). B) Variation of the maximum emission wavelength of Trp187 as a function of the concentration of $CaCl_2$. (●) In the presence of SUV (POPC/POPS 80/20) corresponding to 1.5 mM phospholipids. (○) In the absence of phospholipids. Annexin V concentration: 9.7 $\mu$M (from ref. 61).

tion on the sample containing the same concentration of protein and 1.5 mM phospholipids (L/P = 150) provokes an additional red shift of 5 nm which brings the maximum of emission at 340 nm.

As compared to the spectral shift induced by calcium ions alone, the mid-point of the titration is decreased by about one order of magnitude smaller in the presence of SUV but remains still in the millimolar range of concentrations (Figure 7.14B). One can remark that it is also the range of concentration for the dissociation constant of calcium for PS lipids.[65] This suggests that the protein binds to the preformed Ca-PS complex. The amplitude of the spectral shift induced by binding of annexin V to negatively charged membranes in the presence of calcium is similar to the one provoked by high calcium concentrations in the absence of membranes (Figure 7.3) and by lowering the pH to 4 (Figure 7.9). It reveals likely the existence of a similar conformational change pushing the Trp187 residue out of its hydrophobic pocket inside the protein at neutral pH (Figure 7.4A and B).

#### *7.3.3.2. Conformational Change of Domain III upon Interaction of Annexin V with Phospholipid Membranes: Excited-state Lifetime Distribution*

In the presence of 10-mM calcium, already inducing a shift of the fluorescence emission spectrum by 10 nm (Figure 7.14B), the excited-state lifetime profile is also modified as compared to the protein in the absence of calcium (Figure 7.15B).

Three lifetime populations are still detectable as for the protein in the absence of calcium (Figure 7.15A) but the center of each peak is shifted to longer values and the longest-lived excited state becomes dominant. The interaction of the protein with SUV containing POPS/POPC (20/80) further modifies the excited state lifetime profile as shown on Figure 7.4C–E. The progressive red shift of the Trp187 fluorescence emission spectrum as a function of the L/P ratio, is accompanied by the gradual appearance of a long excited state lifetime of ~6 ns. This lifetime population characterizes the membrane-bound state of the protein. Its existence is in line with the assumed nature of the conformational change of domain III in the protein/membrane complex, bringing the Trp187 indole ring on the protein surface, not in contact with any quenching groups. The larger value of this lifetime as compared to that at pH 4 and in the presence of high calcium concentration may originate from the fact that collisional quenching due to the solvent in the last two cases, is strongly impaired in the membrane hydration layer. The increase of the proportion of this long lifetime begins at the lowest L/P ratio we have tested (L/P = 27) and levels off at L/P ~ 150–200. At the highest L/P ratios, the two longest lifetimes dominate the fluorescence decay (Table 7.1).

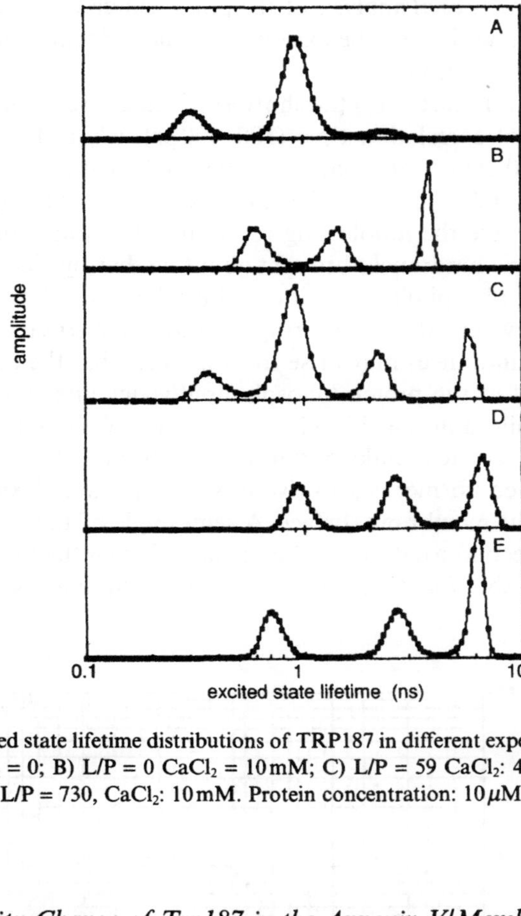


Figure 7.15. Excited state lifetime distributions of TRP187 in different experimental conditions. A) L/P = 0, $CaCl_2$ = 0; B) L/P = 0 $CaCl_2$ = 10 mM; C) L/P = 59 $CaCl_2$: 4 mM; D) L/P = 143, $CaCl_2$: 10 mM; E) L/P = 730, $CaCl_2$: 10 mM. Protein concentration: 10 $\mu$M. Temperature: 20 °C (from ref. 61).

### *7.3.3.3. Mobility Change of Trp187 in the Annexin V/Membrane Complex: Time-resolved Fluorescence Anisotropy Study*

In the presence of POPC/POPS SUV (80/20) at low L/P molar ratios and 10 mM calcium, the fast subnanosecond motion is preserved and an infinite component is observed. The wobbling-in-cone angle of the rotational motion of the indole ring is reduced as compared to the value in the presence of calcium only, but it is much larger than the value in the protein alone (Table 7.1). This wobbling-in-cone angle value is further increased at higher L/P ratios and additional correlation times appear. In all cases the initial anisotropy value is smaller than that expected for an immobile Trp residue at this excitation wavelength.[62] Fast subnanosecond rotational motions are likely not resolved either in the measurements or in the analysis (due to specific couplings between lifetime and correlation times).

The two-dimensional analysis of the polarized decays of the membrane-bound protein shows indeed the existence of such specific coupling between lifetime and correlation times.

At L/P = 730 (Figure 7.16) the shortest lifetimes are associated only with the fast rotational correlation time (300–400 ps), while the long lifetime is coupled both to this fast rotation, to an intermediate motion (5 ns) and to an infinitely long correlation time. The shortest lifetimes correspond therefore to conformers where the indole ring rotation is fast and isotropic, without any apparent steric hindrances brought about neither by elements of protein structure nor by phospholipid moities, while the long lifetime corresponds to a conformer where the indole ring rotation is restricted by steric hindrances. These hindrances may arise from the fact that the aromatic residue is likely confined at the protein/phospholipid/water interface where a large number of mobile amino-acid side chains and of lipid head groups are present. This will create a bulky and flexible environment. In agreement with this picture, this conformer is sensitive to slower protein flexibilities that are still present on the membrane surface. A semi-angle of the wobbling-in-cone of the subnanosecond rotation can be calculated from the $\Gamma(\tau, \theta)$ coefficients summarized in Table 7.4. A value of ~28° for the sub-nanosecond motion is found.

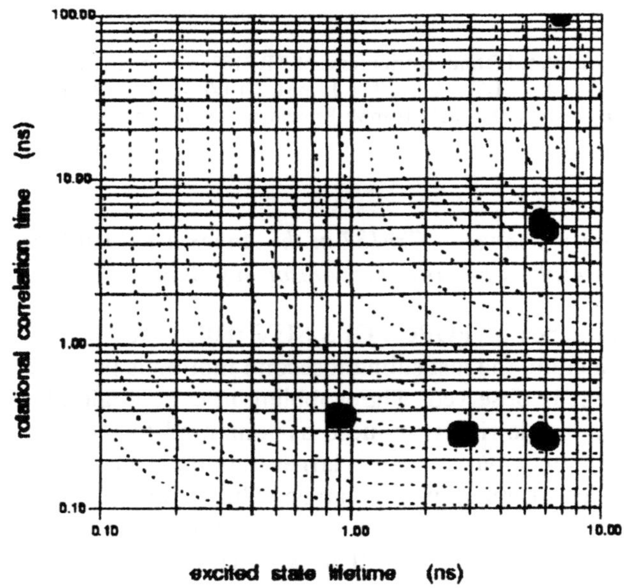


Figure 7.16. *MEM reconstructed $\gamma(\tau, \theta)$ distributions of annexin V bound to membranes.* Lipid/protein molar ratio of 730 (C). Protein concentration: 10 $\mu$M. 10 mM $CaCl_2$ (from ref. 61).

**Table 7.4. Distribution of the $\Gamma(\tau, \theta)$ Parameters of the Polarized Fluorescence Decays of Trp187 at Different Lipid/Protein Molar Ratios (from ref. 61)**

No lipids $CaCl_2$ 10 mM

| $\theta_{(ns)}$ \ $\tau_{(ns)}$ | 0.4 | 1.1 | 2.9 | 5 |
|---|---|---|---|---|
| 0.05–0.5 | 0.42 | — | — | — |
| 1.5–5 | — | 0.41 | 0.05 | — |
| 8–20 | — | — | 0.09 | 0.03 |

L/P = 35

| $\theta_{(ns)}$ \ $\tau_{(ns)}$ | 0.7 | 1.7 | 4.2 | 7.3 |
|---|---|---|---|---|
| <0.02 | 0.22 | — | — | — |
| 0.4–1.8 | 0.21 | 0.23 | 0.08 | 0.06 |
| 10–20 | — | — | 0.07 | — |
| ∞ | 0.08 | — | 0.04 | 0.01 |

L/P = 54

| $\theta_{(ns)}$ \ $\tau_{(ns)}$ | 0.6 | 1.7 | 5.2 |
|---|---|---|---|
| 0.1 | 0.30 | — | — |
| 1.5–4 | — | 0.29 | 0.08 |
| 15 | — | — | 0.18 |
| ∞ | 0.13 | — | 0.02 |

L/P = 342

| $\theta_{(ns)}$ \ $\tau_{(ns)}$ | 0.5 | 1.8 | 5.4 |
|---|---|---|---|
| 0.05 | 0.33 | 0.03 | 0.04 |
| 1.5–2 | — | 0.20 | 0.10 |
| 30 | — | — | 0.19 |
| ∞ | — | — | 0.10 |

L/P = 730

| $\theta_{(ns)}$ \ $\tau_{(ns)}$ | 0.9 | 2.8 | 6.0 |
|---|---|---|---|
| 0.3–0.4 | 0.26 | 0.26 | 0.14 |
| 5–6 | — | — | 0.21 |
| ∞ | — | — | 0.13 |

At low L/P ratios (L/P = 35 and 54), where not enough lipid molecules are present to saturate the protein, the shortest lifetime is also associated with an infinite component of large amplitude (corresponding to a wobbling-in-cone angle of 50°). This is not the case at high L/P ratios where no rotational constraints are sensed by this lifetime as quoted above. The presence of an infinite component indicates the existence of rotational constraints occurring at high protein concentrations on the membrane surface that are released when the protein is more dilute. This suggests the existence of intermolecular contacts between the domain III of adjacent protein molecules at low L/P ratios, when the protein spreads over the membrane surface. The Brownian rotational correlation time is extremely long (infinite value) as compared to the lifetime because the protein is firmly bound to the large rotating body constituted by the lipid vesicle, the size of which corresponds to a microsecond tumbling motion.

The existence of several conformers suggests that domain III remains highly flexible (in a time-scale slower than the fluorescence lifetime) when the protein is bound at the membrane surface. Moreover, in some conformers, the Trp187 residue is moving very fast with a high degree of rotational freedom. Both observations are not compatible with the proposed model of the protein/membrane complex which suggested that the indole ring was inserted into the first methylene region of the phospholipid fatty acid chains.[11]

#### 7.3.3.4. *Accessibility of Trp187 to Acrylamide in the Membrane-bound Protein*

Time-resolved acrylamide quenching experiments were performed at saturating conditions with membranes (L/P = 150). The longest and intermediate lifetimes provide bimolecular quenching constants $k_q$ comparable to that measured for the membrane-free protein but much lower than those measured in the presence of calcium or at pH 4 (Table 7.3). This corresponds to a low accessibility of the indole ring to the quencher. Despite its relatively polar environment, the Trp187 is protected from direct contact with the bulk solvent. This indicates that the indole ring is situated in the water layer covering the membrane surface, which displays a much higher viscosity than the bulk solvent. This hypothesis is in agreement with the experiments in reverse micelle (see below).

### 7.3.4. The Interaction of Annexin V with Reverse Micelles

The binding of annexin V to phospholipid membranes at neutral pH is probably driven mainly by electrostatic interactions involving calcium ions, which maintain the protein at the membrane surface. The protein convex surface is most likely in contact with the first water layers surrounding the phospholipid polar head groups, which constitute the interface of the membrane with the aqueous solvent. In this interfacial water layer region, the proton activity as well as the availability of water molecules for hydration are significantly modified as compared to bulk water.[31,66] We have seen that the interaction of the protein with the membrane surface provoked a conformational change of domain III, in a similar way as high calcium concentrations and mild acidic pH. We have proposed that the driving force in the mechanism of this conformational change was the modifications of specific electrostatic interactions involving acidic residues on the protein surface, which can be modulated by pH and membrane/water interfaces.

Reverse micelles can mimic the interfacial region of membrane with water and its influence on protein conformation and dynamics. These microemulsions can dissolve many proteins of different kinds.[32,34,67] They display convenient properties for optical studies.[68]

#### *7.3.4.1. Modification of the Trp187 Environment in Reverse Micelles: Steady-state Fluorescence Emission Spectrum*

In the micro-molar concentration range, annexin V is soluble in reverse micelles formed by the surfactant AOT in isooctane at a water/surfactant molar ratio ($w_0$) as low as 2.8 as judged by the absorption spectrum which did not exhibit any light scattering.

At this low water content, the fluorescence emission maximum is already red-shifted by about 12 nm with respect to the protein in buffer solution at neutral pH (Figure 7.17).

The fluorescence emission maximum is sensitive to the water content of the micelles. It is more and more shifted to the red when the water content of the reverse micelles is increased (Figure 7.17). Its value culminates at

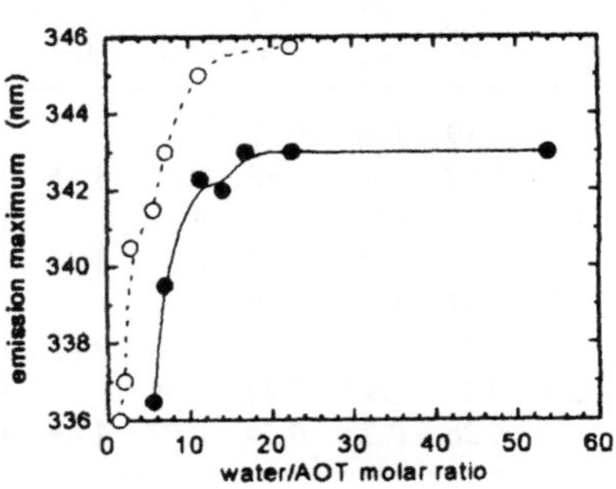


Figure 7.17. *Variation of the emission maximum of W187 of annexin V* (●) and of NATA (○) incorporated into reverse micelles of water/AOT in isooctane as a function of the water/surfactant molar ratio $w_0$. Protein concentration 2.5 $\mu$M for $w_0$ = 2.8 and 10 $\mu$M for the others $w_0$. Excitation wavelength: 295 nm (from ref. 61).

343 nm ($\Delta\lambda_{max} = 18$ nm) for a $w_0$ value of ~15 where after it stays at a constant value. Data for N-acetyl tryptophanamide (NATA) are presented for comparison. The emission maximum of this Trp derivative is always larger than that of the Trp187 in annexin V and it reaches a plateau value at longer emission wavelengths. These observations suggest that the conformational change of domain III of annexin V, which leads to a large exposure of the Trp187 to the aqueous solvent pool of reverse micelles, is occurring in these microemulsions. The difference between the maximum emission wavelength of Trp187 and NATA comes likely from the fact that the effective dielectric constant at the boundary between the protein and water is probably much smaller than in the water core of the reverse micelles.[69]

### *7.3.4.2. Excited State Lifetime Distribution of Trp187: Conformational Change in Reverse Micelles*

The fluorescence emission decay of Trp187 is also modified when the protein is incorporated into reverse micelles. Four well-separated excited state lifetime populations are detected (Figure 7.18). This large heterogeneity suggests the existence in these systems of a large conformational dynamics in the slow time scale with respect to the lifetimes. The proportion of the major lifetime of ~0.9–1 ns, which characterizes the Trp187 emission in the protein in buffer at neutral pH, is strongly decreased when the protein is included into the reverse micelles. The proportion of the 3 ns lifetime is enhanced from few percent to 35–40% whatever the aqueous content of the micelles (Table 7.5).

Table 7.5. Fluorescence intensity and anisotropy decay parameters of Trp187 in annexin V incorporated into reverse micelles as a function of the water/AOT molar ratio ($w_0$). Excitation wavelength: 295 nm (bandwidth 4 nm), emission wavelength: 335 nm (bandwidth 8 nm) (from ref. 61).

| $w_0$ | $\tau_1$ (ns) $C_1$ | $\tau_2$ (ns) $C_2$ | $\tau_3$ (ns) $C_3$ | $\tau_4$ (ns) $C_4$ | $\langle\tau\rangle$ (ns) | $\theta_1$ (ns) $\beta_1$ | $\theta_2$ (ns) $\beta_2$ | $\theta_3$ (ns) $\beta_3$ | $\theta_4$ (ns) $\beta_4$ | $A_{t=0}$ | $S$ | $\omega_{max}$ |
|---|---|---|---|---|---|---|---|---|---|---|---|---|
| 5.6 | 0.19 | 1.28 | 6.12 | | — | — | 1.9 | 61 | 0.143 | 0.76 | 34 | |
| | *0.30* | *0.30* | *0.35* | *0.05* | 1.86 | — | — | *0.027* | *0.116* | | | |
| 11.2 | 0.13 | 0.82 | 2.19 | 4.52 | | 0.1 | 0.8 | 7.1 | 35.6 | 0.206 | 0.58 | 46 |
| | *0.25* | *0.29* | *0.38* | *0.08* | 1.46 | *0.062* | *0.026* | *0.050* | *0.068* | | | |
| 16.8 | 0.41 | 1.47 | 3.30 | — | | 0.17 | 0.7 | 2.6 | 22.3 | 0.193 | 0.61 | 45 |
| | *0.28* | *0.47* | *0.25* | | *1.65* | *0.058* | *0.013* | *0.047* | *0.075* | | | |
| 22.4 | 0.22 | 0.93 | 2.22 | 4.56 | | 0.08 | 0.5 | 2.9 | 26.1 | 0.179 | 0.60 | 46 |
| | *0.26* | *0.28* | *0.39* | *0.07* | 1.48 | *0.027* | *0.023* | *0.058* | *0.071* | | | |

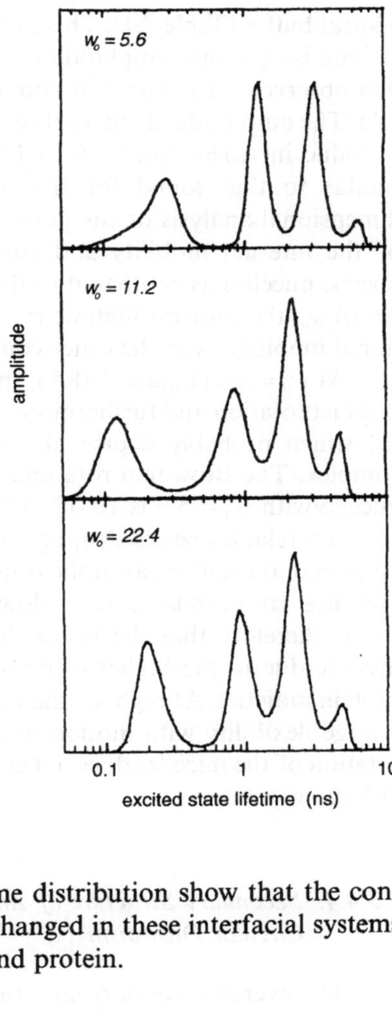


Figure 7.18. *MEM reconstructed excited state lifetime distributions of Trp187 at different water/detergent molar ratios ($w_0$).* Upper panel: $w_0 = 5.6$, middle panel: $w_0 = 11.4$, lower panel: $w_0 = 22.4$. Excitation wavelength: 295 nm (bandwidth 4 nm), emission wavelength: 340 nm (bandwidth 8 nm). Protein concentration: 10 $\mu$M. Temperature: 20 °C.

These modifications of the Trp187 lifetime distribution show that the conformation of domain III is considerably changed in these interfacial systems in a similar way as in the membrane-bound protein.

### *7.3.4.3. Time-resolved Fluorescence Anisotropy Decays*

The sub-nanosecond/nanosecond dynamics of the protein appears to be considerably amplified in reverse micelles. The analysis of the polarized fluorescence decays by the one-dimensional anisotropy model shows that at $w_0 = 5.6$, two correlation times can be obtained in the nanosecond time range (Table 7.5). The longer one displays a very large value, almost infinite as compared to the mean excited state lifetime value. The orientational order parameter $S$ values are significantly lower and the wobbling-in-cone angle ($\omega_{max}$) values are higher at all $w_0$ (Table 7.5) than those calculated for the protein in neutral buffer (Table 7.1). This indicates that the local motion of the Trp187 residue is of larger amplitude in the reverse micelles, a situation similar to that observed for the protein bound to negatively charged membranes (Table 7.1). The amplitude of internal rotation estimated by $\omega_{max}$ increases when the $w_0$ value increases from 5.6 to 11.2 and then remains at a constant value similar to that found for the protein bound to membranes. The two-dimensional analysis of the polarized decays provides confirms the increase of the internal mobility and conformational dynamics of the protein in reverse micelles as compared to the situation in neutral buffer (Figure 7.19). At all $w_0$, the shortest lifetime is associated with a fast subnanosecond rotational mobility as in the other conditions.

At $w_0 = 5.6$ (Figure 7.18A), the major lifetime of 3 ns is also coupled to this fast rotation and furthermore to a long rotational correlation time (12–15 ns) which probably depicts the Brownian rotation of the protein/micelle complex. The Brownian rotational correlation time value for empty reverse micelle with $w_0 = 5.6$ is 10–12 ns.[70] At $w_0 = 22.4$, the diagram displays more cross-correlation peaks. A spot associating the shortest lifetime and the subnanosecond rotation, as in the other conditions, is present. Moreover, longer lifetimes are correlated with slower rotational motions (Figure 7.18B). It seems therefore that the faster the rotation of the indole, the shorter the lifetime due to the higher efficiency of dynamic quenching with proximate protein moieties. At high $w_0$, the domain III of the protein appears to display a large flexibility with motion in the nanosecond time scale. The Brownian rotation of the micelles does not contribute to a large extent to the anisotropy at high $w_0$.

#### *7.3.4.4. Secondary Structure of Annexin V in Reverse Micelles: Circular Dichroism*

The overall secondary structure of the protein is not significantly modified upon incorporation into reverse micelles whatever the water/surfactant molar ratio as assessed from the conservation of the dichroic band characteristic of the $\alpha$-helical structure (not shown).

## 7.4. Discussion

To characterize at the molecular level the mechanism of the potential physiological function(s) of annexins, a number of studies have focused on the understanding of the mechanism of their interaction with pure phos-

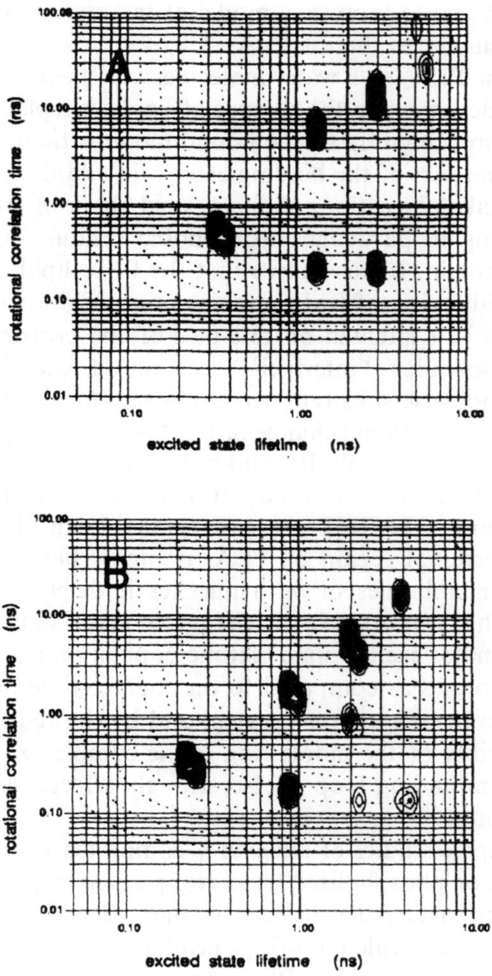


Figure 7.19. *MEM reconstructed $\gamma(\tau, \theta)$ distributions of annexin V in reverse micelles.* A) $w_0 = 5.6$; B) $w_0 = 22.4$. Protein concentration: 10 $\mu$M. Temperature 20 °C.

pholipid model membranes.[71] The knowledge of the 3D-structure of an increasing number of annexin molecules has allowed the comparison with other proteins which require calcium as a co-factor for binding to membranes like phospholipases.[72–73] This led to the observation that the consensus sequences and the structure of the calcium binding sites of annexin V share high similarities with the single calcium site of the calcium binding loop from bovine and porcine pancreas phospholipase $A_2$ ($PLA_2$).[6,72,73] This has prompted the authors to propose a model of interaction of annexin V with phospholipids[11] similar to that of $PLA_2$.[74–76]

In the case of $PLA_2$(s), a water molecule, which occupies one coordination site of the calcium ion in the absence of lipids, is replaced by one oxygen atom of the phosphate group of the bound monomeric phospholipid molecule, providing one part of the binding energy. In addition, many hydrophobic interactions take place along the fatty acid chain of the phospholipidic inhibitor, involving in particular hydrophobic residues in the N-terminal domain of the protein. It was shown that the $PLA_2$ lipid binding site comprises an extended region of hydrophobic amino-acids around the entrance to the active site. It is situated on one face of the protein molecule which constitutes the so-called "interfacial recognition site".[77] The *sn-2* acyl chain of the monomeric substrate penetrates to some extent the interior of $PLA_2$. A large conformational change of the N-terminal region upon binding of the protein to surfactant micelles and membranes, affecting both the exposure to the solvent and local flexibility of Trp3, has been emphasized by fluorescence.[43,78–82] and $^2$D-NMR measurements.[83–84] In the membrane-free protein, the N-terminal region of $PLA_2$ is highly flexible, whereas in the complexes with membranes or with micelles it displays a rigid $\alpha$-helical conformation. This conformational change is made manifest both by the presence of one major excited state lifetime population for Trp3 in the membrane/$PLA_2$ complexes as compared to the four lifetime populations which are present in the membrane-free protein and by the reduction of the subnanosecond mobility of the indole ring in the complexes.[43,79,82] The dynamics of the calcium-binding loop is also strongly affected by the interaction with lipid aggregates.[43,82]

Crystallographic studies of annexin V complexed with small polar molecules, analogues of phospholipid head groups, have suggested that domain III of the protein participated directly in the interaction with acidic phospholipid molecules *via* a calcium bridge involving the calcium binding site in this domain[11] in a similar way as $PLA_2$. In these studies however, only glycerophosphoserine or glycerophosphoethanolamine, without any hydrocarbon chain, were used, unlike the studies with $PLA_2$ which was complexed with monomeric phospholipid analogues bearing a short acyl chain. Although the latter protein is almost inactive on monomeric substrate,[85] the presence of a hydrocarbon chain allows the enzyme to display some affinity for monomeric inhibitors. Annexin V displays also a weak interaction with monomeric phospholipids. The absence of any hydrocarbon chain in the ligand analogues used for annexin V studies reduces the relevance of the data as far as the interaction with true membranes is concerned. Moreover, up to now, no structural evidence has been presented in the case of annexins neither for the existence of an interfacial recognition site nor for the presence of a hydrophobic groove. Another fundamental difference with $PLA_2$ is the absence of stereo-specificity of annexin V for phospholipids[71] whereas $PLA_2$ exhibits a stereo-specificity which is expressed even in its interaction with micelles of surfactant stereo-isomers.[82,83] The mechanisms of binding to membranes of these two classes of proteins seem therefore quite different. It is less demanding in terms of molecular structural adaptation of the ligands for annexins than for $PLA_2$.

Nevertheless, the proposed model of the annexin V/membrane complexes can predict several features which can be tested. First, the hydrophobic interactions between the indole ring and the acyl chains should be substantial. Second, this should produce a large change of polarity of the local environment of Trp187, which should be in turn reflected in its fluorescence properties. Third, if the indole ring is indeed inserted in the membrane bilayer at the level of the first methylene groups of the fatty acid chains, it should restrict their dynamics and modify the orientational order parameter. Finally, the mobility of Trp187 should be strongly affected in the annexin V/membrane complexes. We will discuss these different predictive features in the following paragraphs in the light of time-resolved fluorescence measurements.

### 7.4.1. The Role of the Change of Domain III in the Annexin/Membrane Interactions: Is the Swinging out of Trp187 Crucial for Binding?

It is accepted as a fact, based on many experimental observations, that the major driving forces in the mechanism of interaction of annexin V with membranes are most likely electrostatic interactions. The protein likely presents its convex face to the membrane surface[20,21] where the calcium sites are located. It is however a dogma that the local conformational change of domain III, induced upon protein binding to negatively charged lipid membranes in the presence of calcium, is important for the mechanism of binding. This conformational change probably leads to a swinging out of Trp187 to the protein surface, by analogy to what is occurring in the presence of calcium ions at high concentration.[25–30,86,87] The association of annexin V with negatively charged phospholipidic membranes is nevertheless reversible by EDTA.[26] It is therefore not quite clear whether hydrophobic interactions of the indole ring of Trp187 with phospholipids really exist and participate in the stabilization of the complex on the membrane surface.

The quenching efficiency of the Trp187 fluorescence emission by doxyl-labeled phospholipids using the "parallax method",[88–92] have been interpreted as supporting the existence of hydrophobic interactions involving the indole ring of Trp187 of annexin V and phospholipid molecules. These experiments have suggested however a Trp187 location close to the membrane/water interface.[26] Other experiments performed in mixed micelles of detergent and phospholipids have refined the interpretation in a way that the aromatic side-chain of Trp187 was situated near the first carbon atoms of the *sn*-2 acyl chain.[28,29] These measurements suffer however from known limitations as to their potency to estimate quantitatively the quencher-chromophore inter-distance. Several factors are involved. In particular, these quenching reagents are not strictly contact quenchers but rather short-range quenchers.[88–91] The fluctuations in the vertical direction of the unlabelled and of the doxyl-labeled phospholipids have also to be taken into account.[92] The most severe drawback however, of the estimation of the deepness of fluorophores inside membrane bilayers using these spin-labeled fatty acids or phospholipids, arises from the distortion of the acyl chain conformation by its substitution with the bulky polar doxyl group. Energy minimization calculations show that a kink is formed at the level of the substituent. In the case of the C5 labeled derivative, this leads to a location of the doxyl group at the same depth as the phosphocholine polar head group (not shown). Moreover, careful examination of the experimental data[25] shows a blue-shift of the residual protein fluorescence in the membranes enriched with doxyl-labeled lipids (50% of the phospholipids were doxyl-PC). With the reported L/P ratio of ~180, all annexin V molecules should be bound to the vesicles. Therefore, one cannot exclude the possibility that less protein was bound to the PS/doxyl-PC vesicles than to the PS/PC vesicles in these experiments, which will therefore decrease the fluorescence signal and lead to an overestimation of the quenching efficiency. This can occur especially in the case of the C5-doxyl derivative, which is the most bilayer disturbing probe of the series.

The results reported in mixed micelles should also be taken with caution since they may reflect the specific case of the host micelles of the surfactant C12E8 that has been used for these studies.[28–30] Surfactant micelles in water are not quite similar to bilayer membranes especially concerning their dynamics. The packing forces of the polar head groups are weaker in the surfactant micelle systems than in the phospholipid bilayer. The ordering of the acyl chain is not as high as it can be in phospholipid bilayers. This may favor the ability of surfactant molecules to penetrate protein crevices more readily than phospholipids. This could lead to artifactual interactions of the protein with surfactant molecules and host lipids.[93] Therefore, to our opinion, the importance of the swinging out of the Trp187 residue in the binding process of annexin V to the phospholipidic membranes was not demonstrated by these experiments.

### 7.4.2. The Location of Trp187 at the Membrane/Protein/Water Interface

We have reexamined therefore in more detail the former interpretation of the steady-state fluorescence data available in the literature in the light of the time-resolved data, in order to get a more satisfying description of the importance of domain III in the interaction of annexin V with membranes. Moreover, we have estimated the protein local dynamics by fluorescence anisotropy since the effect of protein binding to the membrane should affect strongly the indole motion in the frame of the proposed model of interaction.[11] Indeed, the insertion of the indole ring between the first methylene groups of the fatty acyl chains or between the glycerol moieties should lead to a strong immobilization of the aromatic ring since this membrane region possesses the strongest molecular packing.

The red-shift of the fluorescence emission spectrum by ~15 nm, observed so far upon binding of the protein to the membranes, corresponds to a stabilization energy of the Trp187 excited state of ~4 kcal.mol$^{-1}$ with respect to the buried conformation inside the protein. This spectral shift to lower energies could be due to dipolar interactions with polar groups in the close environment of the indole ring.[36] The energy balance is therefore in favor of stronger dipolar interactions of the Trp187 excited state, or larger local electric field[94,95] in the membrane-bound conformation of domain III as compared to the free protein at neutral pH. The comparison of the steady-state fluorescence data of annexin V bound to membranes with those obtained on transmembrane helix-forming peptides bearing a Trp residue at different positions on the amino-acid sequence, shows that when the Trp residue was located on position 1, the fluorescence emission maximum was at around 338 nm. This indicates a surface location with no hydrophobic contact.[96] The spectral blue shift with respect to water is due to slower dielectric relaxation of the surrounding dipoles at the membrane/water interface. By contrast, when the Trp was placed at position 6 (~2 helix turns within the membrane bilayer), the emission maximum was located at 322 nm.[96] In the case of Trp187 of membrane-bound annexin V, the maximum of emission is clearly at 340 nm, which is not in agreement with hydrophobic interactions between the indole ring and the acyl chains of the phospholipids.

The increase in quantum yield, early observed in the literature when annexin V binds to the membrane, has been assigned to the transfer of the Trp187 residue from a less to a more hydrophobic environment.[28] There is however no obvious correlation between high quantum yields values and blue fluorescence emission or conversely between low quantum yield and red emission of indole. While the energy distribution of the photons (the emission spectrum) is ruled by molecular intrinsic properties of the chromophore and by the sensitivity of the excited state to dipolar interactions or to local electric field), the quantum yield depends on the respective efficiencies of the non-radiative processes and of the emission of photon. This balance is also ruled by intramolecular and intermolecular factors. Intermolecular quenching efficiency is increased by the presence in the close neighborhood of the chromophore of electron scavengers such as disulfide bridges, peptide bonds or protonated amino-acid side chains.[59,60,97–109] There is no contradiction, in principle, between a high quantum yield value and a red shift of the fluorescence emission spectrum. In the case of multi-tryptophan containing proteins, the "blue" emitting Trp(s) are often associated with short lifetimes whereas the "red" ones emit with long lifetimes (see ref.[35] for an illustrative example).

Our interpretation of the change of the fluorescence parameters upon membrane binding in the specific case of annexin V is that the quenching interactions involving the Thr224 peptide bond are released in the protein membrane-bound form (like in the calcium-bound form in domain III) due to the swinging out of Trp187 on the protein surface in a polar environment. The major conformer corresponding to the long-lived excited state does not share any contact with quenching moieties belonging to neither the protein nor lipid molecules. In this location, the Trp187 residue evidences however an heterogeneity of conformations which points to the preservation of the flexibility of domain III in the membrane-bound form of the protein.

We suggest therefore that the Trp187 is probably not inserted into the membrane bilayer but remains in the hydration water layer of the membrane which extends some 5–6 Å from the molecular membrane surface, *i.e.* the thickness of two–three water molecules. The less efficient dipolar relaxation of local dipoles would explain the partial shift of the fluorescence spectrum (340 nm) with respect to that expected for bulk water (356 nm). This hypothesis implies that domain III is not rigidly anchored by Trp insertion in the membrane and that the conformational change of this domain leading to the exposure of the Trp187 on the protein surface is not necessary for the binding of the protein to the membrane. It is rather a consequence of and not a prerequisite for the binding, although it can play a role in the stability of the final complex. The interactions of domains I and II of annexin V appear stronger than those of domain III and IV according to recent experiments involving single point mutations[110] and also by parallel studies of binding enthalpies and intrinsic fluorescence changes.[111] Insertion of a Trp residue in each of the calcium site of the protein might allow exploring the effect of the protein binding to the membrane on the local dynamics and conformation of the other domains.

The influence of the Trp187 residue on the binding of the protein to lipidic membranes was recently checked with a mutant lacking this residue.[114] The effect of the mutant protein on the self-quenching of NBD-PS, used as a binding test, appeared less pronounced than that of the wild type protein. This could however be interpreted either by a weaker binding of the protein or by a lower number of affected PS molecules. The second binding test, involving a blood coagulation assay, did not show any significant difference between the wild type protein and the Trp187 mutant.[112] This suggests that the aromatic residue is not strongly involved in the binding of the protein to the membrane.

The effects of annexin V on the organization and dynamics of phospholipid bilayers are also in agreement with an interaction of the protein only with the membrane surface. Annexin V binding has been shown to provoke weak perturbations of the lipid molecular packing and of the acyl chain flexibility as evidenced by $^2$H-NMR and by fluorescence anisotropy measurements with 1,6-diphenyl-1,3,5-hexatriene. It reduces considerably the lipid lateral diffusion as evaluated by fluorescence-recovery-after-photobleaching (FRAP) experiments or by excimer formation and the mobility of the phosphatidylserine head groups, as shown by $^2$H-NMR.[113,114] These observations are compatible with a model of surface adsorption of the protein which assumes no significant insertion of the protein in the membrane.

### 7.4.3. The Mechanism of the Conformational Change on the Membrane Surface

The Trp187 environment in terms of polarity, dynamics and interaction with specific groups, resembles that observed in reverse micelles, in mild acidic pH conditions and in the presence of high calcium concentrations. This suggests a common global structure or folding of domain III in these different experimental conditions. This common structure is likely close to that observed in the crystals of the P1 form obtained in the presence of high calcium concentration[9] which shows the movement of loops IIIE-D and IIIA-B and also of Trp187 (Figure 7.4B). The effect of pH is particularly meaningful.[58,115] It suggests that the mechanism of this conformational change involves the breakage of few specific electrostatic interactions, important in the folding and dynamics of the A form without calcium in domain III.[9] Molecular modeling suggest the important participation of Asp226 which plays a role in the pathway that leads to the opening of the calcium binding (Sopkova *et al.*, in preparation). These interactions may be weakened in mild acidic pH conditions, at the membrane surface and in reverse micelles. In the latter system, the proton activity and therefore the *pKs* of this residue can be modified by several units.[66] It is worth to remark that residue Asp226 is replaced in annexin III by a Lys residue. In this last protein, the conformation of domain III is similar to that of the "calcium" form of annexin V with its Trp191 exposed to the solvent.[14] A recent construction of the mutant Asp226-Lys of annexin V led to the solvent exposure of the Trp187 residue in the absence of calcium at neutral pH (Sopkova *et al.*, in preparation). This suggests that among other residues which are exchanged between the two proteins, Asp226 plays a specific role.

#### 7.4.4. What Could be the Role of the Conformational Change of Domain III of Annexin V in the Formation of the Trimeric Complexes at the Membrane Surface?

A large conformational flexibility of domain III of annexin V is a characteristic feature of this protein. It plays probably a role in the adaptability of the protein to changes in the physico-chemical properties of the environment at the membrane surface. It was believed that the swinging out of the Trp187 residue was important to stabilize the complex of the protein on the membrane surface. However, it seems from these and other recent data, that this hypothesis is not supported by all the data. It appears that domain III of the protein is probably not in strong interaction with the membrane. Recent Atomic Force Microscopy results suggest that domain III is located some 6 Å apart from the membrane surface.[116]

The conformational change of domain III may however facilitate the formation of the protein network which is observed in supported bilayers.[116] This network is probably one important aspect of the mechanism of action of annexin V to protect the membrane from lipolysis by $PLA_2$. This protein network is organized on the basis of a trimeric unit.[117,118] We suggest that the basic trimer could be stabilized by interactions between on one side domains I and II of one molecule and domain III of the "calcium" structure on the other. In this organization, the domains III will stand on the exterior of the trimer.

## Acknowledgments

We are very grateful to Dr. I. Maurer-Fogy (Bender and Co., Vienna, Austria) for a generous gift of pure recombinant human annexin V. Dr. M. Takahashi is acknowledged for the CD measurements. Dr. A. Lewit-Bentley is gratefully acknowledged for continuous support to this work and valuable discussions. We thank Pr. A. P. Demchenko for helpful criticism of the manuscript. This work has been supported in part by a grant from EC (n° ERBBIO4CT960083). J. S. is the recipient of a post-doctoral support from this grant. The technical staff of LURE is acknowledged for running the synchrotron machine Super-ACO during the beam sessions. M. V. wishes to thank INSERM for continuous financial support.

## References

1. Meers, P. In *Annexins: Molecular Structure to Cellular Function*, B. A. Seaton, ed. Chapman and Hall, New York, 1996.
2. Gerke, V. and Moss, S. E. Annexins and membrane dynamics. *Biochim. Biophys. Acta* 1997; 135; 129–154.
3. Raynal, P. and Pollard, H. B. Annexins: the problem of assessing the biological role for a gene family of multifunctional calcium- and phospholipid-binding proteins. *Biochim. Biophys. Acta* 1994; 1197; 63–93.
4. Huber, R., Römisch, J. and Paques, E. P. The crystal and molecular structure of human annexin V, an anticoagulant protein that binds to calcium and membranes. *EMBO J.* 1990; 9; 3867–3874.
5. Huber, R., Schneider, M., Mayr, I., Römisch, J. and Paques, E. P. The calcium binding sites in human annexin V by crystal structure analysis at 2.0 A resolution. Implications for membrane binding and calcium channel activity. *FEBS Lett.* 1990; 275; 15–21.
6. Huber, R., Berendes, R., Burger, A., Schneider, M., Karshikov, A., Luecke, H., Romisch, J. and Paques, E. P. Crystal and molecular structure of human annexin V after refinement. Implications for structure, membrane binding and ion channel formation of the annexin family of proteins. *J. Mol. Biol.* 1992; 223; 683–704.
7. Lewit-Bentley, A., Morera, S., Huber, R. and Bodo, R. The effect of metal binding on the structure of annexin V and implications for membrane binding. *Eur. J. Biochem.* 1992; 210; 73–77.
8. Concha, N. O., Head, J. F., Kaetzel, M. A., Dedman, J. R. and Seaton, B. A. Rat annexin V crystal structure: Ca(2+)-induced conformational changes. *Science* 1993; 261; 1321–1324.
9. Sopkova, J., Renouard, M. and Lewit-Bentley, A. The crystal structure of a new high-calcium form of annexin V. *J. Mol. Biol.* 1993; 234; 816–825.
10. Lewit-Bentley, A., Bentley, G. A., Favier, B., L'Hermite, G. and Renouard, M. The interaction of metal ions with annexin V: a crystallographic study. *FEBS Lett.* 1994; 345; 38–42.
11. Swairjo, M. A., Concha, N. O., Kaetzel, M. A., Dedman, J. R. and Seaton, B. A. Ca(2+)-bridging mechanism and phospholipid head group recognition in the membrane-binding protein annexin V. *Nature Struct. Biol.* 1995; 2; 968–974.
12. Weng, X., Luecke, H., Song, I. S., Kang, D. S., Kim, S. H. and Huber, R. Crystal structure of human annexin I at 2.5 A resolution. *Protein Sci.* 1993; 2; 448–458.
13. Luecke, H., Chang, B. T., Maillard, W. S., Schlaepfer, D. D. and Haigler, H. T. Crystal structure of the annexin XII hexamer and implications for bilayer insertion. *Nature* 1995; 378; 512–515.
14. Favier-Perron, B., Lewit-Bentley, A. and Russo-Marie, F. The high-resolution crystal structure of human annexin III shows subtle differences with annexin V. *Biochemistry* 1996; 35; 1740–1744.
15. Burger, A., Berendes, R., Liemann, S., Benz, J., Hofmann, A., Göttig, P., Huber, R., Gerke, V., Thiel, C., Römisch, J. and Weber, J. The crystal structure and ion channel

activity of human annexin II, a peripheral membrane protein. *J. Mol. Biol.* 1996; 257; 839–847.

16. Benz, J., Bergner, A., Hofmann, A., Demange, P., Göttig, P., Liemann, S., Huber, R. and Voges, D. The structure of recombinant human annexin VI in crystals and membrane-bound. *J. Mol. Biol.* 1996; 260; 638–643.
17. Kawasaki, H., Avilasakar, A., Creutz, C. E. and Kretzinger, R. H. The crystal structure of annexin VI indicates relative rotation of the two lobes upon membrane binding. *Biochim. Biophys. Acta* 1996; 1313; 277–282.
18. Zanotti, G., Malpeli, G., Gliubich, F., Folli, C., Stoppini, M., Olivi, L., Savoia, A. and Berni, R. Structure of the trigonal crystal form of bovine annexin IV. *Biochem. J.* 1998; 329; 101–106.
19. Brisson, A., Mosser, G. and Huber, R. Structure of soluble and membrane-bound human annexin V. *J. Mol. Biol.* 1991; 220; 199–203.
20. Pigault, C., Follenius-Wund, A., Schmutz, M., Freyssinet, J.-M. and Brisson, A. Formation of two-dimensional arrays of annexin V on phosphatidylserine-containing liposomes. *J. Mol. Biol.* 1994; 236; 199–208.
21. Voges, D., Berendes, R., Burger, A., Demange, P., Baumeister, W. and Huber, R. Three-dimensional structure of membrane-bound annexin V. A correlative electron microscopy-X-ray crystallography study. *J. Mol. Biol.* 1994; 238; 199–213.
22. Kretsinger, R. H. Classification and evolution of EF-hand proteins. *Biometals.* 1998; 11; 277–295.
23. Dijkstra, B. W., Renetseder, R., Kalk, K. H., Hol, W. G. J. and Drenth, J. Structure of porcine pancreatic phospholipase A2 at 2.6 A resolution and comparison with bovine phospholipase A2. *J. Mol. Biol.* 1983; 168; 163–179.
24. Berendes, R., Voges, D., Demange, P., Huber, R. and Burger, A. Structure-function analysis of the ion channel selectivity filter in human annexin V. *Science* 1993; 262; 427–430.
25. Sopkova, J., Gallay, J., Vincent, M., Pancoska, P. and Lewit-Bentley, A. The dynamic behavior of annexin V as a function of calcium ion binding: a circular dichroism, UV absorption, and steady-state and time-resolved fluorescence study. *Biochemistry* 1994; 33; 4490–4499.
26. Meers, P. Location of tryptophans in membrane-bound annexins. *Biochemistry* 1990; 29; 3325–3330.
27. Meers, P., Bentz, J., Alford, D., Nir, S., Paphadjopoulos, D. and Hong, K. Synexin enhances the aggregation rate but not the fusion rate of liposomes. *Biochemistry* 1988; 27; 4430–4439.
28. Meers, P. and Mealy, T. R. Relationship between annexin V tryptophan exposure, calcium, and phospholipid binding. *Biochemistry* 1993; 32; 5411–5418.
29. Meers, P. and Mealy, T. R. Calcium-dependent annexin V binding to phospholipids: stoichiometry, specificity, and the role of negative charge. *Biochemistry* 1993; 32; 11711–11721.
30. Meers, P. and Mealy, T. R. Phospholipid determinants for annexin V binding sites and the role of tryptophan 187. *Biochemistry* 1994; 33; 5829–5837.
31. Luisi, P. L. and Magid, L. J. Solubilization of enzymes and nucleic acids in hydrocarbon micellar solutions. *Crit. Rev. Biochem.* 1986; 20; 409–474.
32. Nicot, C. and Waks, M. Proteins as invited guests of reverse micelles: conformational effects, significance, applications. *Biotechnol. Genet. Eng. Rev.* 1996; 13; 267–314.
33. Maurer-Fogy, I., Reutelingsperger, C. P. M., Peiters, J., Bodo, G., Stratowa, C. and Hauptman, R. Cloning and expression of cDNA for human vascular anticoagulant, a $Ca^{2+}$-dependent phospholipid-binding protein. *Eur. J. Biochem.* 1988; 174; 585–592.

34. Gallay, J., Vincent, M., Nicot, C. and Waks, M. Conformational aspects and rotational dynamics of synthetic adrenocorticotropin-(1–24) and glucagon in reverse micelles. *Biochemistry* 1987; 26; 5738–5747.
35. Rouvière, N., Vincent, M., Craescu, C. T. and Gallay, J. Immunosuppressor binding to the immunophilin FKBP59 affects the local structural dynamics of a surface beta-strand: a time-resolved fluorescence study. *Biochemistry* 1997; 36; 7339–7352.
36. Vincent, M., Gallay, J. and Demchenko, A. P. Solvent relaxation around the excited state of indole: analysis of fluorescence lifetime distribution and time-dependent spectral shift. *J. Phys. Chem.* 1995; 99; 14931–14941.
37. Livesey, A. K., Licinio, P. and Delaye, M. Maximum entropy analysis of dynamic parameters via the Laplace transform *J. Chem. Phys.* 1986; 84; 5102–5107.
38. Livesey, A. K. and Brochon, J. C. Analyzing the distribution of decay constants in pulse-fluorimetry using the maximum entropy method. *Biophys. J.* 1987; 52; 693–706.
39. Brochon, J.-C. Maximum entropy method of data analysis in time-resolved spectroscopy. *Methods in Enzymol.* 1994; 240; 262–311.
40. Vincent, M., Brochon, J.-C., Mérola, F., Jordi, W. and Gallay, J. Nanosecond dynamics of horse heart apocytochrome c in aqueous solution as studied by time-resolved fluorescence of the single tryptophan residue (Trp-59). *Biochemistry* 1988; 27; 8752–8761.
41. Mérola, F., Rigler, R., Holmgren, A. and Brochon, J.-C. Picosecond tryptophan fluorescence of thioredoxin: evidence for discrete species in slow exchange. *Biochemistry* 1989; 28; 3393–3398.
42. Gentin, M., Vincent, M., Brochon, J. C., Livesey, A. K., Cittanova, N. and Gallay, J. Time-resolved fluorescence of the single tryptophan residue in rat alpha-fetoprotein and rat serum albumin: analysis by the maximum-entropy method. *Biochemistry* 1990; 29; 10405–10412.
43. Kuipers, O. P., Vincent, M., Brochon, J.-C., Verheij, H. M., de Haas, G. H. and Gallay, J. Insight into the conformational dynamics of specific regions of porcine pancreatic phospholipase A2 from a time-resolved fluorescence study of a genetically inserted single tryptophan residue. *Biochemistry* 1991; 30; 8771–8785.
44. Vincent, M., Li de la Sierra, I. M., Berberan-Santos, M. N., Diaz, A., Diaz, M., Padron, G. and Gallay, J. Time-resolved fluorescence study of human recombinant interferon alpha 2. Association state of the protein, spatial proximity of the two tryptophan residues. *Eur. J. Biochem.* 1992; 210; 953–961.
45. Gallay, J., Vincent, M., Li de la Sierra, I. M., Alvarez, J., Ubieta, R., Madrazo, J. and Padrón, G. Protein flexibility and aggregation state of human epidermal growth factor. A time-resolved fluorescence study of the native protein and engineered single-tryptophan mutants. *Eur. J. Biochem.* 1993; 211; 213–219.
46. Bouhss, A., Vincent, M., Munier, H., Gilles, A.-M., Takahashi, M., Bârzu, O., Danchin, A. and Gallay, J. Conformational transitions within the calmodulin-binding site of Bordetella pertussis adenylate cyclase studied by time-resolved fluorescence of Trp242 and circular dichroism. *Eur. J. Biochem.* 1996; 237; 619–628.
47. Jaynes, E. T. *Papers on Probability Statistics and Statistical Physics* (Rosenkrantz, R. D., Ed.) Reidel, Dordrecht, The Netherlands. 1983.
48. Livesey, A. K. and Skilling, J. Maximum entropy theory. *Acta Crystallogr., Sect. A: Found. Crystallogr* 1985; A41; 113–122.
49. Wahl, Ph. Analysis of fluorescence anisotropy decay by a least square method. *Biophys. Chem.* 1979; 10; 91–104.
50. Vincent, M. and Gallay, J. The interactions of horse heart apocytochrome c with phospholipid vesicles and surfactant micelles: time-resolved fluorescence study of the single tryptophan residue (Trp-59). *Eur. J. Biophys.* 1991; 20; 183–191.

51. Blandin, P., Mérola, F., Brochon, J.-C., Trémeau, O. and Menez, A. Dynamics of the active loop of snake toxins as probed by time-resolved polarized tryptophan fluorescence. *Biochemistry* 1994; 33; 2610–2619.
52. Ichiye, T. and Karplus, M. Fluorescence depolarization of tryptophan residues in proteins: a molecular dynamics study. *Biochemistry* 1983; 22; 2884–2893.
53. Levy, R. M. and Szabo, A. *J. Amer. Chem. Soc.* 1982; 104; 2073–2075.
54. Kinosita, K. Jr, Kawato, S. and Ikegami, A. On the wobbling-in-cone analysis of fluorescence anisotropy decay. *Biophys. J.* 1977; 20; 289–305.
55. Malinowski, E. R. in *Factor Analysis in Chemistry*, John Willey and Sons, Second Edition, New York, 1991, p. 109.
56. Striekland, E. H. *Crit. Rev. in Biochem. and Molec. Biol.* 1974; 2; 113.
57. Kraulis, P. J. MOLSCRIPT: a program to produce both detailed and schematic plots of protein structures. *Appl. Crystallogr.* 1991; 24; 946–950.
58. Sopkova, J., Vincent, M., Takahashi, M., Lewit-Bentley, A. and Gallay, J. Conformational flexibility of domain III of annexin V studied by fluorescence of tryptophan 187 and circular dichroism: the effect of pH. *Biochemistry* 1998; 37; 11962–11970.
59. Chen, Y., Liu, B., Hong-Tao, Y. and Barkley, M. D. The Peptide Bond Quenches Indole Fluorescence. *J. Am. Chem. Soc.* 1996; 118; 9271-9278.
60. Chen, Y. and Barkley, M. D. Toward understanding tryptophan fluorescence in proteins. *Biochemistry* 1998; 37; 9976–9982.
61. Sopkova, J., Vincent, M., Takahashi, M., Lewit-Bentley, A. and Gallay, J. Conformational flexibility of domain III of annexin V at membrane/water interfaces. *Biochemistry* 1999; 38; 5447–5458.
62. Valeur, B. and Weber, G. Resolution of the fluorescence excitation spectrum of indole into the 1La and 1Lb excitation bands. *Photochem. Photobiol.* 1977; 25; 441–444.
63. Eftink, M. in *Topics in Fluorescence Spectroscopy* vol. 2 Principles (J. R. Lakowicz, ed.) chap. 2, pp. 53–126, Plenum Press, New York London. 1991.
64. Johnson, D. A. and Yguerabide, J. Solute accessibility to N epsilon-fluorescein isothiocyanate-lysine-23 cobra alpha-toxin bound to the acetylcholine receptor. A consideration of the effect of rotational diffusion and orientation constraints on fluorescence quenching. *Biophys. J.* 1985; 48; 949–955.
65. Newton, C. Pangborn, W. Nir, S. Papahadjopoulos, D. Specificity of Ca2+ and Mg2+ binding to phosphatidylserine vesicles and resultant phase changes of bilayer membrane structure. *Biochim Biophys Acta* 1978; 506; 281–287.
66. El Seoud, O. A. (1984) in *Reverse Micelles* (P. L. Luisi and B. E. Straub, eds.), pp. 81–93, Plenum Press, New York and London.
67. Marzola, P. and Gratton, E. Hydration and protein dynamics: frequency domain fluorescence spectroscopy on proteins in reverse micelles. *J. Phys. Chem.* 1991; 95; 9488–9495.
68. Visser, A. J. W. G. Time-resolved fluorescence on self-assembly membranes. *Curr. Opin. Coll. and Interface Scie.* 1997; 2; 27–36.
69. Wong, M., Thomas, J. K. and Grätzel, M. Fluorescence probing of inverted micelles. The state of solubilized water clusters in alkane/diisooctyl sulfosuccinate (Aerosol OT) solution. *J. Am. Chem. Soc.* 1976; 98; 2391–2397.
70. Keh, E. and Valeur, B. Investigation of water-containing inverted micelles by fluorescence polarization. Determination of size and internal fluidity. *J. Coll. Interface Sci.* 1981; 79; 465–478.
71. Meers, P. in *Annexins: Molecular Structure to Cellular Function* (B. A. Seaton, ed.) Chapman and Hall, New York. 1996.
72. Dijkstra, B. W., Kalk, K. H., Hol, W. G. J. and Drenth, J. Structure of bovine pancreatic phospholipase $A_2$ at 1.7 Å resolution. *J. Mol. Biol.* 1981; 147; 97–123.

73. Thunissen, M. M. G. M., Kalk, K. H., Drenth, J., Dijkstra, B. W., Kuipers, O. P., Dijkman, R., de Haas, G. H. and Verheij, H. M. X-ray structure of phospholipase A2 complexed with a substrate-derived inhibitor. *Nature* 1990; 374; 689–691.
74. Scott, D. L., Otwinowski, Z., Gelb, M. H. and Sigler, P. Crystal structure of bee-venom phospholipase A2 in a complex with a transition-state analogue. *Science* 1990; 250; 1563–1566.
75. Scott, D. L., Otwinowski, Z., Yuan, K., Gelb, M. H. and Sigler, P. Interfacial catalysis: the mechanism of phospholipase A2. *Science* 1990; 250; 1541–1546.
76. White, S. P., Scott, D. L., Otwinowski, Z., Gelb, M. H. and Sigler, P. Crystal structure of cobra-venom phospholipase A2 in a complex with a transition-state analogue. *Science* 1990; 250; 1560–1563.
77. Volwerk, J. J. and de Haas, G. H. in *Molecular Biology of Lipid-Protein interactions* (eds. Griffith, O. H. and Jost, P. C.) pp. 69–149, Wiley, New York. 1982.
78. van Dam-Mieras, M. C. E., Slotboom, A. J., Pieterson, W. A. and de Haas, G. H. The interaction of phospholipase A2 with micellar interfaces. The role of the N-terminal region. *Biochemistry* 1975; 14; 5387–5394.
79. Ludescher, R. D., Johnson, I. D., Volwerk, J. J., de Haas, G. H., Jost, P. and Hudson, B. Rotational dynamics of the single tryptophan of porcine pancreatic phospholipase $A_2$, its zymogen, and an enzyme/micelle complex. A steady-state and time-resolved anisotropy study. *Biochemitry* 1988; 27; 6618–6628.
80. Ludescher, R. D., Volwerk, J. J., de Haas, G. H. and Hudson, B. Complex photophysics of the single tryptophan of porcine pancreatic phospholipase $A_2$, its zymogen, and an enzyme/micelle complex. *Biochemistry* 1985; 24; 7240–7249.
81. Jain, M. H. and Maliwal, B. P. Spectroscopic properties of the states of pig pancreatic phospholipase $A_2$ at interfaces and their possible molecular origin. *Biochemistry* 1993; 32; 11838–11846.
82. Vincent, M., Deveer, A.-M., de Haas, G. H., Verheij, H. M. and Gallay, J. Stereospecificity of the interaction of porcine pancreatic phospholipase A2 with micellar and monomeric inhibitors. A time-resolved fluorescence study of the tryptophan residue. *Eur. J. Biochem.* 1993; 215; 531–539.
83. van den Berg, B., Tessari, M., Boelens, R., Dijkman, R., de Haas, G. H., Kaptein, R. and Verheij, H. M. NMR structures of phospholipase $A_2$ reveal conformational changes during interfacial activation. *Nature Struct. Biol.* 1995; 2; 402–406.
84. Verger, R. and de Haas, G. H. Interfacial enzyle kinetics of lipolysis. *Annu. Rev. Biophys. Bioeng.* 1976; 5; 77–119.
85. Deveer, A. M. T. J., den Ouden, A. T., Vincent, M., Gallay, J., Verger, R., Egmont, M. R., Verheij, H. M. and de Haas, G. H. Competitive inhibition of lipolytic enzymes. VIII: Inhibitor-induced aggregation of porcine pancreatic phospholipase A2. *Biochim. Biophys. Acta* 1992; 1126; 95–104.
86. Sopkova, J. PhD thesis, Universities of Prague and Orsay. 1994.
87. Follenius-Wund, A., Piémont, E, Freyssinet, J.-M., Gerard, D. and Pigault, C. Conformational adaptation of annexin V upon binding to liposomes: a time-resolved fluorescence study. *Biochem. Biophys. Res. Comm.* 1997; 234; 111–116.
88. London, E. and Feigenson, G. W. Fluorescence quenching in model membranes. 1. Characterization of quenching caused by a spin-labeled phospholipid. *Biochemistry* 1981; 20; 1932–1938.
89. Chattopadhyay, A. and London, E. Parallax method for direct measurement of membrane penetration depth utilizing fluorescence quenching by spin-labeled phospholipids. *Biochemistry* 1987; 26; 39–45.

90. Yeager, M. D. and Feigenson, G. W. Fluorescence quenching in model membranes: phospholipid acyl chain distributions around small fluorophores. *Biochemistry* 1990; 29; 4380–4392.
91. Ladokhin, A. S. and Holloway, P. W. Fluorescence of membrane-bound tryptophan octyl ester: a model for studying intrinsic fluorescence of protein-membrane interactions. *Biophys. J.* 1995; 69; 506–517.
92. Ladokhin, A. S. Analysis of protein and peptide penetration into membranes by depth-dependent fluorescence quenching: theoretical considerations. *Biophys. J.* 1999; 76; 946–955.
93. Tanford, C. and Reynolds, J. A. Characterization of membrane proteins in detergent solutions. *Biochim. Biophys. Acta* 1976; 457; 133–170.
94. Callis, P. R. and Burgess, B. K. Tryptophan Fluorescence Shifts in Proteins from Hybrid Simulations: An Electrostatic Approach. *J. Phys. Chem.* 1997; 101; 9429–9432.
95. Callis, P. R. [1]La and [1]Lb transitions of tryptophan: applications of theory and experimental observations to fluorescence of proteins. *Methods in Enzymology* 1997; 278; 113–150.
96. Voges, K.-P., Jung, G. and Sawyer, W. H. Depth-dependent fluorescent quenching of a tryptophan residue located at defined positions on a rigid 21-peptide helix in liposomes. *Biochim. Biophys. Acta* 1987; 896; 64–76.
97. Cowgil, R. W. Fluorescence and protein structure X. Reappraisal of solvent and structural effect. *Biochim. Biophys. Acta* 1967; 133; 6–18.
98. Russell, E. C. and Cowgill, R. W. Fluorescence and protein structure. 13. Further effects of side-chain groups. *Biochim. Biophys. Acta* 1968; 154; 231–233.
99. Steiner, R. F. and Kirby, E. P. The interaction of the ground and excited states of indole derivatives with electron scavengers. *J. Phys. Chem.* 1969; 73; 4130–4135.
100. Bushueva, T. L., Busel, E. P. and Burstein, E. A. The interaction of protein functional groups with indole chromophores III. Amine, amide and thiol groups. *Stud. Biophys.* 1975; 52; 41–52.
101. Bushueva, T. L., Busel, E. P., Bushueva, V. N. and Burstein, E. A. The interaction of protein functional groups with indole chromophore I. Imidazole group. *Stud. Biophys.* 1974; 44; 129–139.
102. Ricci, R. W. and Nesta, J. M. Inter- and intramolecular quenching of indole fluorescence by carbonyl compounds. *J. Phys. Chem.* 1976; 80; 974–980.
103. Petrich, J. W., Chang, M. C., McDonald, D. M. and Fleming, G. R. On the origin of nonexponential fluorescence decay in tryptophan and its derivatives. *J. Am. Chem. Soc.* 1983; 105; 3824–3832.
104. Shizuka, H., Scrizawa, M., Shimo, T., Saito, I. and Matsuura, T. Fluorescence-quenching mechanism of tryptophan. Remarkably efficient internal proton-induced quenching and charge-transfer quenching. *J. Am. Chem. Soc.* 1988; 110; 1930–1934.
105. Tilstra, L., Sattler, M. C., Cherry, W. R. and Barkley, M. D. Fluorescence of a rotationally constrained tryptophan derivative, 3-carboxy-1,2,3,4-tetrahydro-2-carboline. *J. Am. Chem. Soc.* 1990; 112; 554–563.
106. Szabo, A. G. and Rayner, D. M. Fluorescence decay of tryptophan conformers in aqueous solution. *J. Am. Chem. Soc.* 1980; 102; 554–563.
107. McMahon, L. P., Colucci, W. J., McLaughlin, M. L. and Barkley, M. D. Deuterium isotope effects in constriend tryptophan derivatives: implications for tryptophan photophysics. *J. Am. Chem. Soc.* 1992; 114; 8442–8448.
108. Boens, N., Janssens, L. D., van Dommelen, L., de Schryver, F. C. and Gallay, J. Photophysics of tryptophan: global analysis of the fluorescence decay surface as a function of ph, temperature, quencher concentration, excitation and emission wavelengths, timing

calibration and deuterium isotope effect. Time-Resolved Laser Spectroscopy in Biochemistry III, *SPIE Proceedings*. 1992; 140; 58–69.

109. Vos, R. and Engelborghs, Y. A fluorescence study of tryptophan-histidine interactions in the peptide anantin and in solution. *Photochem. Photobiol.* 1994; 60; 24–32.

110. Cordier-Ochsenbein, F. PhD thesis, University Paris-Sud. 1997.

111. Plager, D. A. and Nestuelen, G. L. Direct enthalpy measurements of the calcium-dependent interaction of annexins V and VI with phospholipid vesicles. *Biochemistry* 1994; 33; 13239–13249.

112. Campos, B., Mo, Y. D., Mealy, T. R., Swairjo, M. A., Balch, C., Head, J. F., Retzinger, G., Dedman, J. R. and Seaton, B. A. Mutational and crystallographic analyses of interfacial residues in annexin V suggest direct interactions with phospholipid membrane components. *Biochemistry* 1998; 37; 8004–8010.

113. Saurel, O., Cezanne, L., Milon, A., Tocanne, J. F. and Demange, P. Influence of annexin V on the structure and dynamics of phosphatidylcholine/phosphatidylserine bilayers: a fluorescence and NMR study. *Biochemistry* 1998; 37; 1403–1410.

114. Cezanne, L., Lopez, A., Loste, F., Parnaud, G., Saurel, O., Demange, P. and Tocanne, J. F. Organization and dynamics of the proteolipid complexes formed by annexin V and lipids in planar supported lipid bilayers. *Biochemistry* 1999; 38; 2779–2786.

115. Beermann, Br. B., Hinz, H.-J., Hofmann, A. and Huber, R. Acid induced equilibrium unfolding of annexin V wild type shows two intermediate states. *FEBS Lett.* 1998; 423; 265–269.

116. Reviakine, I., Bergma-Schutter, W. and Brisson, A. Growth of Protein 2-D Crystals on Supported Planar Lipid Bilayers Imaged in Situ by AFM. *J. Struct. Biol.* 1998; 121; 356–361.

117. Concha, N. O., Head, J. F., Kaetzel, M. A., Dedman, J. R. and Seaton, B. A. Annexin V forms calcium-dependent trimeric units on phospholipid vesicles. *FEBS Lett.* 1992; 314; 159–162.

118. Brisson, A. and Lewit-Bentley, A. in *Annexins: Molecular Structure to Cellular Function* (B. A. Seaton, ed.), chap. 4, pp. 43–52, Chapman and Hall, New York. 1996.

# 8

# Tryptophan Calmodulin Mutants

Jacques Haiech and Marie-Claude Kilhoffer

## 8.1. Introduction

Calcium is one of the most important second messengers in eukaryotic cells and may also play a role in prokaryotic cells[1–4] although less convincing evidence has been reported in these systems. Pioneered by the work of Ringer at the end of the last century, one had to wait until the sixties to start to get some insight into the intracellular molecular mechanisms underlying calcium signalling. The development and use of calcium chelators on the one hand,[5–8] and the purification and characterization of membrane fractions able to accumulate calcium against a calcium concentration at the expense of ATP hydrolysis[9–11] on the other hand, constituted the first steps along the road that led us to the understanding of the role of calcium inside cells. The next milestone, at the beginning of the seventies, was the discovery of eukaryotic calcium binding proteins belonging to a unique evolutionary family and the description of their multidomain structure.[12–25] The end of the seventies and early eighties were marked by the description of calcium channels.[26–29] These were extensively investigated thanks to patch clamp and molecular biology resulting in a fine classification of the different types of calcium channels and the development of useful pharmacological tools, some of which pursued a career as drug stars in the eighties;[30–34] for review see.[35–37]

In cells, the different elements ($Ca^{2+}$ ATPases, $Ca^{2+}$ channels and $Ca^{2+}$ binding proteins) are combined in order to:

*Jacques Haiech and Marie-Claude Kilhoffer* • Pharmacologie et Physico-Chimie des Interactions Cellulaires et Moléculaires, UMR CNRS 7034, Université Louis Pasteur de Strasbourg, Faculté de Pharmacie, 67401 Illkirch France.

*Topics in Fluorescence Spectroscopy, Volume 6: Protein Fluorescence*, edited by Joseph R. Lakowicz. Kluwer Academic / Plenum Publishers, New York, 2000

- maintain a calcium gradient between two cellular compartments (intracellular vs. extracellular compartments or two intracellular compartments),
- trigger, upon specific stimuli, transient and localized increases in the cytosolic calcium concentration,
- detect localized calcium transients and transduce them into cellular events.

Good knowledge of the individual components of the calcium machinery has been obtained in the two last decades. However, the detailed molecular mechanisms explaining how a localized calcium transient leads to a given physiological response is still unclear. In addressing this question, Haiech and Demaille, in 1980, proposed the concept of the calcisome.[38] The calcisome was defined as "a specific assembly of calcium sensors, target enzymes and inhibitory proteins, associated with or in close vicinity to the membrane which contains $Ca^{2+}$ pumps and channels and able to respond specifically to a transient and localized rise in $Ca^{2+}$ concentration". The extraordinary development of cell imaging, multiphoton microscopy and of microspectrophotometric techniques will undoubtedly bring exciting information and shed new light onto how calcium signalling is deciphered in living cells.

One of the key events in calcium signal transduction is the detection of calcium transients. Since the early seventies, the intracellular eukaryotic calcium binding proteins appear to be the main calcium detectors (for reviews see[39–44]). Most of these proteins belong to the EF-hand domain protein family and present in their structure the canonical EF-hand domain, constituted by two 12 residue-long alpha helices surrounding a 12 residue-long calcium binding loop,[19,21,45] suggesting their probable evolution from a single EF-hand domain by duplication. The prototype of this family is calmodulin, a four EF-hand domain protein.[22,23,46–50] Whereas most of the calcium binding proteins are specifically localized and are representative of a given cellular state, calmodulin appears to be ubiquitous, present in all eukaryotic species and involved in a multitude of calcium dependent cellular events, through its interaction with various target enzymes. Therefore, numerous studies were undertaken in order to obtain detailed mechanistic insight into calcium binding to this fascinating protein.

Calmodulin was identified as an activator of cyclic nucleotide phosphodiesterase in 1970 by Cheung and Kakiuchi.[13,14] The biological activity of the protein was investigated during the seventies,[51–59] but the complete amino acid sequence appeared only in 1980.[60,61] The protein sequence was confirmed by DNA sequencing.[62–64] Calmodulin crystallization was difficult and the first 3D-structure was released in 1985 and refined in 1988.[65,66] In its crystal form, the protein appeared as a dumbbell, composed of two lobes linked by a long alpha helix. Each lobe is composed of two EF-hand calcium binding domains (Figure 8.1).

Calmodulin sequences from different organisms are very similar and based on these sequences, a synthetic hybrid calmodulin gene was created in 1985[67] (Figure 8.2). This gene allowed to produce in *Escherichia coli* the first recombinant calmodulin, initially termed VU-1 and then SynCaM (for synthetic calmodulin). The protein, which presented all the activation properties of natural calmodulin including activation of plant NAD-kinase, was used as a standard of comparison.

In 1985, when we started our work with synthetic calmodulin, the dogma that prevailed in the calmodulin field were the following:

- calmodulin presents four calcium binding sites on two independent lobes,
- calmodulin interacts with numerous target proteins in a similar way, schematically depicted as follows: upon calcium binding, calmodulin exposes (a) hydrophobic patch(es), which constitute(s) the area of interaction with the various target proteins. Therefore, any mutation in calmodulin will appear upon calcium binding, being either neutral or able to block most if not all interactions with target proteins and subsequent cellular events.

In order to challenge such a view, calmodulin mutations were performed along two lines. The first, aimed to modify the electrostatic potential of SynCaM, gave rise to the family of electrostatic mutants.[68,69] The use of such mutants clearly showed that calmodulin interacts differently with its various target proteins. The second line was aimed to introduce a reporter group in the protein in order to follow the protein structural changes induced by ligand binding. This led us to develop the family of tryptophan containing calmodulin mutants.

The present chapter will deal exclusively with the latter strategy. The main results obtained using tryptophan containing calmodulin mutants are

Figure 8.1. 3D structure of calcium-loaded SynCaM. The cylinders correspond to the $\alpha$-helices, the spheres to the $Ca^{2+}$ ions.

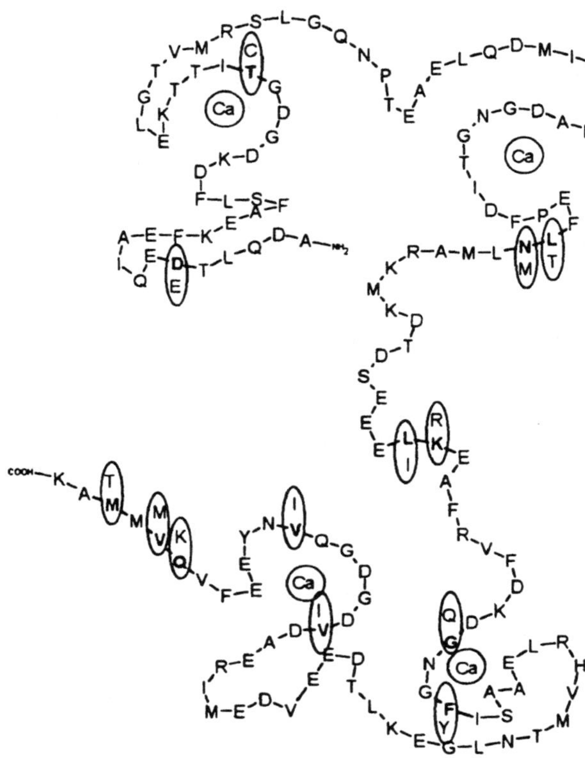


Figure 8.2. Synthetic calmodulin sequence compared to spinach calmodulin and mammalian calmodulin sequences.

presented with emphasis on how they allowed to progress in the understanding of the protein structure and function.

## 8.2. Building Tryptophan Containing Calmodulin Mutants

The primary structure of SynCaM contains a single tyrosyl residue (Tyr138) and no tryptophan. This latter residue has been shown to be an important probe in studies of protein structures and dynamics. Indeed, its spectral characteristics (namely maximum of emission wavelength and quantum yield) are very sensitive to the chromophore surrounding. We therefore decided to introduce a single tryptophan residue in different areas of the protein, in order to monitor the surrounding of the inserted residue and the changes in its surrounding upon calcium binding. It was therefore of paramount importance that the tryptophan mutation did not induce any major structural or functional modification of the synthetic calmodulin. In other words, the newly designed mutants have to be isofunctional.

### 8.2.1. Where to Insert the Tryptophanyl Residue?

At first, to find amino acid positions which can be substituted by tryptophanyl residues, the primary structures of all known EF hand domains were compared in order to detect the positions exhibiting the lowest level of constraints, in other words, the positions where the residue conservation is extremely weak. The $7^{th}$ position in the calcium loop appeared to fulfill this criteria. It is interesting to note that in calmodulin this position is often occupied by an aromatic residue. In a second step, the effect of tryptophan substitution was analyzed on computer-generated models. Starting from the crystal structure of rat calmodulin, the SynCaM structure was computed. This requires 10 conservative mutations. In the SynCaM structure thus obtained, an amino acid at a selected position was replaced by a tryptophanyl residue and the structure was minimized by molecular mechanics. In 1986, when the present work was started, different force fields were used to generate the energy-minimized model structures. All led to similar results. Currently, with the increase in computer power and the ease to do energy minimization and dynamic computation, we generate computer models with a more general and complete strategy (Figure 8.3) using SwissModel tools (www.expasy.ch). This strategy can be applied to any protein as soon as a 3D structure is available. For calmodulin, three 3D structures are available: apocalmodulin,[70,71] calcium saturated calmodulin[66,72] and calcium-calmodulin—target peptide complex.[73,74] The insertion of the tryptophan residue can thus be checked on all three structures.

In 1986, we decided to make five tryptophan containing calmodulin mutants, four of them with a tryptophan mutation in one of the four calcium binding loops (SYNCAM-32 (T26W), SYNCAM-33 (T62W), SYNCAM-9 (F99W) and SYNCAM-34 (Q135W)) and a fifth with a single tryptophan mutation in the central helix that links the two lobes of the protein together (SYNCAM-31 (S81W)). The five mutants are presented in Figure 8.4.

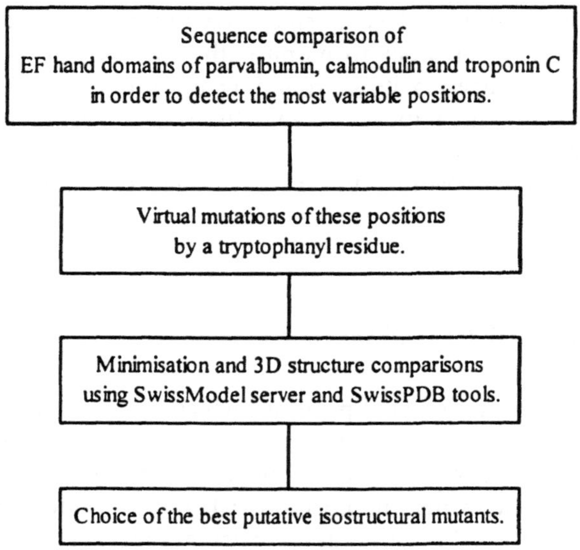


Figure 8.3. Schematic representation of the strategy now used to introduce a tryptophanyl residue in SynCaM. URL for Swissmodel: www.expasy.ch

### 8.2.2. How to Insert Tryptophan?

The synthetic calmodulin gene with its unique, regularly spaced endonuclease restriction sites was designed also to facilitate cassette-based site-specific mutagenesis.[67,75,76] Using the appropriate restriction endonucleases, a specific gene segment containing the codon to be substituted was removed and replaced by a new gene segment with the desired tryptophan codon.[76,77] Nowadays, PCR techniques would allow to build and to modify (by using recombinant PCR[78,79]) synthetic genes in a simpler way.

### 8.2.3. Expression, Purification and Characterization of the Tryptophan Containing Mutants

The gene encoding SynCaM and by extension all the derived mutants including tryptophan mutants were cloned into the expression vector pKK223-3[67] under control of a hybrid *trp-lac* promoter (P*tac*) which allows IPTG-induced high-level expression of proteins in *E. coli*. Stronger promoters can be found in different commercialized expression vectors (e.g. T7

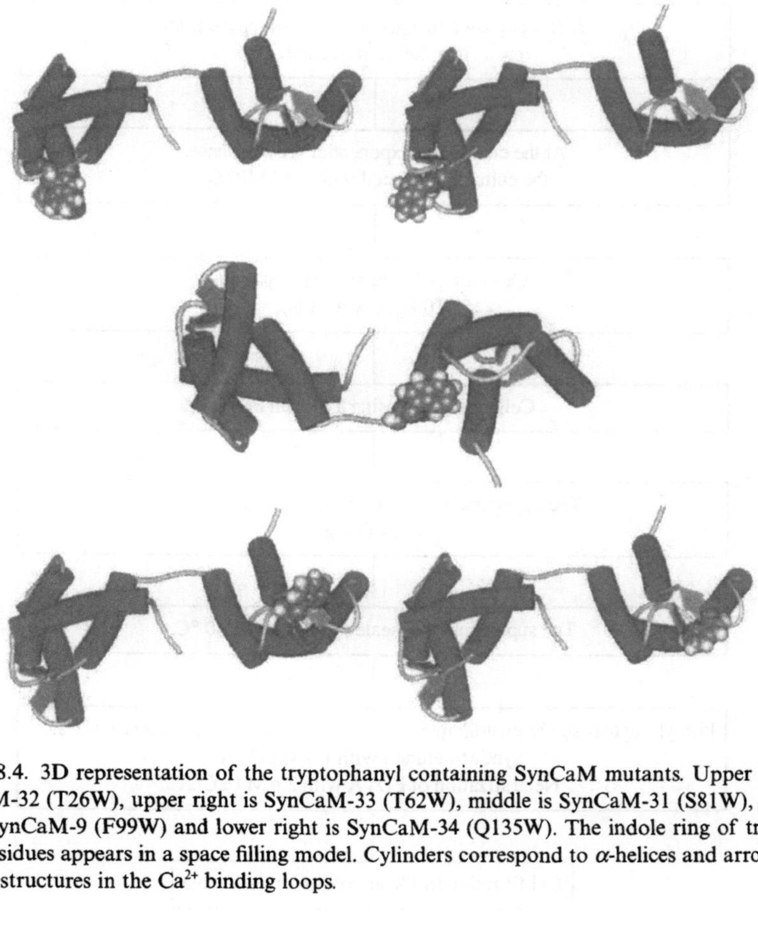

Figure 8.4. 3D representation of the tryptophanyl containing SynCaM mutants. Upper left is SynCaM-32 (T26W), upper right is SynCaM-33 (T62W), middle is SynCaM-31 (S81W), lower left is SynCaM-9 (F99W) and lower right is SynCaM-34 (Q135W). The indole ring of tryptophan residues appears in a space filling model. Cylinders correspond to $\alpha$-helices and arrows to $\beta$-sheet structures in the $Ca^{2+}$ binding loops.

promoter). However, very strong promoters are sometimes detrimental by overwhelming the metabolism of the host cell, thus generating a heterogeneous protein population. A common example is the decrease in the amino terminal methionine cleavage efficiency, leading to a mixture of proteins carrying or not this residue. The P*tac* promoter appeared as an excellent compromise, allowing both qualitative (homogeneous proteins) and quantitative recombinant protein production.

The general recombinant protein purification scheme is described in Figure 8.5. Five hundred milligrams to one gram of proteins are obtained from a 200 L. fermentor using a growth medium where glucose was replaced by glycerol. Each batch of protein was characterized by

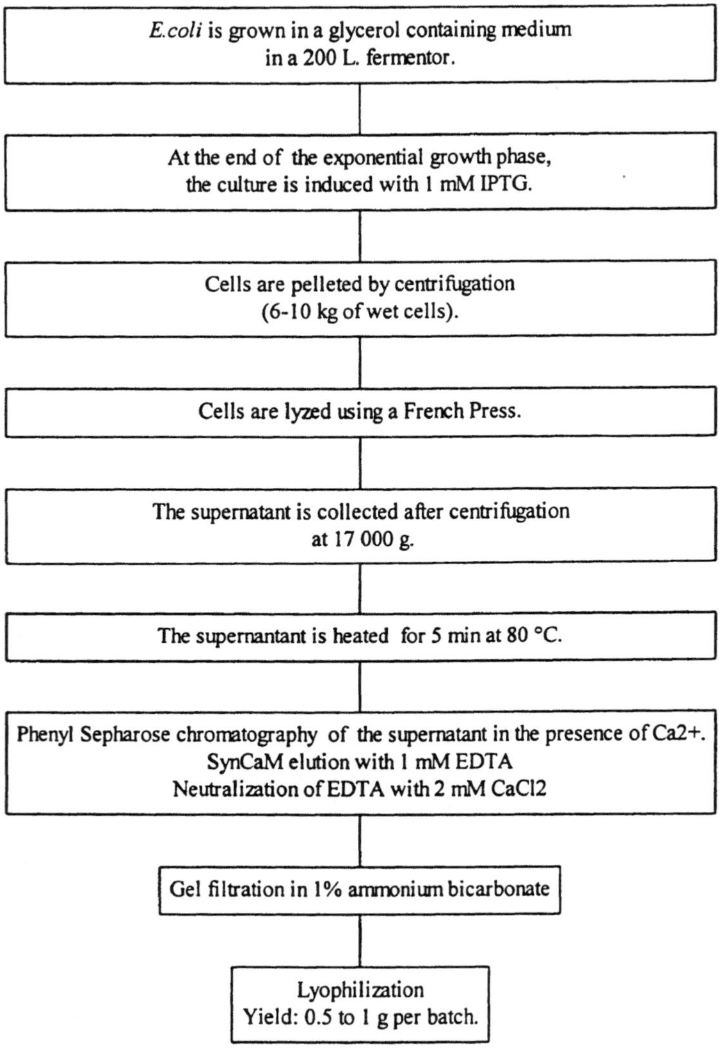


Figure 8.5. Schematic representation of SynCaM purification scheme.

electrophoresis in SDS gels, HPLC chromatography, amino acid analysis, sequencing of the mutated peptide and mass spectrometry. Our different studies showed that, in order to get a homogeneous batch of protein, long exposure to any calcium chelator had to be avoided. SynCaM and its derivatives were kept lyophilized at −20 °C and have proved to be stable for more than 10 years.

## 8.3. Analysis of the Tryptophan Containing Calmodulin Mutants

The isofunctionality of the different mutants was checked with respect to their enzyme activation abilities. Four calmodulin dependent enzymes were used for the assay (myosin light chain kinase, adenylate cyclase, NAD kinase and cyclic nucleotide phosphodiesterase). The five tryptophan containing mutants activated these enzymes in a manner similar to SynCaM. Nevertheless, the presence of small differences cannot be completely ruled out and it is probably wise to think than any change in calmodulin induces small perturbations in its properties. Therefore, an absolute isofunctionality of the different tryptophan mutants appears unlikely.

### 8.3.1. The Mutants Have to Be Isostructural

Ideally, the structure of each mutant has to be as similar as possible to the structure of SynCaM, the synthetic standard calmodulin. Checking this point in a strict manner would require either crystallization of each mutant or a NMR study. We used a lower level of constraints and performed CD spectra in order to check that, under similar experimental conditions, each mutant presents a level of alpha helix content similar to that of SynCaM. It is important to note that the cacodylate buffer used in the crystallization experiments promotes helix formation.[80,81] Therefore, it is not surprising that the alpha helix content computed from CD spectra is not the same obtained from the 3D structure of SynCaM.

### 8.3.2. The Mutants Have to Be Similar to SynCaM in their Calcium Binding Properties

Calcium binding to these mutants was performed using flow dialysis under different experimental conditions (Table 8.1). Under all conditions tested, the binding curves obtained were indistinguishable.

Taking together the different observations, we were fairly confident that the 3D structure was conserved in the five tryptophan containing mutants.

Table 8.1. Calcium Binding Parameters of the Different Calmodulin Tryptophan Containing Mutants in the Absence or Presence of Ethylene Glycol (EG)

| Mutants | EG % | $K_d$ (μM) | $n$ |
|---|---|---|---|
| T26W | 0 | 3.85 | 4.4 |
| | 20 | 0.63 | 4 |
| | 40 | 0.27 | 3.5 |
| T62W | 0 | 2.5 | 3.7 |
| | 20 | — | — |
| | 40 | 0.28 | 3.6 |
| S81W | 0 | 2.2 | 3.8 |
| | 20 | 0.68 | 3.7 |
| | 40 | 0.28 | 3.8 |
| F99W | 0 | 2.56 | 4.4 |
| | 20 | 1.1 | 3.8 |
| | 40 | 0.53 | 4 |
| Q135W | 0 | 2.9 | 3.9 |
| | 20 | 1.02 | 3.9 |
| | 40 | 0.48 | 4.2 |

Experiments were performed in 50 mM Hepes, pH 7.5, in the absence or presence of EG at the concentrations indicated. $K_d$, the $Ca^{2+}$ dissociation constant, and $n$, the number of $Ca^{2+}$ binding sites, were obtained from flow dialysis experiments and determined using the Scatchard equation: $v = nKx/1 + Kx$, where $K = 1/K_d$ is the $Ca^{2+}$ association constant and $v$ and $x$ are the concentrations of bound and free $Ca^{2+}$, respectively. Reprinted with permission from (82). 

## 8.4. Using Tryptophan Containing Calmodulin Mutants as Tools to Obtain Deeper Insight into the Structure and Calcium Binding Mechanism of Calmodulin

Calcium binding to calmodulin is one of the first steps in the expression of the protein activity. Understanding the mechanism of $Ca^{2+}$ binding to calmodulin is thus of paramount importance for understanding the relationship between an increase in the intracellular $Ca^{2+}$ concentration and the physiological cell response. Various approaches and techniques have been used to obtain insight into the mechanism of $Ca^{2+}$ binding to calmodulin. Total $Ca^{2+}$ binding to the protein was investigated using equilibrium dialysis, flow dialysis and $Ca^{2+}$-sensitive electrodes.[83–89] Conformational changes associated with $Ca^{2+}$ binding to the protein were followed using changes in the fluorescence properties of Tyr 99 and Tyr 138 (native calmodulin), absorption properties, circular dichroism, NMR and small angle neutron diffraction.[83,87,90–110] Finally, kinetics of $Ca^{2+}$ dissociation have been measured using tyrosine fluorescence as a internal probe or Quin II as an external probe.[111,112] Several models have been proposed to explain the results obtained. The first model, based on the quasi-linearity of the Scatchard plots considers calmodulin as a protein with four identical $Ca^{2+}$ binding sites.[113] This model did not take into consideration results obtained from conformational studies, where changes in the properties of a probe located either in the COOH or the $NH_2$ terminal half of the molecule were recorded for molar ratios of $Ca^{2+}$ to calmodulin of zero to two, or two to four, respectively. This led to a second model in which calmodulin was considered as a protein with two high affinity and two low affinity sites, the sites being independent and the difference in $K_d$ being about two orders of magnitude. This model took into account NMR-, fluorescence- and CD data, but was not in agreement with the binding isotherm. The third model introduced cooperativity in the $Ca^{2+}$ binding mechanism.[84,114–116] One variant of this model assumes calmodulin to exhibit positive cooperativity between the two sites of a given lobe, the two lobes (COOH and $NH_2$ lobes) being independent. In addition, in this model, the mean $Ca^{2+}$ affinity of the COOH terminal lobe is 6–10 times greater than the mean $Ca^{2+}$ affinity of the $NH_2$ lobe. In the second variant of the cooperative model, termed sequential model or perhaps more appropriately the preferential pathway binding model, not only do the two $Ca^{2+}$ binding sites in a given lobe interact, but so do also the two lobes of the protein. The preferential binding pathway means that $Ca^{2+}$ binding sites of calmodulin are occupied in an ordered manner, binding of calcium to one specific site facilitating $Ca^{2+}$ binding to the next specific site. This latter model took into consideration all the data from the literature (binding isotherms, spectroscopic data, kinetic data). For a while, the four models coexisted in the calmodulin field. In this chapter, we will show how the tryptophan mutants helped in the understanding of the molecular mechanism of $Ca^{2+}$ binding to calmodulin.

### 8.4.1. Fluorescent Properties of the Tryptophan Containing SynCaM Mutants

As indicated above, tryptophanyl residues in the calcium binding loops all occupy the same position of the loop ($7^{th}$ position in the 12 residue-long $Ca^{2+}$ binding loop). It was therefore interesting to investigate the steady state fluorescent properties of the different mutants in the absence or presence of $Ca^{2+}$. Results are shown in Table 8.2.

Table 8.2. Spectroscopic Properties of Tryptophan Containing SynCaM Mutants and Tryptophan Model Compounds

| Parameters / SynCaM | Conditions | $\varepsilon_M$ ($M^{-1}$ $cm^{-1}$) | $\lambda_{max}$ (nm) | $\phi_{297}$ |
|---|---|---|---|---|
| T26W | $-Ca^{2+}$ | 6150 | 345 | 0.14 |
| | $+Ca^{2+}$ | 6350 | 354 | 0.31 |
| | 6M Gua | — | 353 | 0.11 |
| | 70% EG, $-Ca^{2+}$ | — | 341 | 0.21 |
| | 70% EG, $+Ca^{2+}$ | — | 353 | 0.31 |
| T62W | $-Ca^{2+}$ | 7400 | 343 | 0.12 |
| | $+Ca^{2+}$ | 7400 | 346 | 0.13 |
| | 6M Gua | — | 353 | 0.12 |
| F99W | $-Ca^{2+}$ | 7400 | 348 | 0.19 |
| | $+Ca^{2+}$ | 7400 | 348 | 0.15 |
| | 6M Gua | — | 352 | 0.11 |
| Q135W | $-Ca^{2+}$ | 6150 | 350 | 0.23 |
| | $+Ca^{2+}$ | — | 355 | 0.29 |
| S81W | $-Ca^{2+}$ | 6150 | 348 | 0.18 |
| | $+Ca^{2+}$ | 6000 | 342 | 0.2 |
| | 6M Gua | — | 352 | 0.12 |
| L-Trp or NATA | in $H_2O$ | 5450 | 352 | 0.14 |
| | in 50 mM Hepes, 70%EG, pH 7.5 | — | 350 | 0.21 |

*$\varepsilon_M$ corresponds to the molar extinction coefficient of the protein at the absorption maximum $\lambda_{max}$ corresponds to the maximum of emission (±1 nm) for excitation at 297 nm. $\phi_{297}$ is the protein quantum yield (±0.01) for excitation at 297 nm. $-Ca^{2+}$ and $+Ca^{2+}$ stand for the protein in the absence of $Ca^{2+}$ and at a saturating $Ca^{2+}$ concentration, respectively. Experiments were performed in 50 mM Hepes, pH 7.5, unless indicated otherwise. NATA stands for N-acetyltryptophanamide.

For the different mutants the emission maxima range between 341 and 355 nm. In the absence of $Ca^{2+}$, tryptophans in the $NH_2$ terminal half of the molecule exhibit similar fluorescence properties ($\lambda_{max}$ and quantum yields) as do tryptophans in the COOH terminal half. Emission maxima of Trp 99 and Trp 135 are slightly red-shifted compared to those of Trp 26 and Trp 62, although their quantum yields are higher. In the presence of $Ca^{2+}$, tryptophan fluorescence properties ($\lambda_{max}$ and quantum yields) regroup differently in that Trp 26 properties resemble those of Trp 135 and Trp 62 properties are similar to those of Trp 99. In contrast to SynCaM T26W, SynCaM T62W, SynCaM F99W and SynCaM Q135W where $Ca^{2+}$ binding to the protein induces either no change in the emission maxima of the tryptophanyl residues or a red-shift, SynCaM S81W, with the tryptophanyl residue in the central helix, shows a 6 nm blue shift. Results presented here for the $Ca^{2+}$ loaded proteins are similar to those obtained by Chabbert *et al.*[77] The emission maxima and quantum yields of the tryptophan containing mutants in 6 M guanidinium chloride were close to the values commonly observed in denatured proteins ($\phi = 0.11$ and $\lambda_{max} \cong 352$ nm). Emission wavelength maxima of the different tryptophans strongly suggest that the residues are well exposed (in the case of $Ca^{2+}$ loaded SynCaM T26W, apo- and $Ca^{2+}$ loaded SynCaM Q135W, apo –SynCaM S81W) or rather well exposed (in the case of apo-SynCaM T26W, apo- and $Ca^{2+}$ loaded SynCaM T62W, apo- and $Ca^{2+}$ loaded SynCaM F99W and $Ca^{2+}$ loaded SynCaM S81W) to the aqueous medium. Most of these mutants have quite unusual high quantum yields, the most striking being $Ca^{2+}$ loaded SynCaM T26W and SynCaM Q135W with quantum yields around 0.3, associated with the most red shifted emission maxima (354 and 355 nm, respectively). For SynCaM T26W, $Ca^{2+}$ binding induces a 120% increase in the quantum yield from 0.14 to 0.31 with a simultaneous red shift of the emission maximum from 345 nm to 354 nm. ApoSynCaM Q135W exhibits a quantum yield of 0.23 associated with an emission maximum at 350 nm. Addition of $Ca^{2+}$ leads to a quantum yield of 0.29 associated with an emission maximum at 355 nm. The properties of these tryptophanyl residues were at odds with the commonly held concepts which governed analysis of protein fluorescence parameters. Indeed, tryptophan containing proteins usually were classified according to their quantum yields and maximum emission wavelengths, with blue-shifted spectra ($\lambda_{max} <$ 330 nm) going along with high quantum yields ($\phi > 0.20$) and red-shifted spectra associated with quantum yields below 0.14.

For SynCaM T26W, quenching experiments were undertaken in order to acquire more information on the degree of exposure of the tryptophanyl residue. Ionic quenchers ($I^-$ and $Cs^+$) and the neutral acrylamide quencher were used (Figure 8.6). Stern-Volmer plots of SynCaM T26W acrylamide quenching show an upward curvature pointing to a static quenching component. Such static quenching has been observed also for SynCaM F99W[117] and other single tryptophan containing proteins (for review and discussion see[118]). Experiments performed with $I^-$ and $Cs^+$ led to linear quenching curves. Acrylamide quenching rates of SynCaM T26W (Table 8.3) in the absence or presence of $Ca^{2+}$ are characteristic of partially buried tryptophan residues[119] (fully exposed tryptophanyl residues have quenching rate constants close to $4 \times 10^9\,M^{-1}\,s^{-1}$ and buried tryptophanyl residues have $k_q$ values close to $0.5 \times 10^9\,M^{-1}\,s^{-1}$). Looking at different single tryptophan containing proteins, Eftink showed a dependence of the acrylamide quenching rates on the emission maximum.[118] Emission maximum of Trp26, for apo- and $Ca^{2+}$ loaded SynCaMT26W are completely off this curve indicating that the chromophores, although partially buried remain in a totally polar environment. Binding of $Ca^{2+}$ did not significantly affect acrylamide exposure of Trp26. Exposure of Trp 26 to ionic quenchers (both positively and negatively

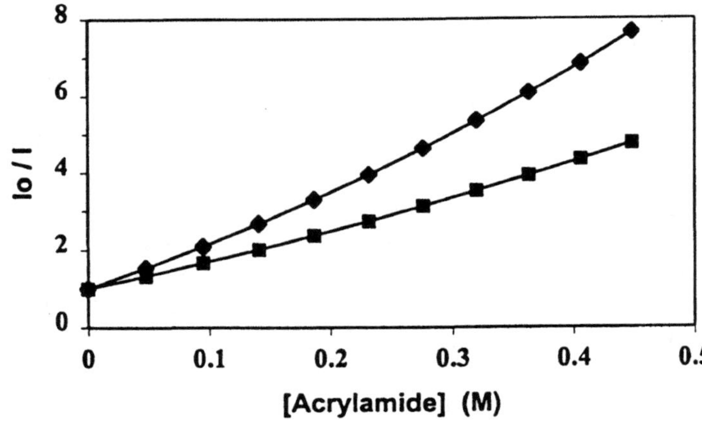


Figure 8.6. Stern-Volmer plot of SynCaM T26W quenching by acrylamide. Quenching experiments were performed in 50 mM Hepes, pH 7.5 in the absence of $Ca^{2+}$ (■) or at saturating $Ca^{2+}$ concentration (◆). Excitation wavelength was set at 297 nm. Io and I correspond to the area under the emission spectrum in the absence and presence of acrylamide, respectively. Aliquots of a stock solution of acrylamide were added to the protein solution. Data were corrected for the dilution and the screening effect due to the absorption of acrylamide at 297 nm.[117] SynCaM T26W concentration was $3 \times 10^{-5}$ M. Solid lines correspond to the data fitted according to the modified Stern-Volmer relationship[119] $I_o/I = (1 + K_{sv}\,[Q])\,e^{V[Q]}$, where [Q] stands for the quencher concentration and $K_{sv}$ for the collisional quenching constant.

Table 8.3. Fluorescence Quenching Parameters of SynCaM T26W and SynCaM S81W

| Quencher | SynCaM | Conditions | Ksv ($M^{-1}$) | V ($M^{-1}$) | $k_q$ ($M^{-1}\,s^{-1}$) |
|---|---|---|---|---|---|
| Acrylamide | T26W | $-Ca^{2+}$ | 6.05 | 0.55 | $1.5 \times 10^9$ |
| | | $+Ca^{2+}$ | 10.1 | 0.74 | $1.6 \times 10^9$ |
| KI | T26W | $-Ca^{2+}$ | 1 | 0 | $0.25 \times 10^9$ |
| | | $+Ca^{2}$ | 2.4 | 0 | $0.38 \times 10^9$ |
| CsCl | T26W | $-Ca^{2+}$ | 2.5 | 0 | $0.62 \times 10^9$ |
| | | $+Ca^{2}$ | 2.8 | 0 | $0.44 \times 10^9$ |
| Acrylamide | S81W* | $+Ca^{2+}$* | 5.9* | 1.2* | $2 \times 10^9$ |

Experiments were performed in 50 mM Hepes, pH 7.5 in the absence or presence of 1 mM $Ca^{2+}$. Quenching data were analyzed according to the Stern-Volmer equation (119): $I_o/I = (1 + K_{sv}\,[Q])\,e^{V[Q]}$, where $I_o$ and I are the fluorescence intensities at an appropriate wavelength in the absence ($I_o$) and presence (I) of a given quencher concentration [Q], $K_{sv}$ is the collisional quenching constant which corresponds to the product of the bimolecular quenching rate constant ($k_q$) and the average value of the lifetimes $\bar{\tau} = \Sigma\alpha_i\tau_i$ in the absence of quencher (Table 8.4) and V is the static quenching parameter. * data from (77) charged) is hindered, reflecting electrostatic repulsion by both positive and negative charges close to the chromophore.

$Ca^{2+}$ binding slightly increases the exposure to $I^-$ and decreases the exposure to $Cs^+$. This could be attributed to negative charge neutralization by $Ca^{2+}$ or to $Ca^{2+}$ induced change in charge partition around Trp26. Trp 81 in the central helix of $Ca^{2+}$ loaded SynCaM S81W showed an exposure to acrylamide close to that of Trp 26. Its emission maximum at 342 nm is blue-shifted compared to that of Trp 26, and becomes closer to the values expected for partially buried tryptophans. Extensive studies of all tryptophan containing mutants has not been performed so far and much is still to be done in order to understanding precisely the physical meaning underlying fluorescence properties in proteins.

### 8.4.2. Calcium Titration of the Mutants: A Probe of the Sequential $Ca^{2+}$ Binding Mechanism

The five tryptophan mutants all exhibit four $Ca^{2+}$ binding sites (with a mean dissociation constant of $4 \times 10^{-6}$ M) and show $Ca^{2+}$ binding isotherms indistinguishable from that of SynCaM, the standard of comparison calmodulin (Table 8.1). In order to further investigate the $Ca^{2+}$ binding mechanism, $Ca^{2+}$ binding to the five mutants was studied using fluorescence spectroscopy by monitoring changes in the single tryptophanyl residue located in a given calcium binding site or in the central helix. In addition, fluorescence stopped–flow experiments allowed to determine the kinetics of $Ca^{2+}$ removal from the mutants and to refine the model of $Ca^{2+}$ binding to calmodulin. Experiments were performed either in 50 mM Hepes, pH 7.5, or in 50 mM Hepes, pH 7.5 with 40% ethylene glycol. Ethylene glycol (EG) was first chosen as an anti-freeze agent in order to perform stopped-flow studies at low temperatures,[120,121] but in addition it allowed to dissect the different steps underlying $Ca^{2+}$ binding. Concerning the relevance of using EG, it should be remembered that one of the effects of EG is to decrease the molarity of water. At 40% EG, the concentration of water is ~33 M. This value is close to the water concentration probably prevailing in the cytoplasm of living cells.

#### *8.4.2.1. $Ca^{2+}$ Titrations in the Absence of Ethylene Glycol*

Changes in fluorescence polarization, fluorescence intensities and fluorescence lifetimes upon $Ca^{2+}$ binding to the five tryptophan containing mutants were analyzed.

All mutants, except for SynCaM S81W, exhibit fluorescence polarization changes when $Ca^{2+}$ was added to the medium. For SynCaM T26W and SynCaM T62W, changes took place between 2 and 4 $Ca^{2+}$ bound per protein, whereas for SynCaM F99W and SynCaM Q135W changes took place between 0 and 2 $Ca^{2+}$ bound per protein. This is consistent with a non equivalence of the $Ca^{2+}$ binding sites and is in agreement with the generally accepted model that $Ca^{2+}$ first binds to the COOH terminal half of the molecule and then to the $NH_2$ terminal half of the molecule. Tryptophan polarization thus appears to report local conformational changes taking place in the $Ca^{2+}$ binding site where the chromophore is located.

Changes in fluorescence intensities (Figure 8.7) of the mutants appeared more informative in that they reported local and remote conformational changes. For example, SynCaM Q135W exhibited 25% of its fluorescence change when binding the two first $Ca^{2+}$, the remaining 75% change occurring when the last two $Ca^{2+}$ bound. Trp135 in domain IV is thus sensitive to $Ca^{2+}$

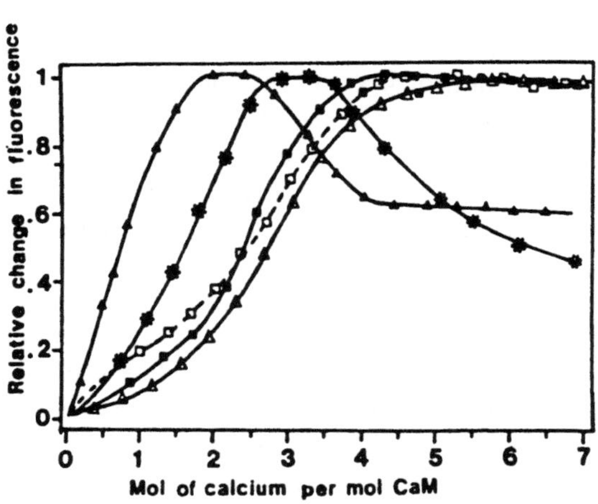


Figure 8.7. Fluorescence intensity changes of the different tryptophan containing calmodulin mutants as a function of $Ca^{2+}$ added to the proteins. $Ca^{2+}$ bound to the proteins was calculated by taking into account the $Ca^{2+}$ affinity constants obtained from $Ca^{2+}$ binding studies. Under the conditions used, $Ca^{2+}$ added corresponds to $Ca^{2+}$ bound up to 4 moles of $Ca^{2+}$ added per mole of protein. Fluorescence intensities correspond to the area under the fluorescence spectra, and changes (which can correspond to an increase or a decrease) are expressed as the percentage of the maximum change. Experiments were performed in 50 mM Hepes, pH 7.5. Protein concentrations ranged between $2.5 \times 10^{-5}$ and $3.5 \times 10^{-5}$ M. Mutants: (▲) SynCaM S81W (5186-4786); (*) SynCaM F99W (4908-5841); (□) SynCaM Q135W (3894-5324); (△) SynCaM T26W (4394-11 028); (■) SynCaM T62W (2985-3435). For each mutant, the areas under the fluorescence spectra in the absence of $Ca^{2+}$ and at the maximum of the change, respectively, are indicated in parentheses (values are in arbitrary units). Reprinted with permission from.[82] Copyright 1999 American Chemical Society.

binding to the COOH terminal lobe where Trp135 is located, but also to $Ca^{2+}$ binding to the $NH_2$ terminal lobe. In the context where $Ca^{2+}$ is presumed to bind first to the sites in the COOH terminal lobe and then to the sites in the $NH_2$ terminal lobe, this observation is a strong argument in favor of a communication between the two lobes of calmodulin. More evidence comes from SynCaM F99W, where Trp99 fluorescence (in the COOH terminal lobe) is affected by $Ca^{2+}$ binding to the sites in the opposite lobe.

### 8.4.2.2. *$Ca^{2+}$ Titrations in the Presence of Ethylene Glycol*

In the presence of EG, calmodulin still binds four $Ca^{2+}$ (Table 8.1). However, the binding isotherms are shifted to lower $Ca^{2+}$ concentrations (pointing to an increase in the mean $Ca^{2+}$ affinity), but their shape remains similar to those obtained in the absence of EG.[82] Scatchard representations of the data are linear.

When $Ca^{2+}$ binding to the five mutants was monitored by the spectral changes of the tryptophans, the results were quite interesting (Figure 8.8).

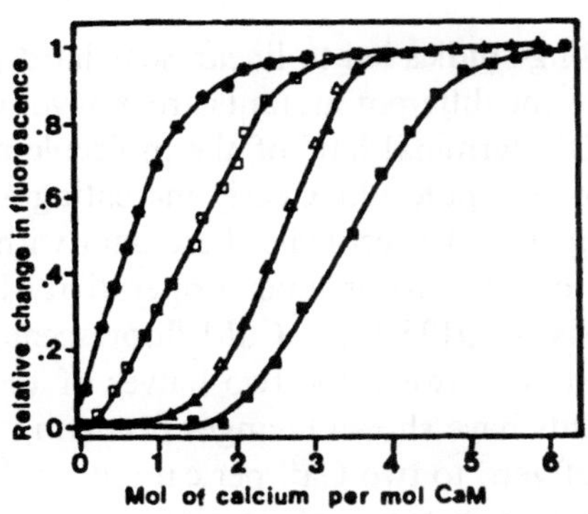


Figure 8.8. Fluorescence intensity changes of the different tryptophan containing calmodulin mutants in 40% EG as a function of $Ca^{2+}$ added to the proteins. $Ca^{2+}$ bound to the proteins was calculated by taking into account the $Ca^{2+}$ affinity constants obtained from $Ca^{2+}$ binding studies. Under the conditions used, $Ca^{2+}$ added corresponds to $Ca^{2+}$ bound up to 4 moles of $Ca^{2+}$ added per mole of protein. Changes were expressed as percentage of the maximum change. Protein concentrations ranged between $2.5 \times 10^{-5}$ and $3.5 \times 10^{-5}$ M. Buffer conditions: 50 mM Hepes, 40% EG, pH 7.5. Mutants: (*) SynCaM F99W ($I_{344}$ 57-33); (□) SynCaM Q135W ($I_{344}$ 76-59); (□) SynCaM S81W ($I_{343}$ 52-47); (△) SynCaM T26W ($I_{353}$ 104-202); (■) SynCaM T62W ($I_{339}$ 87-65). Excitation was set at 297 nm. For each mutant, intensities measured at a given wavelength in the absence of $Ca^{2+}$ and at the maximum of the change induced by $Ca^{2+}$binding, respectively, are indicated in parentheses (values are in arbitrary units). Reprinted with permission from.[82] Copyright 1999 American Chemical Society.

As $Ca^{2+}$ was added, a change in Trp99 fluorescence appeared first, followed by changes in Trp135 and Trp81, then in Trp26 and finally by that of Trp62. This indicates that in EG, $Ca^{2+}$ binding to SynCaM still follows a preferential pathway. Moreover, in this experiment, the sequence of $Ca^{2+}$ binding can be clearly seen and can be given as follows: $Ca^{2+}$ first binds to domain III, the second $Ca^{2+}$ binds to domain IV, the third to domain I and the fourth to domain II ($Ca^{2+}$ binding domains are numbered I to IV, starting from the $NH_2$ terminal part of the molecule; site 1, which is the $Ca^{2+}$ binding site first occupied by $Ca^{2+}$, corresponds to domain III). In these experiments, EG acted as a perturbing agent which enabled to dissect the overall $Ca^{2+}$ binding process into its individual components. A similar pathway of binding was suggested when $Ca^{2+}$ binding was investigated by analyzing tryptophan fluorescence decays of the mutants in buffer without EG.

### *8.4.2.3. Comments*

Results obtained from the study of $Ca^{2+}$ binding to the different tryptophan containing mutants can be summarized as follows:

- direct $Ca^{2+}$ binding studies led to linear Scatchard plots
- $Ca^{2+}$ titrations of the different mutants are in favor of binding occurring first in the COOH terminal half of the molecule and then in the $NH_2$ terminal half of the molecule, clearly indicating a non equivalence of the $Ca^{2+}$ binding sites. To reconcile these data with the linearity of the Scatchard plot, one has to assume cooperativity between the sites. In addition, analysis of Q135W SynCaM fluorescence points to the presence of a cross-talk between the two halves of the molecule. Previous NMR experiments have shown changes occurring in the $NH_2$ terminal over the range of zero to two $Ca^{2+}$ per camodulin,[98,106,122] but the results were never discussed in terms of possible interaction between the two calmodulin lobes.
- EG, which apparently did not qualitatively alter the mechanism of $Ca^{2+}$ binding *per se*, pinpoints the preferential pathway of $Ca^{2+}$ binding and allows to precise the different steps in this pathway.

In the last years, elegant studies performed on recombinant calmodulin by approaches different from the ones we used, reinforce the view of a cross-talk between the two lobes of calmodulin, with conformational change propagation occurring between the COOH terminal and $NH_2$ terminal halves of the molecule.[123–125] In addition, using quantitative proteolytic footprinting of whole calmodulin on the one hand, and separated COOH terminal and $NH_2$ terminal parts on the other hand, Sorensen and Shea[126] showed that the $NH_2$ domain had a higher affinity for $Ca^{2+}$ when it was an isolated peptide, clearly suggesting that whole calmodulin is not the sum of its parts as was proposed earlier. Quantitative proteolytic footprinting also suggested that domain III had a higher $Ca^{2+}$ affinity than does domain IV. At present, the preferential pathway $Ca^{2+}$ binding model still appears to be the only model able to integrate all the data published so far.

#### *8.4.2.4. Fluorescence Stopped-flow as a Probe of a Limiting Step in the Kinetics of $Ca^{2+}$ Binding to Calmodulin*

Steady state fluorescence studies of the five mutants in EG have shown that Trp135, Trp99 and Trp81 are reporter groups for binding of the two first $Ca^{2+}$ ions whereas Trp62 and Trp26 are reporter groups for binding of the last two $Ca^{2+}$ ions. Stopped-flow fluorescence studies of the five mutants in 40% EG were therefore performed in order to obtain insight into the kinetics of $Ca^{2+}$ removal from calmodulin.

$Ca^{2+}$ was removed from the proteins using both EDTA and Quin 2.[82] In the first case, changes in the tryptophan fluorescence is monitored, whereas in the second case, changes in Quin 2 fluorescence upon $Ca^{2+}$ binding to this compound is recorded. When $Ca^{2+}$ was removed with EDTA, relaxation of Trp62 and Trp26 appeared to follow fast kinetics ($k \sim 100\,s^{-1}$ at 25 °C), whereas relaxation of Trp135, Trp99 and Trp81 followed slow kinetics ($k \sim 1\,s^{-1}$ at 25 °C). On the other hand, kinetics of $Ca^{2+}$ removal with Quin 2 is biphasic with a rapid phase and a slow phase differing by two orders of magnitude, each associated with removal of two $Ca^{2+}$ ions. To reconcile steady state data, linear Scatchard plots and kinetic data (biphasic kinetics of $Ca^{2+}$ removal with the two phases differing by two orders of magnitude) the $Ca^{2+}$ binding model proposed earlier[76,84] was refined[82] assuming that a conformational step is associated with each calcium binding step. The slow phase in the $Ca^{2+}$ removal kinetics was therefore reported to be associated with a conformational change taking place after the removal of the first two $Ca^{2+}$ ions. The sequence of $Ca^{2+}$ removal would thus be: a fast removal of two $Ca^{2+}$ from domains I and II in the $NH_2$ terminal part, followed by a slow and kinetically limiting conformational change and finally a $Ca^{2+}$ removal from domains III and IV in the COOH terminal lobe of the protein (the kinetics of this step being at most 1 order of magnitude faster then the removal from domains I and II). This limiting step which also occurs during the $Ca^{2+}$ binding process, is not seen by steady state studies and accounts for the absence of a break in the Scatchard plots.

### 8.4.3. Fluorescence Lifetimes of Tryptophan Mutants

#### *8.4.3.1. Time Domain Lifetimes*

Fluorescence decay experiments were performed with the pulse fluorometry technique as described previously.[127] Excitation was set at 297 nm to excite selectively tryptophan and avoid complication due to eventual Tyr138 → tryptophan energy transfer. Decay data were analyzed as sums of exponentials

$$I(t,\lambda) = \sum \alpha_i(\lambda) e^{-t/\tau_i(\lambda)} \quad \text{with} \quad \sum \alpha_i(\lambda) = 1 \tag{8.1}$$

by using an iterative, non linear, least-squares convolution based on the Marquart algorithm.

For all tryptophan containing mutants, the optimal fit of the decay data (judged by the reduced $\chi^2$ value ranging between 1.1 and 1.3, the randomness of the weighted residuals and the autocorrelation function) was obtained by using a tri-exponential analysis (Table 8.4). The amplitudes of the third component for the different mutants ranged between 0.01 and 0.19 and the lifetimes between 0.1 and 0.4 ns; this component is lowest and almost negligible for the $Ca^{2+}$ saturated form of SynCaM T26W and for SynCaM Q135W apo- and $Ca^{2+}$ loaded form. Fluorescence decay parameters were investigated as a function of wavelength for SynCaM T26W (Figure 8.9) and SynCaM F99W[127] in the absence or presence of $Ca^{2+}$. For SynCaM T26W, under both

Table 8.4. Single Photon Counting Fluorescence Lifetimes Analysis for the Different Tryptophan Containing Syncams

| SynCaM | Conditions | $\tau_1$(ns) | $\alpha_1$ | $\tau_2$(ns) | $\alpha_2$ | $\tau_3$(ns) | $\alpha_3$ | $\bar{\tau}$(ns) |
|---|---|---|---|---|---|---|---|---|
| T26W | $-Ca^{2+}$ | 5.1 | 0.7 | 1.9 | 0.24 | 0.40 | 0.06 | 4 |
| | $+Ca^{2+}$ | 6.9 | 0.87 | 2.6 | 0.12 | 0.1 | 0.01 | 6.3 |
| T62W | $-Ca^{2+}$ | 3.74 | 0.34 | 1.73 | 0.5 | 0.55 | 0.16 | 2.2 |
| | $+Ca^{2+}$ | 4.98 | 0.34 | 1.29 | 0.39 | 0.42 | 0.27 | 2.3 |
| F99W | $-Ca^{2+}$ | 5.65 | 0.46 | 2.18 | 0.43 | 0.33 | 0.11 | 3.6 |
| | $+Ca^{2+}$ | 7.67 | 0.2 | 1.8 | 0.7 | 0.29 | 0.1 | 2.8 |
| Q135W | $-Ca^{2+}$ | 6.69 | 0.71 | 2.45 | 0.27 | 0.38 | 0.02 | 5.4 |
| | $+Ca^{2+}$ | 7.43 | 0.83 | 2.95 | 0.16 | 0.17 | 0.01 | 6.6 |
| S81W | $-Ca^{2+}$ | 4.77 | 0.48 | 2 | 0.44 | 0.41 | 0.08 | 3.2 |
| | $+Ca^{2+}$ | 5.38 | 0.4 | 1.8 | 0.41 | 0.32 | 0.19 | 3 |

Experiments were performed in 50 mM Hepes, pH 7.5 without $Ca^{2+}$ ($-Ca^{2+}$) or in the presence of a saturating $Ca^{2+}$ concentration ($+Ca^{2+}$). The average of the lifetimes is defined by $\bar{\tau} = \Sigma\alpha_i\tau_i$. Excitation and emission wavelengths were set at 297 nm and 350 nm, respectively.

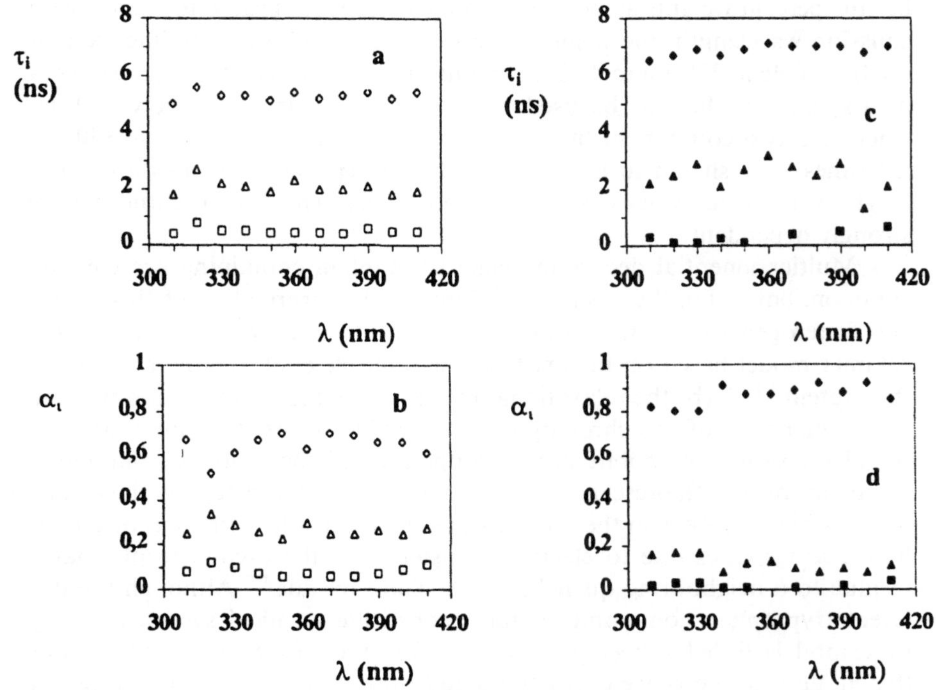


Figure 8.9. Fluorescence decay parameters of SYNCAM T26W as a function of emission wavelength in the absence of $Ca^{2+}$ (a and b) or in the presence of saturating $Ca^{2+}$ concentrations (c and d). Lifetimes $\tau_i$ are given in (a) and (c) where ◇ and ◆ stand for $\tau_1$, △ and ▲ for $\tau_2$, □ and ■ for $\tau_3$. The weighted preexponential factors $\alpha_i$, are given in (b) and (d), where ◇ and ◆ stand for $\alpha_1$, and ▲ for $\alpha_2$, □ and ■ for $\alpha_3$. Excitation wavelength was set at 297 nm.

conditions, lifetimes and amplitudes were approximately independent of the wavelength. In the absence of $Ca^{2+}$ average values (over all wavelengths) for the three lifetimes were 5.3 ns, 2.1 ns and 0.49 ns, with amplitudes of 65%, 27% and 8% respectively. At saturating $Ca^{2+}$ concentrations, the third component is either absent or very low (lifetime average value of 0.3 ns with a percent weight of 2%). The average values of the two other components are $\tau_1 = 6.9$ ns and $\tau_2 = 2.5$ ns with respective average weights of 86% and 12%. The decay parameters of SynCaM T26W in the presence of $Ca^{2+}$ closely resembles those of $Ca^{2+}$ loaded SynCaM Q135W[77] and echoe the similarity of their steady-state fluorescence properties. A similar study was performed previously for SynCaM F99W.[127] For the apo-protein, decay data were fit by a bi-exponential function (with $\chi^2$ around 1.8). Lifetimes were approximately independent of the emission wavelength and averaged around 5.3 and 1.3 ns, but the percent weights associated with the lifetimes appeared to change with emission wavelength, the weight associated with the longer lifetime increasing from 0.46 at 310 nm to 0.63 at 400 nm. In the presence of $Ca^{2+}$, a tri-exponential analysis had to be used for emission wavelengths below 340 nm, whereas a two component fit was acceptable at wavelengths above 340 nm. Lifetimes were shown to be approximately independent of emission wavelength with mean values of 7, 1.7 and 0.6 ns, whereas amplitudes were strongly dependent.

Multiexponential decays in single tryptophan containing proteins are common, but so far, there is no straightforward interpretation of this apparent "heterogeneity". Different models have been proposed including (a) the rotamer model based on the existence of multiple conformational states of the protein,[128–130] (b) the relaxational model where the process takes place in the excited state of the chromophore and which is based on the reorientational relaxation of mobile polar groups around the indole chromophore occurring on the fluorescence decay time scale[131–136] and (c) the dark state model which involves, in the excited state, the reversible formation of a non-fluorescent species due to electron transfer from the excited tryptophanyl residue to a neighboring quenching side chain residue.[137] Although the different tryptophan containing mutants constitute an ideal system to study these models, little has been performed so far, except for SynCaM F99W. For this mutant, fluorescence decay times and time resolved spectra best agreed with the existence of two conformers, characterized by a different lifetime value and different emission spectra.[127]

From the mutants quantum yields (Table 8.2) and decay parameters (Table 8.4), one can estimate the radiative decay rates, $k_r$ with $\phi = k_r\ (\Sigma\alpha_i\tau_i)$ and non radiative decay rates $k_{nr}$ ($k_{nr} = (1 - \phi)/\ \Sigma\alpha_i\tau_i$. $k_r$ values for the different mutants (Table 8.5) with the exception of apo SynCaM T26W are close to those of tryptophan or NATA in $H_2O$. This suggests that the high quantum yield observed for $Ca^{2+}$ saturated SynCaM T26W and Q135W should be related to the decreased non radiative decay rates compared to tryptophan or NATA in $H_2O$. Further studies would be required to understand the mechanism underlying the fluorescent features of these mutants tryptophanyl residues.

#### *8.4.3.2. Time Resolved Spectra: A Probe of the Selection of Conformation Upon Calcium Binding*

A time-resolved fluorescence study of SynCaM F99W in the absence and presence of various molar ratios of $Ca^{2+}$ to protein was performed by Chabbert and coworkers.[127] Time resolved spectra show a clear time dependent

Table 8.5. Radiative and Non Radiative Decay Rates for Tryptophans in the Different Proteins

| SynCaM | Conditions | Lifetimes $\bar{\tau}$ (ns) | Radiative decay rate $k_f$ ($\times 10^9 s^{-1}$) | Non radiative decay rate $k_{nr}$ ($\times 10^9 s^{-1}$) |
|---|---|---|---|---|
| T26W | $-Ca^{2+}$ | 4 | 0.035 | 0.21 |
| | $+Ca^{2+}$ | 6.3 | 0.05 | 0.11 |
| T62W | $-Ca^{2+}$ | 2.2 | 0.054 | 0.4 |
| | $+Ca^{2+}$ | 2.3 | 0.056 | 0.38 |
| F99W | $-Ca^{2+}$ | 3.6 | 0.053 | 0.21 |
| | $+Ca^{2+}$ | 2.8 | 0.054 | 0.28 |
| Q135W | $-Ca^{2+}$ | 5.4 | 0.042 | 0.13 |
| | $+Ca^{2+}$ | 6.6 | 0.044 | 0.12 |
| S81W | $-Ca^{2+}$ | 3.2 | 0.056 | 0.26 |
| | $+Ca^{2+}$ | 3 | 0.067 | 0.27 |
| L-Trp or NATA | $H_2O$ | 2.6 | 0.05 | 0.33 |

relaxation with an isosbestic point at 345 nm ± 1 nm. The time-dependent spectral shift was attributed to the existence of conformers with different spectra that convert into each other with rates slower than or on the time scale of fluorescence emission. Data were interpreted in terms of a two state model, one blue-shifted and the other red-shifted. In the absence of $Ca^{2+}$, the relaxation of the spectrum was completed within 5 ns and its amplitude was small (140 $cm^{-1}$ shift in the emission wavenumber barycenter). $Ca^{2+}$ addition increases the amplitudes of the wavenumber barycenter relaxation (up to 600 $cm^{-1}$) and decreases the relaxation rate which did not reach completion after 20 ns. Data analysis suggests that $Ca^{2+}$ binding changes both the ground state equilibrium (with a complete shift towards the blue shifted state in the presence of $Ca^{2+}$) and the interconversion rates between the two excited states pointing to an increase in the rigidity of the protein. The phenomenon was dependent upon the average number of $Ca^{2+}$ bound to the protein until half saturation of the $Ca^{2+}$ binding sites was attained. Taking into account results from other studies (Ref. 82), it can be assumed that time resolved fluorescence data of Trp99 report changes occurring in the protein structure when $Ca^{2+}$ binds to the COOH terminal half of the protein.

Steady state fluorescence spectra did not show a big change between the apo and $Ca^{2+}$ loaded form of the protein (maximum close to 348 nm in both cases). Time resolved fluorescence clearly indicated that the polarity of the environment was changed under the two conditions, the similarity of emission being the result of a least two opposite effects.

These results suggest that calmodulin explores different sets of conformational states upon $Ca^{2+}$ binding. From a physiological point of view, only the $Ca^{2+}$ loaded form of calmodulin was considered to be relevant. Nevertheless, it has been shown that the apoform may interact with calmodulin target structures[138,139] and may therefore contribute to the diversity and specificity of calmodulin associated responses. For some target structures, complex formation in the absence of $Ca^{2+}$ involves the COOH terminal part of calmodulin.[138] It now appears important to analyze the role of other regions of calmodulin in the interaction with target peptides. This could be done using tryptophan containing mutants.

### 8.4.4. Measurements of Distances by Radiationless Energy Transfer

Tryptophan mutants all contain one tryptophan and one tyrosine (Tyr138) located in the fourth $Ca^{2+}$ binding loop. These mutants thus constitute ideal systems to perform fluorescence resonance energy transfer (FRET) measurements in order to evaluate the distances between the chromophores and get an idea on the structure and changes in structure of the protein upon $Ca^{2+}$ binding. Calmodulin, in its crystal form is a dumbbell shaped protein with two lobes connected by a long central helix composed of seven $\alpha$-helical turns. Energy-transfer measurements between Tyr138 and one of the five tryptophanyl residues of the different SYNCAM mutants should in theory bring information of the distances between residues located either (a) in the same $Ca^{2+}$ binding domain (Tyr138 → Trp 135), (b) in different $Ca^{2+}$ binding domains of the same lobe (Tyr138 → Trp99), (c) in different lobes (Tyr138 → Trp62; Tyr138 → Trp26) and (d) in one $Ca^{2+}$ binding domain and the central helix (Tyr138 → Trp26). This was especially interesting for calmodulin in the context of debate concerning the flexibility of the central helix in solution allowing the two domains to adopt various relative orientations and separations in solution.[110,140–142]

Energy transfer efficiency $\eta$ is related to the distance ($R$) between the chromophores according to the relationship

$$R = \left[(1-\eta)-1\right]^{1/6} R_0 \tag{8.2}$$

where $R_0$ is the Förster critical distance for a given donor and acceptor pair.

Static energy transfer measurements are available for SynCaM T26W and SynCaM F99W. $\eta$ was calculated according to equation

$$\phi_P^{280}(Trp) = \phi_P^{297}(f_{Trp}^{280} + \eta\, f_{Tyr}^{280})$$

where $\phi_P^{297}$ is the quantum yield of the protein excited at 297 nm and $\phi_P^{280}(Trp)$, the fractional quantum yield of the tryptophan residue in the protein excited at 280 nm, determined as indicated previously.[117] Values of $\phi_P^{297}$ for the mutants are given in Table 8.2. $f_{Tyr}^{280}$ is the fractional absorption of Tyr138 at 280 nm. Its value (Table 8.6) was determined from the knowledge of the molar extinction coefficient of a given mutant at 280 nm and the molar extinction coefficient of SynCaM, which contains only Tyr138 ($\varepsilon_{M,280}$ = 1500 $M^{-1}\,cm^{-1}$).

For the tyrosine—tryptophan pair, $R_0$ was calculated according to the method of Eisinger *et al.*[143]

$$R_0^{\,6} = (8.79x10^{-25})\kappa^2 n^{-4} \phi_D J_{AD}\ (cm^6) \qquad (8.3)$$

where $J_{AD} = 4.8 \times 10^{-16}\,M^{-1}\,cm^6$ and $n$, the refractive index equals 1.335. $\phi_D$, the quantum yield of the donor corresponds, in the case of the tryptophan mutants, to the quantum yield of Tyr138. The value of this quantum yield, obtained from the study of SynCaM was found to be equal to 0.031 and 0.061 in the absence and presence of $Ca^{2+}$, respectively.[117] $\kappa^2$, the orientation factor, varies from 0 to 4. The minimum values are obtained when the donor emission and the acceptor absorption dipoles are perpendicular to each other, and the maximum value corresponds to parallel and aligned dipoles. If dipoles sample all orientations during the interval of the excited state, $\kappa^2 = 2/3$.[144] Taking a value of $2/3$ for $\kappa^2$, the calculated distance $R_0$ was 11.8 Å for the apo mutants and 13.2 Å for the $Ca^{2+}$ loaded forms.

Table 8.6. Fluorescence Energy Transfer Measurements

| SynCaM | Conditions | $f_{Tyr}^{280}$ | $\eta$ | R (Å) |
|---|---|---|---|---|
| T26W | $-Ca^{2+}$ | 0.24 | 0.192 | 15 |
| | $+Ca^{2+}$ | 0.24 | <0.1 | >20 |
| S81W | $-Ca^{2+}$ | 0.24 | 0.097 | 17 |
| | $+Ca^{2+}$ | 0.25 | 0.35 | 15 |
| F99W | $-Ca^{2+}$ | 0.202 | 0.3 | 13.5 |
| | $+Ca^{2+}$ | 0.202 | <0.1 | >20 |

$f_{Tyr}^{280}$ stands for the fractional absorption of Tyr138 at 280 nm, $\eta$ to energy transfer efficiency and R to the distance between Tyr138 and one of the tryptophans. Reported values were obtained taking $\kappa^2 = 2/3$.

From the knowledge of the three crystal structures of calmodulin and the assumption, corroborated by modelisation, that the tryptophanyl residue does not modify the overall structure of the protein, distances between Tyr138 and tryptophanyl residues were computed and compared to distances measured by steady state energy transfer (Table 8.6). There is no correlations whatsoever. For SynCaM T26W and SynCaM F99W a value around 14 Å and a distance greater than 20 Å (corresponding to the absence of meaningful measurable energy transfer) was found for the apoprotein and the $Ca^{2+}$ saturated proteins, respectively. If the central helix of calmodulin in solution exists in its extended form, one does not expect any energy transfer between Tyr138 and Trp26. This is the case for holo SynCaM T26W, but not for the apoform. Whatever the discussion about orientation factor could be, the energy transfer measured in the latter case constitutes a strong argument for flexibility in the protein tether allowing Tyr138 and Trp26 to come closer together, at least in part of the protein population. The distance between Tyr138 and Trp26 should not be dramatically altered by $Ca^{2+}$ binding to the protein. $Ca^{2+}$ induced rigidification of the protein, altering the value of $\kappa^2$, is probably responsible for the difference observed. Intradomain distance measurements, using SynCaM F99W, is also error prone due to the difficulty in estimating $\kappa^2$.[117,145] Only values obtained for SynCaM S81W are close to expected values. Therefore, the use of static efficiency to analyze the separation between two residues has to be taken "*cum grano saltis*". For one of the mutant (SynCaM F99W99), the distribution of separations between Trp99 and Tyr138 after nitrosylation of this residue were measured using time-domain dynamic fluorescence measurements of energy transfer.[145] This method appears to be much more accurate and yields more information about the segmental flexibility of the protein. Similar studies on the other tryptophan mutants should be performed in order to validate such techniques.

Preliminary results where the anisotropy time decays of the 5 tryptophan containing mutants have been measured as a function of calcium occupancy of the different $Ca^{2+}$ binding sites appear to support segmental flexibility modification of SynCaM as a function of calcium binding.

## 8.5. Perspectives and Open Questions

The ease and power of protein engineering enable insertion of single tryptophanyl resides in any soluble protein at specific locations. Fluorescence from this reporter group then allows to follow interactions between the mutated proteins and ligands, and to analyze the local conformational changes upon ligand binding. On the other hand, analysis of the fluorescence properties of the single tryptophanyl residue might help in getting insight into the chromophore surroundings. However, in the absence of a precise link between the measured fluorescence characteristics of a tryptophanyl residue and the physical structure of the residue (e.g. interpretation of multiexponential decays of single tryptophan containing proteins), this latter type of investigation remains unsuccessful and often frustrating. A clear theory of tryptophan fluorescent decay times in proteins is required in order to use the fluorescent characteristics to get information on the local structure of the protein.

Although interpretation of anisotropy decay times appear more straightforward, precise and reproducible values are not easily obtained. Improvement in the techniques and in the homogeneity of the proteins are required before information on the proteins local flexibility can be obtained from anisotropy decay times. Fluorescence energy transfer (FRET) seems to be the most promising technique. Again, this has to be coupled to molecular dynamics and a better knowledge of fluorescence time decay in order to refine our interpretation of FRET results. The use of fluorescent reporter groups in the analysis of the structure-function relationship of proteins is becoming a central strategy. Our next goal is to be able to use these techniques directly inside cellular compartments as easily as in a test tube.

## References


1. Herbaud, M. L., Guiseppi, A., Denizot, F., Haiech, J., and Kilhoffer, M. C., Calcium signalling in Bacillus subtilis, *Biochim Biophys Acta, 1448*, 212–226 (1998).
2. Smith, R. J., Calcium and bacteria, *Adv Microb Physiol, 37*, 83–133 (1995).
3. Youatt, J., Calcium and microorganisms, *Crit Rev Microbiol, 19*, 83–97 (1993).
4. Onek, L. A., and Smith, R. J., Calmodulin and calcium mediated regulation in prokaryotes, *J Gen Microbiol, 138*, 1039–1049 (1992).
5. Schwartzenbach, G., Senn, H., and Anderegg, G., Komplexone XXIV. Ein grosser chelateffekt besonderer art, *Helv Chem Acta, 40*, 1886 (1957).
6. Tsien, R. Y., New calcium indicators and buffers with high selectivity against magnesium and protons: design, synthesis, and properties of prototype structures, *Biochemistry, 19*, 2396–2404 (1980).
7. Tsien, R. Y., Pozzan, T., and Rink, T. J., Calcium homeostasis in intact lymphocytes: cytoplasmic free calcium monitored with a new, intracellularly trapped fluorescent indicator, *J Cell Biol, 94*, 325–334 (1982).
8. Grynkiewicz, G., Poenie, M., and Tsien, R. Y., A new generation of $Ca^{2+}$ indicators with greatly improved fluorescence properties, *J Biol Chem, 260*, 3440–3450 (1985).
9. Ebashi, S., A granule-bound relaxation factor in skeletal muscle, *Archiv Biochem Biophys, 76*, 410–423 (1958).

10. Hasselbach, W., and Makinose, M., Die calciumpumpe der "erschlaffungsgrana" des muskels und ihre abhängigkeit von der ATP-spaltung, *Biochem Z*, *333*, 518–528 (1961).
11. Hasselbach, W., and Makinose, M., ATP and active transport, *Biochem Biophys Acta*, *7*, 132–136 (1962).
12. Ebashi, S., and Kodama, A., Native tropomyosin-like action of troponin on trypsin-treated myosin B, *J Biochem (Tokyo)*, *60*, 733–734 (1966).
13. Cheung, W. Y., Cyclic 3′,5′-nucleotide phosphodiesterase. Demonstration of an activator, *Biochem Biophys Res Commun*, *38*, 533–538 (1970).
14. Kakiuchi, S., and Yamazaki, R., Calcium dependent phosphodiesterase activity and its activating factor (PAF) from brain studies on cyclic 3′,5′-nucleotide phosphodiesterase (3), *Biochem Biophys Res Commun*, *41*, 1104–1110 (1970).
15. Benzonana, G., Capony, J. P., and Pechere, J. F., The binding of calcium to muscular parvalbumins, *Biochim Biophys Acta*, *278*, 110–116 (1972).
16. Teo, T. S., and Wang, J. H., Mechanism of activation of a cyclic adenosine 3′:5′-monophosphate phosphodiesterase from bovine heart by calcium ions. Identification of the protein activator as a $Ca^{2+}$ binding protein, *J Biol Chem*, *248*, 5950–5955 (1973).
17. Weeds, A. G., and McLachlan, A. D., Structural homology of myosin alkali light chains, troponin C and carp calcium binding protein, *Nature*, *252*, 646–649 (1974).
18. Moews, P. C., and Kretsinger, R. H., Refinement of the structure of carp muscle calcium-binding parvalbumin by model building and difference Fourier analysis, *J Mol Biol*, *91*, 201–225 (1975).
19. Kretsinger, R. H., Hypothesis: calcium modulated proteins contain EF hands. *In: Carafoli E, et al., Ed. Calcium transport in contraction and secretion. Amsterdam, North-Holland*, pp. 469–478 (1975).
20. Kretsinger, R. H., Evolution and function of calcium-binding proteins, *Int Rev Cytol*, *46*, 323–393 (1976).
21. Kretsinger, R. H., The informational role of calcium in the cytosol, *Adv Cyclic Nucleotide Res*, *11*, 1–26 (1979).
22. Watterson, D. M., and Vincenzi, F. F., Calmodulin and cell functions, Vol. 356, The New York Academy of Sciences, New York (1980).
23. Klee, C. B., Crouch, T. H., and Richman, P. G., Calmodulin, *Annu Rev Biochem*, *49*, 489–515 (1980).
24. Demaille, J. G., Haiech, J., and Goodmann, M., Calcium binding proteins: from diversification of structure to diversification of function, *Protides of the biological fluids*, *28*, 95–98 (1980).
25. Cheung, W. Y., Calmodulin, in *Calcium and cell function*, Vol. I, Cheung, W. Y., Ed., Academic Press, New York, London, Toronto, Sydney, San Francisco (1980).
26. Hagiwara, S., Ozawa, S., and Sand, O., Voltage clamp analysis of two inward current mechanisms in the egg cell membrane of a starfish, *J Gen Physiol*, *65*, 617–644 (1975).
27. Reuter, H., Calcium movements in excitable cells, Baker, P. F. and Reuter, H., Ed. Pergamon, New-York pp. 55–67 (1975).
28. Llinas, R., and Yarom, Y., Electrophysiology of mammalian inferior olivary neurones in vitro. Different types of voltage-dependent ionic conductances, *J Physiol (Lond)*, *315*, 549–567 (1981).
29. Llinas, R., and Sugimori, M., Electrophysiological properties of in vitro Purkinje cell dendrites in mammalian cerebellar slices, *J Physiol (Lond)*, *305*, 197–213 (1980).
30. Reuter, H., Stevens, C. F., Tsien, R. W., and Yellen, G., Properties of single calcium channels in cardiac cell culture, *Nature*, *297*, 501–504 (1982).
31. Lux, H. D., and Hofmeier, G., Properties of a calcium- and voltage-activated potassium current in Helix pomatia neurons, *Pflugers Arch*, *394*, 61–69 (1982).

32. Brown, A. M., Camerer, H., Kunze, D. L., and Lux, H. D., Similarity of unitary $Ca^{2+}$ currents in three different species, *Nature, 299*, 156–158 (1982).
33. Tsien, R. W., Calcium channels in excitable cell membranes, *Annu Rev Physiol, 45*, 341–358 (1983).
34. Reuter, H., Ion channels in cardiac cell membranes, *Annu Rev Physiol, 46*, 473–484 (1984).
35. Lux, H. D., Observations on single calcium channels: an overview, in *Single channel recording*, Sakman, B., and Neher, E., Ed., Plenum Press, New York and London (1983).
36. Wray, D. W., Norman, R. I., and Hess, P., Calcium channels: structure and function., Vol. 560, The New York Academy of Sciences, New York (1989).
37. Hille, B., Calcium channel, in *Ionic channels of excitable membranes*, Second edition, Hille, B., Ed., Sinauer Associates Inc, Sunderland, Massachusetts (1992).
38. Haiech, J., and Demaille, J. G., Supramolecular organization of regulatory proteins into calcisomes: a model of the concerted regulation by calcium ions and cyclic adenosine 3′:5′-monophosphate in eukaryotic cells, in *Metabolic Interconversion of Enzymes. Proc. in Life Sciences*, Springer Verlag, Berlin, pp. 303–313 (1981).
39. Wasserman, R. H., Corradino, R. A., Carafoli, E., Kretsinger, R. H., MacLennan, D. H., and Siegel, F. L., Calcium-binding binding proteins and calcium function, North Holland, New York, Amsterdam, Oxford (1977).
40. Goodman, M., Pechere, J. F., Haiech, J., and Demaille, J. G., Evolutionary diversification of structure and function in the family of intracellular calcium-binding proteins, *J Mol Evol, 13*, 331–352 (1979).
41. Pochet, R., Lawson, D. E. M., and Heizmann, C. W., Calcium binding proteins in normal and transformed cells, in *Advances in experimental medicine and biology*, Vol. 269, Plenum Press, New York, London (1989).
42. Moncrief, N. D., Kretsinger, R. H., and Goodman, M., Evolution of EF-hand calcium-modulated proteins. I. Relationships based on amino acid sequences, *J Mol Evol, 30*, 522–562 (1990).
43. Heizmann, C. W., Novel calcium binding proteins. Fundamentals and clinical implications, Springer—Verlag, Berlin, Heidelberg, New York, London, Paris, Tokyo, Hong Kong, Barcelona, Budapest (1991).
44. Gerke, V., Heizmann, C. W., and Krebs, J., Special Issue. 5th European symposium on calcium binding proteins in normal and transformed cells, in *Biochem. Biophys. Acta. Molecular cell research*, Vol. 1148 (2), Avruch, J., Ed., Elsevier, Amsterdam, Lausanne, New York, Oxford, Shannon, Tokyo (1998).
45. Kretsinger, R. H., and Nockolds, C. E., Carp muscle calcium-binding protein. II. Structure determination and general description, *J Biol Chem, 248*, 3313–3326 (1973).
46. Cheung, W. Y., Calmodulin plays a pivotal role in cellular regulation, *Science, 207*, 19–27 (1980).
47. Marx, J. L., Calmodulin: a protein for all seasons, *Science, 208*, 274–276 (1980).
48. Cox, J. A., Comte, M., Malnoe, A., and Stein, E. A., Mode of action of the regulatory protein calmodulin, in *Metal ions in biological systems*, Vol. 17, Sigel, H., Ed., Marcel Dekker, New York, Basel, pp. 215–273 (1984).
49. Watterson, D. M., Burgess, W. H., Lukas, T. J., Iverson, D. Marshak, D. R., Schleicher, M., Erikson, B. W., Fok, K.-F., and Van Eldik, L. J., Towards a molecular and atomic anatomy of calmodulin and calmodulin-binding proteins, in *Advances in cyclic nucleotide and protein phosphorylation research*, Strada, S. J., and Thomson, W. J., Ed., Raven Press, New York, pp. 205–226 (1984).
50. Cohen, P., and Klee, C. B., Calmodulin, in *Molecular aspects of cellular regulation*, Vol. 5, Cohen, P., Ed., Elsevier, Amsterdam, New York, Oxford (1988).

51. Anderson, J. M., and Cormier, M. J., Calcium-dependent regulation of NAD kinase, *Biochem Biophys Res Commun*, *84*, 595–602 (1978).
52. Brostrom, C. O., Huang, Y. C., Breckenridge, B. M., and Wolff, D. J., Identification of a calcium-binding protein as a calcium-dependent regulator of brain adenylate cyclase, *Proc Natl Acad Sci USA*, *72*, 64–68 (1975).
53. Cohen, P., Burchell, A., Foulkes, J. G., and Cohen, P. T., Identification of the $Ca^{2+}$-dependent modulator protein as the fourth subunit of rabbit skeletal muscle phosphorylase kinase, *FEBS Lett*, *92*, 287–293 (1978).
54. Dabrowska, R., Aromatorio, D., Sherry, J. M., and Hartshorne, D. J., Composition of the myosin light chain kinase from chicken gizzard, *Biochem Biophys Res Commun*, *78*, 1263–1272 (1977).
55. Gopinath, R. M., and Vincenzi, F. F., Phosphodiesterase protein activator mimics red blood cell cytoplasmic activator of ($Ca^{2+}$-$Mg^{2+}$)ATPase, *Biochem Biophys Res Commun*, *77*, 1203–1209 (1977).
56. Jarrett, H. W., and Penniston, J. T., Partial purification of the $Ca^{2+}$-$Mg^{2+}$ ATPase activator from human erythrocytes: its similarity to the activator of 3′:5′—cyclic nucleotide phosphodiesterase, *Biochem Biophys Res Commun*, *77*, 1210–1216 (1977).
57. Stewart, A. A., Ingebritsen, T. S., Manalan, A., Klee, C. B., and Cohen, P., Discovery of a $Ca^{2+}$- and calmodulin-dependent protein phosphatase: probable identity with calcineurin (CaM-BP80), *FEBS Lett*, *137*, 80–84 (1982).
58. Schulman, H., and Greengard, P., Stimulation of brain membrane protein phosphorylation by calcium and an endogenous heat-stable protein, *Nature*, *271*, 478–479 (1978).
59. Yazawa, M., and Yagi, K., Calcium-binding subunit of myosin light chain kinase, *J Biochem (Tokyo)*, *82*, 287–289 (1977).
60. Watterson, D. M., Sharief, F., and Vanaman, T. C., The complete amino acid sequence of the $Ca^{2+}$-dependent modulator protein (calmodulin) of bovine brain, *J Biol Chem*, *255*, 962–975 (1980).
61. Watterson, D. M., Iverson, D. B., and Van Eldik, L. J., Spinach calmodulin: isolation, characterization, and comparison with vertebrate calmodulins, *Biochemistry*, *19*, 5762–5768 (1980).
62. Putkey, J. A., Ts'ui, K. F., Tanaka, T., Lagace, L., Stein, J. P., Lai, E. C., and Means, A. R., Chicken calmodulin genes. A species comparison of cDNA sequences and isolation of a genomic clone, *J Biol Chem*, *258*, 11864–11870 (1983).
63. Lagace, L., Chandra, T., Woo, S. L., and Means, A. R., Identification of multiple species of calmodulin messenger RNA using a full length complementary DNA, *J Biol Chem*, *258*, 1684–1688 (1983).
64. Simmen, R. C., Tanaka, T., Ts'ui, K. F., Putkey, J. A., Scott, M. J., Lai, E. C., and Means, A. R., The structural organization of the chicken calmodulin gene [published erratum appears in J Biol Chem 1987 Apr 5;262(10):4928], *J Biol Chem*, *260*, 907–912 (1985).
65. Babu, Y. S., Sack, J. S., Greenhough, T. J., Bugg, C. E., Means, A. R., and Cook, W. J., Three-dimensional structure of calmodulin, *Nature*, *315*, 37–40 (1985).
66. Babu, Y. S., Bugg, C. E., and Cook, W. J., Structure of calmodulin refined at 2.2 A resolution, *J Mol Biol*, *204*, 191–204 (1988).
67. Roberts, D. M., Crea, R., Malecha, M., Alvarado-Urbina, G., Chiarello, R. H., and Watterson, D. M., Chemical synthesis and expression of a calmodulin gene designed for site-specific mutagenesis, *Biochemistry*, *24*, 5090–5098 (1985).
68. Craig, T. A., Watterson, D. M., Prendergast, F. G., Haiech, J., and Roberts, D. M., Site-specific mutagenesis of the alpha-helices of calmodulin. Effects of altering a charge cluster in the helix that links the two halves of calmodulin, *J Biol Chem*, *262*, 3278–3284 (1987).

69. Lukas, T. J., Craig, T. A., Roberts, D. M., Watterson, D. M., Haiech, J., and Prendergast, F. G., An interdisciplinary approach to the molecular mechanisms of calmodulin action: comparative biochemistry, site specific mutagenesis and protein engineering, in *Calcium binding proteins in health and disease*, Norman, A. W., Vanaman, T. C., and Means, A. R., Ed., Academic Press, Inc, pp. 533–543 (1987).
70. Zhang, M., Tanaka, T., and Ikura, M., Calcium-induced conformational transition revealed by the solution structure of apo calmodulin, *Nat Struct Biol, 2*, 758–767 (1995).
71. Kuboniwa, H., Tjandra, N., Grzesiek, S., Ren, H., Klee, C. B., and Bax, A., Solution structure of calcium-free calmodulin, *Nat Struct Biol, 2*, 768–776 (1995).
72. Chattopadhyaya, R., Meador, W. E., Means, A. R., and Quiocho, F. A., Calmodulin structure refined at 1.7 A resolution, *J Mol Biol, 228*, 1177–1192 (1992).
73. Meador, W. E., Means, A. R., and Quiocho, F. A., Target enzyme recognition by calmodulin: 2.4 A structure of a calmodulin-peptide complex, *Science, 257*, 1251–1255 (1992).
74. Ikura, M., Clore, G. M., Gronenborn, A. M., Zhu, G., Klee, C. B., and Bax, A., Solution structure of a calmodulin-target peptide complex by multidimensional NMR, *Science, 256*, 632–638 (1992).
75. Roberts, D. M., Zimmer, W. E., and Watterson, D. M., The use of synthetic oligodeoxyribonucleotides in the examination of calmodulin gene and protein structure and function, *Methods Enzymol, 139*, 290–303 (1987).
76. Kilhoffer, M. C., Roberts, D. M., Adibi, A. O., Watterson, D. M., and Haiech, J., Investigation of the mechanism of calcium binding to calmodulin. Use of an isofunctional mutant with a tryptophan introduced by site-directed mutagenesis, *J Biol Chem, 263*, 17023–17029 (1988).
77. Chabbert, M., Lukas, T. J., Watterson, D. M., Axelsen, P. H., and Prendergast, F. G., Fluorescence analysis of calmodulin mutants containing tryptophan: conformational changes induced by calmodulin-binding peptides from myosin light chain kinase and protein kinase II, *Biochemistry, 30*, 7615–7630 (1991).
78. Martin, A., Toselli, E., Rosier, M. F., Auffray, C., and Devignes, M. D., Rapid and high efficiency site-directed mutagenesis by improvement of the homologous recombination technique, *Nucleic Acids Res, 23*, 1642–1643 (1995).
79. Jones, D. H., and Howard, B. H., A rapid method for recombination and site-specific mutagenesis by placing homologous ends on DNA using polymerase chain reaction, *Biotechniques, 10*, 62–66 (1991).
80. Bayley, P. M., and Martin, S. R., The alpha-helical content of calmodulin is increased by solution conditions favouring protein crystallisation, *Biochim Biophys Acta, 1160*, 16–21 (1992).
81. Torok, K., Lane, A. N., Martin, S. R., Janot, J. M., and Bayley, P. M., Effects of calcium binding on the internal dynamic properties of bovine brain calmodulin, studied by NMR and optical spectroscopy, *Biochemistry, 31*, 3452–3462 (1992).
82. Kilhoffer, M. C., Kubina, M., Travers, F., and Haiech, J., Use of engineered proteins with internal tryptophan reporter groups and pertubation techniques to probe the mechanism of ligand-protein interactions: investigation of the mechanism of calcium binding to calmodulin, *Biochemistry, 31*, 8098–8106 (1992).
83. Crouch, T. H., and Klee, C. B., Positive cooperative binding of calcium to bovine brain calmodulin, *Biochemistry, 19*, 3692–3698 (1980).
84. Haiech, J., Klee, C. B., and Demaille, J. G., Effects of cations on affinity of calmodulin for calcium: ordered binding of calcium ions allows the specific activation of calmodulin-stimulated enzymes, *Biochemistry, 20*, 3890–3897 (1981).

85. Keller, C. H., Olwin, B. B., LaPorte, D. C., and Storm, D. R., Determination of the free-energy coupling for binding of calcium ions and troponin I to calmodulin, *Biochemistry*, *21*, 156–162 (1982).
86. Kilhoffer, M. C., Haiech, J., and Demaille, J. G., Ion binding to calmodulin. A comparison with other intracellular calcium-binding proteins, *Mol Cell Biochem*, *51*, 33–54 (1983).
87. Burger, D., Cox, J. A., Comte, M., and Stein, E. A., Sequential conformational changes in calmodulin upon binding of calcium, *Biochemistry*, *23*, 1966–1971 (1984).
88. Ogawa, Y., and Tanokura, M., Calcium binding to calmodulin: effects of ionic strength, $Mg^{2+}$, pH and temperature, *J Biochem (Tokyo)*, *95*, 19–28 (1984).
89. Iida, S., and Potter, J. D., Calcium binding to calmodulin. Cooperativity of the calcium-binding sites, *J Biochem (Tokyo)*, *99*, 1765–1772 (1986).
90. Dedman, J. R., Potter, J. D., Jackson, R. L., Johnson, J. D., and Means, A. R., Physicochemical properties of rat testis $Ca^{2+}$-dependent regulator protein of cyclic nucleotide phosphodiesterase. Relationship of $Ca^{2+}$-binding, conformational changes, and phosphodiesterase activity, *J Biol Chem*, *252*, 8415–8422 (1977).
91. Drabikowski, W., Kuznicki, J., and Grabarek, Z., Similarity in $Ca^{2+}$-induced changes between troponic-C and protein activator of 3′:5′-cyclic nucleotide phosphodiesterase and their tryptic fragments, *Biochim Biophys Acta*, *485*, 124–133 (1977).
92. Klee, C. B., Conformational transition accompanying the binding of $Ca^{2+}$ to the protein activator of 3′,5′-cyclic adenosine monophosphate phosphodiesterase, *Biochemistry*, *16*, 1017–1024 (1977).
93. Wolff, D. J., Poirier, P. G., Brostrom, C. O., and Brostrom, M. A., Divalent cation binding properties of bovine brain $Ca^{2+}$-dependent regulator protein, *J Biol Chem*, *252*, 4108–4117 (1977).
94. Richman, P. G., and Klee, C. B., Specific perturbation by $Ca^{2+}$ of tyrosyl residue 138 of calmodulin, *J Biol Chem*, *254*, 5372–5376 (1979).
95. Drabikowski, W., Brzeska, H., Kuznicki, J., and Grabarek, Z., Studies on structure and function of calmodulin, *Ann NY Acad Sci*, *356*, 374–375 (1980).
96. Kilhoffer, M. C., Demaille, J. G., and Gerard, D., Tyrosine fluorescence of ram testis and octopus calmodulins. Effects of calcium, magnesium, and ionic strength, *Biochemistry*, *20*, 4407–4414 (1981).
97. Seamon, K. B., and Moore, B. W., Octopus calmodulin. Structural comparison with bovine brain calmodulin, *J Biol Chem*, *255*, 11644–11647 (1980).
98. Seamon, K. B., Calcium- and magnesium-dependent conformational states of calmodulin as determined by nuclear magnetic resonance, *Biochemistry*, *19*, 207–215 (1980).
99. Seamon, K., NMR studies on tyrosine-138 of calmodulin, *Ann NY Acad Sci*, *356*, 433–434 (1980).
100. Andersson, T., Drakenberg, T., Forsen, S., and Thulin, E., Characterization of the $Ca^{2+}$ binding sites of calmodulin from bovine testis using 43Ca and 113Cd NMR, *Eur J Biochem*, *126*, 501–505 (1982).
101. Wallace, R. W., Tallant, E. A., Dockter, M. E., and Cheung, W. Y., Calcium binding domains of calmodulin. Sequence of fill as determined with terbium luminescence, *J Biol Chem*, *257*, 1845–1854 (1982).
102. Andersson, A., Forsen, S., Thulin, E., and Vogel, H. J., Cadmium-113 nuclear magnetic resonance studies of proteolytic fragments of calmodulin: assignment of strong and weak cation binding sites, *Biochemistry*, *22*, 2309–2313 (1983).
103. Klevit, R. E., Spectroscopic analyses of calmodulin and its interactions, *Methods Enzymol*, *102*, 82–104 (1983).

104. Minowa, O., and Yagi, K., Calcium binding to tryptic fragments of calmodulin, *J Biochem (Tokyo)*, *96*, 1175–1182 (1984).
105. Yazawa, M., Kawamura, E., Minowa, O., Yagi, K., Ikura, M., and Hikichi, K., N-terminal region (domain 1) of calmodulin is the low affinity site for $Ca^{2+}$. A 13C NMR study of S-cyanocalmodulin, *J Biochem (Tokyo)*, *95*, 443–446 (1984).
106. Klevit, R. E., Dalgarno, D. C., Levine, B. A., and Williams, R. J., 1H-NMR studies of calmodulin. The nature of the $Ca^{2+}$-dependent conformational change, *Eur J Biochem*, *139*, 109–114 (1984).
107. Thulin, E., Andersson, A., Drakenberg, T., Forsen, S., and Vogel, H. J., Metal ion and drug binding to proteolytic fragments of calmodulin: proteolytic, cadmium-113, and proton nuclear magnetic resonance studies, *Biochemistry*, *23*, 1862–1870 (1984).
108. Martin, S. R., and Bayley, P. M., The effects of $Ca^{2+}$ and $Cd^{2+}$ on the secondary and tertiary structure of bovine testis calmodulin. A circular-dichroism study, *Biochem J*, *238*, 485–490 (1986).
109. Trewhella, J., Liddle, W. K., Heidorn, D. B., and Strynadka, N., Calmodulin and troponin C structures studied by Fourier transform infrared spectroscopy: effects of $Ca^{2+}$ and $Mg^{2+}$ binding, *Biochemistry*, *28*, 1294–1301 (1989).
110. Heidorn, D. B., and Trewhella, J., Comparison of the crystal and solution structures of calmodulin and troponin C, *Biochemistry*, *27*, 909–915 (1988).
111. Bayley, P., Ahlstrom, P., Martin, S. R., and Forsen, S., The kinetics of calcium binding to calmodulin: Quin 2 and ANS stopped-flow fluorescence studies, *Biochem Biophys Res Commun*, *120*, 185–191 (1984).
112. Martin, S. R., Andersson Teleman, A., Bayley, P. M., Drakenberg, T., and Forsen, S., Kinetics of calcium dissociation from calmodulin and its tryptic fragments. A stopped-flow fluorescence study using Quin 2 reveals a two-domain structure, *Eur J Biochem*, *151*, 543–550 (1985).
113. Milos, M., Schaer, J. J., Comte, M., and Cox, J. A., Calcium-proton and calcium-magnesium antagonisms in calmodulin: microcalorimetric and potentiometric analyses, *Biochemistry*, *25*, 6279–6287 (1986).
114. Linse, S., Helmersson, A., and Forsen, S., Calcium binding to calmodulin and its globular domains, *J Biol Chem*, *266*, 8050–8054 (1991).
115. Wang, C. L., A note on $Ca^{2+}$ binding to calmodulin, *Biochem Biophys Res Commun*, *130*, 426–430 (1985).
116. Yazawa, M., Ikura, M., Hikichi, K., Ying, L., and Yagi, K., Communication between two globular domains of calmodulin in the presence of mastoparan or caldesmon fragment. $Ca^{2+}$ binding and 1H NMR, *J Biol Chem*, *262*, 10951–10954 (1987).
117. Kilhoffer, M. C., Roberts, D. M., Adibi, A., Watterson, D. M., and Haiech, J., Fluorescence characterization of VU-9 calmodulin, an engineered calmodulin with one tryptophan in calcium binding domain III, *Biochemistry*, *28*, 6086–6092 (1989).
118. Eftink, Florescence quenching: Theory and applications, in *Topics in Fluorescence Spectroscopy*, Vol. 2, Lakowicz, J. R., Ed., Plenum Press, New York and London, pp. 53–126 (1991).
119. Eftink, M. R., and Ghiron, C. A., Exposure of tryptophanyl residues in proteins. Quantitative determination by fluorescence quenching studies, *Biochemistry*, *15*, 672–680 (1976).
120. Biosca, J. A., Barman, T. E., and Travers, F., Transient kinetics of the binding of ATP to actomyosin subfragment 1: evidence that the dissociation of actomyosin subfragment 1 by ATP leads to a new conformation of subfragment 1, *Biochemistry*, *23*, 2428–2436 (1984).

121. Biosca, J. A., Travers, F., Hillaire, D., and Barman, T. E., Cryoenzymic studies on myosin subfragment 1: perturbation of an enzyme reaction by temperature and solvent, *Biochemistry*, *23*, 1947–1955 (1984).
122. Ikura, M., Hiraoki, T., Hikichi, K., Mikuni, T., Yazawa, M., and Yagi, K., Nuclear magnetic resonance studies on calmodulin: calcium-induced conformational change, *Biochemistry*, *22*, 2573–2579 (1983).
123. Shea, M. A., Verhoeven, A. S., and Pedigo, S., Calcium-induced interactions of calmodulin domains revealed by quantitative thrombin footprinting of Arg37 and Arg106, *Biochemistry*, *35*, 2943–2957 (1996).
124. Pedigo, S., and Shea, M. A., Quantitative endoproteinase GluC footprinting of cooperative $Ca^{2+}$ binding to calmodulin: proteolytic susceptibility of E31 and E87 indicates interdomain interactions, *Biochemistry*, *34*, 1179–1196 (1995).
125. Pedigo, S., and Shea, M. A., Discontinuous equilibrium titrations of cooperative calcium binding to calmodulin monitored by 1-D 1H-nuclear magnetic resonance spectroscopy, *Biochemistry*, *34*, 10676–10689 (1995).
126. Sorensen, B. R., and Shea, M. A., Interactions between domains of apo calmodulin alter calcium binding and stability, *Biochemistry*, *37*, 4244–4253 (1998).
127. Chabbert, M., Kilhoffer, M. C., Watterson, D. M., Haiech, J., and Lami, H., Time-resolved fluorescence study of VU-9 calmodulin, an engineered calmodulin possessing a single tryptophan residue, *Biochemistry*, *28*, 6093–6098 (1989).
128. Royer, C. A., Understanding fluorescence decay in proteins, *Biophys J*, *65*, 9–10 (1993).
129. Kim, S. J., Chowdhury, F. N., Stryjewski, W., Younathan, E. S., Russo, P. S., and Barkley, M. D., Time resolved fluorecence of the single tryptophan of Bacillus stearothermophilus phosphofructokinase, *Biophys J*, *065*, 215–226 (1993).
130. Brown, M. P., and Royer, C., Fluorescence spectroscopy as a tool to investigate protein interactions, *Curr Opin Biotechnol*, *8*, 45–49 (1997).
131. Callis, P. R., 1La and 1Lb transitions of tryptophan: applications of theory and experimental observations to fluorescence of proteins, *Methods Enzymol*, *278*, 113–150 (1997).
132. Callis, P. R. a. B., B. K., Tryptophan fluorescence shifts in proteins from hybrid simulations—An electrostatic approach, *J Phys Chem*, *101*, 9429–9432 (1997).
133. Demchenko, A. P., Red edge fluorescence spectroscopy of single tryptophan proteins, *Eur Biophys*, *16*, 121–129 (1988).
134. Demchenko, A. P., Dielectric behavior of proteins, *J Mol Liq*, *56*, 127–139 (1993).
135. Grinvald, A., and Steinberg, I. Z., Fast relaxation processes in a protein revealed by the decay kinetics of tryptophan fluorescence, *Biochemistry*, *13*, 5170–5178 (1974).
136. Lakowicz, J. R., and Cherek, H., Dipolar relaxation in proteins on the nanosecond timescale observed by wavelength-resolved phase fluorometry of tryptophan fluorescence, *J Biol Chem*, *255*, 831–834 (1980).
137. Hudson, B. S., Huston, J. M., and Soto-Campos, G., A reversible "dark state mechanism for complexity of the fluorescence of tryptophan in proteins., *J Phys Chem*, *103*, 2227–2234 (1999).
138. Tsvetkov, P. O., Protasevich, I. I., Gilli, R., Lafitte, D., Lobachov, V. M., Haiech, J., Briand, C., and Makarov, A. A., Apocalmodulin binds to the myosin light chain kinase calmodulin target site, *J Biol Chem*, *274*, 18161–18164 (1999).
139. Kilhoffer, M. C., Lukas, T. J., Watterson, D. M., and Haiech, J., The heterodimer calmodulin: myosin light-chain kinase as a prototype vertebrate calcium signal transduction complex, *Biochim Biophys Acta*, *1160*, 8–15 (1992).
140. Barbato, G., Ikura, M., Kay, L. E., Pastor, R. W., and Bax, A., Backbone dynamics of calmodulin studied by 15N relaxation using inverse detected two-dimensional NMR spectroscopy: the central helix is flexible, *Biochemistry*, *31*, 5269–5278 (1992).

141. Matsushima, N., Izumi, Y., Matsuo, T., Yoshino, H., Ueki, T., and Miyake, Y., Binding of both $Ca^{2+}$ and mastoparan to calmodulin induces a large change in the tertiary structure, *J Biochem (Tokyo)*, *105*, 883–887 (1989).
142. Seaton, B. A., Head, J. F., Engelman, D. M., and Richards, F. M., Calcium-induced increase in the radius of gyration and maximum dimension of calmodulin measured by small-angle X-ray scattering, *Biochemistry*, *24*, 6740–6743 (1985).
143. Eisinger, J., Feuer, B., and Lamola, A. A., Intramolecular singlet excitation transfer. Applications to polypeptides, *Biochemistry*, *8*, 3908–3915 (1969).
144. Cheung, H. C., Resonance Energy Transfer, in *Topics in fluorescence Spectroscopy*, Vol. 2, Lakowicz, J. R., Ed., Plenum Press, New York and London, pp. 127–176 (1991).
145. Steiner, R. F., Albaugh, S., and Kilhoffer, M. C., Distribution of separations between groups in an engineered calmodulin, *J Fluorescence*, *1*, 15–22 (1991).

# 9

# Luminescence Studies with *trp* Aporepressor and Its Single Tryptophan Mutants

Maurice R. Eftink

## 9.1. Introduction

To fully exploit the power of fluorescence techniques to study protein structure, function and dynamics, it is important to be able to assign the fluorescence properties of individual tryptophan residues. The problem lies in the fact that the fluorescence of tryptophan residues is highly overlapped. By a combination of time-resolved and quenching-resolved techniques, the fluorescence properties of individual tryptophan residues can sometimes be assigned (in cases where there are only two or perhaps three such residues). However, the most useful strategy for resolving the properties of individual tryptophan residues is site-directed mutagenesis, to selectively remove the residues, thus simplifying the spectroscopic data.

This strategy has been applied to numerous proteins. This chapter will summarize studies with the homodimeric protein *trp* aporepressor, the wild type form of which contains two tryptophans per subunit. Luminescence studies with *trp* aporepressor mutants will show that trp → trp energy transfer occurs in the wild type protein. The $R_0^{2/3}$ for 50% energy transfer between tryptophan residues has been estimated to be in the range of 10 Å. Though it is likely that such energy transfer between tryptophan residues occurs in proteins, it is very difficult to find experimental evidence for its existence. Yet, identifying the existence of trp → trp energy transfer is necessary when attempting to interpret fluorescence decay parameters. The studies summarized here with *trp* aporepressor illustrate how such evidence can be obtained.

---

*Maurice R. Eftink* • Department of Chemistry, The University of Mississippi, Oxford, MS 38677.

*Topics in Fluorescence Spectroscopy, Volume 6: Protein Fluorescence*, edited by Joseph R. Lakowicz. Kluwer Academic / Plenum Publishers, New York, 2000

*Trp* aporepressor from *E. coli* is a homodimeric protein having a subunit molecular weight of 12.5 kDa. *Trp* repressor is a DNA binding regulatory protein. Upon binding the co-repressor L-tryptophan, the *trp* aporepressor-tryptophan complex binds to the operator that controls the transcription of enzymes involved in the biosynthesis of tryptophan. The crystal structure of *trp* repressor shows it to have the helix-turn-helix DNA-binding motif and to undergo a change in the positioning of its "reading heads" upon binding tryptophan.[1] A very interesting feature of *trp* aporepressor is that the two identical subunits are highly intertwined. There are two tryptophan residues, Trp-19 and Trp-99, in each subunit. From the crystal structure, Trp-19 is located near the subunit interface, while Trp-99 is located in a more accessible position as part of the C-terminal $\alpha$-helix.

Two single tryptophan mutants, W19F and W99F, are available, as is a tryptophan-less mutant, W19F/W99F.[2]

## 9.2. Fluorescence Studies with Wild Type and Mutant Forms of *trp* Aporepressor

*Steady-state data:* Since *trp* aporepressor has two types of intrinsic tryptophan residues, there is a challenge of dissecting and assigning the fluorescence contribution of the two type of residues. Shown in Figure 9.1 is the

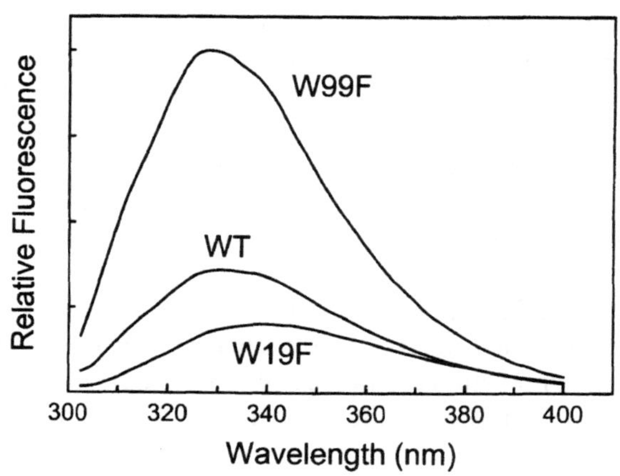


Figure 9.1. Fluorescence emission spectra of wild type, W19F, and W99F *trp* aporepressor, at 20 °C, pH 7.5, with excitation at 295 nm. The spectra are for solutions having the same absorbance at the excitation wavelength. This plot was reproduced from Eftink *et al.* (1993) with permission from the American Chemical Society.

steady-state emission of wild type *trp* aporepressor and the W19F and W99F mutants. Whereas wild type protein has an emission maximum of 332 nm, Trp-19 in W99F has a bluer maximum of 328 nm and a higher quantum yield, and Trp-99 of W19F has a redder maximum of 339 nm and a lower quantum yield.[3]

Acrylamide quenching data (not shown) indicate that Trp-99 of W19F is the more exposed ($k_q = 4.5 \times 10^9 M^{-1} s^{-1}$) than is Trp-19 of W99F ($k_q = 1.0 \times 10^9 M^{-1} s^{-1}$). The wild type protein gives an apparent $k_q$ value that is between the values for the two mutants, as expected.[3]

*Time-resolved fluorescence data:* Frequency domain fluorescence lifetime studies of wild type and the two mutants have been carried out by Royer[4] and by Eftink and coworkers,[3] with congruent results. All three proteins show a non-exponential fluorescence decay, with the decay profile of the wild type protein being something of an average between that of the two single tryptophan type proteins.

Shown in Figure 9.2 are frequency domain data, with the resulting decay times given in Table 9.1.[3] Trp-19 of W99F has the longer mean decay time of ~3 ns, whereas Trp-99 of W19F has a mean decay time of ~1 ns. The mean decay time of the wild type is intermediate, 1.4 ns. We obtained satisfactory fits to our data with bi-exponential decay laws. The fluorescence of Trp-19 of W99F is dominated by a ~4 ns long lifetime component. The decay of Trp-99 of W19F has nearly equal contributions from a ~3 ns and a ~0.5 ns component.

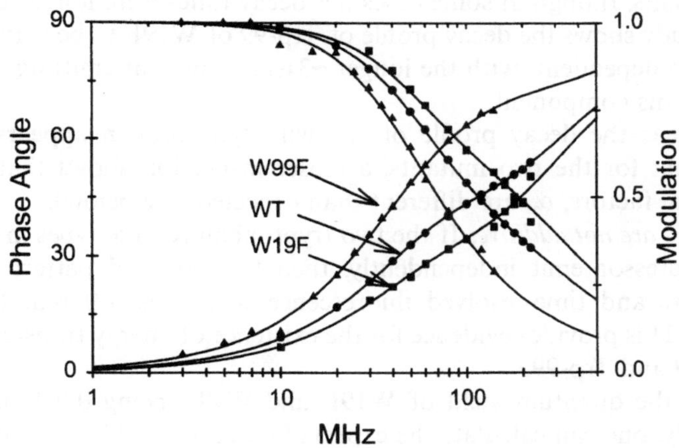


Figure 9.2. Frequency domain phase/modulation fluorescence lifetime data for wild type (●), W19F (■), and W99F (▲) *trp* aporepressor at 20 °C, pH 7.5, in 0.15 M KCl and 0.02 M potassium phosphate. The solid curves are fits of a bi-exponential intensity decay law with the parameters given in Table 9.1. This plot was reproduced from Eftink *et al.* (1993) with permission from the American Chemical Society.

Table 9.1. Luminescence Properties of Wild Type *trp* Aporepressor and Its W19F and W99F Mutants

| Property | WT | W19F | W99F |
|---|---|---|---|
| **Fluorescence** | | | |
| $\lambda_{max}$ (nm) | 332 | 339 | 328 |
| Quantum Yield | 0.063 | 0.039 | 0.164 |
| Lifetimes | | | |
| $\tau_1$ (ns) | 3.36 | 3.14 | 2.94 |
| $\alpha_1$ | 0.30 | 0.17 | 0.72 |
| $f_1$ | 0.716 | 0.526 | 0.962 |
| $\tau_2$ (ns) | 0.57 | 0.58 | 0.40 |
| $\alpha_2$ | 0.70 | 0.83 | 0.28 |
| $f_2$ | 0.284 | 0.474 | 0.038 |
| mean $\langle\tau\rangle$ (ns) | 1.41 | 1.01 | 2.95 |
| **Phosphorescence** | | | |
| 0–0 transition (nm) | 407<br>415 | 415 | 407 |

Parameters from Eftink *et al.* (3).

Royer[4] performed decay measurements as a function of emission wavelength to construct decay associated spectra. Her global analysis yielded similar results, though in some cases her decay times were longer than ours. Royer's study shows the decay profile of Trp-99 of W19F to be very emission wavelength dependent, with the longer ~3 ns component emitting to the red of the ~0.5 ns component.

Whereas the decay profile of the wild type protein appears to be a sum of that for the two mutants, a closer inspection shows that the preexponential factors, $\alpha_i$, are different than expected (see below).

*Things are not additive:* If the two tryptophan residue types in wild type *trp* aporepressor emit independently, then the expected pattern for both steady-state and time-resolved fluorescence would be different from that observed. This provides evidence for the existence of energy transfer between the Trp-19 and Trp-99.

With the quantum yield of W19F and W99F being 0.039 and 0.164, respectively, one can calculate the expected quantum yield of $\Phi$ for the wild type protein, if emission were independent. (This calculation takes into consideration the slight difference in the absorption spectrum of the two tryptophans.) That is, the quantum yield is expected to be

$$\Phi = \alpha_{19}\Phi_{19} + \alpha_{99}\Phi_{99} \tag{9.1}$$

where the $\Phi_i$ are the quantum yields of the individual residues and the $\alpha_i$ are the normalized relative absorbances of the individual residues at the excitation wavelength (these $\alpha_i$ are analogous to the $\alpha_i$ pre-exponential factors in a bi-exponential decay law). If there is no energy transfer between residues, the above equation predicts that the observed $\Phi$ would be 0.091 (with experimentally estimated $\alpha_{19} = 0.42$ and $\alpha_{99} = 0.58$ at the excitation wavelength).

However, the observed quantum yield is 0.063, which is closer than expected to the quantum yield of Trp-99 in W19F. If there is energy transfer from the bluer Trp-19 to the redder Trp-99, with transfer efficiency $E_T$, then the quantum yield should be

$$\Phi = \alpha_{19}\Phi_{19}\cdot(1 - E_T) + \alpha_{99}\Phi_{99}\cdot(1 + E_T\alpha_{19}/\alpha_{99}) \tag{9.2}$$

Substituting the experimentally observed $\Phi = 0.063$ into this equation, we can calculate that the efficiency of energy transfer is $E_T = 0.54$. This analysis assumes that the $\Phi_{19}$ and $\Phi_{99}$ determined from the mutants applies to the same tryptophans in the wild type. Since the thermodynamic stability of the mutants is not too much different than the wild type,[5] we have no reason to suspect that the environments of the tryptophan residues has changed by the mutations. This energy transfer model enables us to explain the fact that the quantum yield of the wild type is lower than expected from consideration of the two mutants.

The time-resolved fluorescence data also show evidence for Trp-19 to Trp-99 energy transfer, but a complete analysis of these data is more difficult. If there is energy transfer, then we would expect the decay time of the donor to be reduced and the pre-exponential for the acceptor to be increased. Referring to Scheme 9.1 for the case of Trp-19 to Trp-99 energy transfer, the

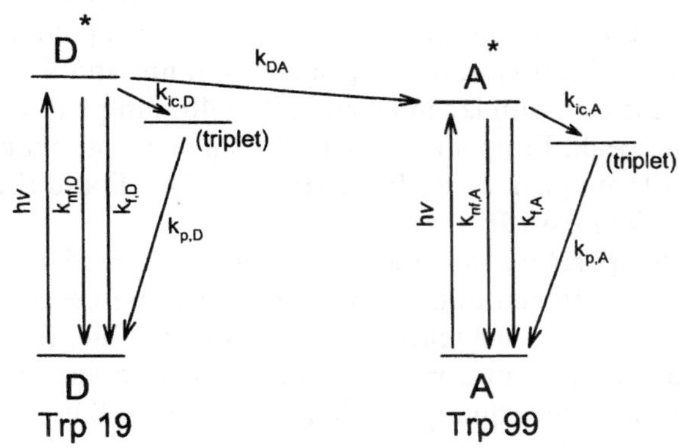


Scheme 9.1.

decay time of donor, $\tau_D$, will be $1/\tau_D = k_{f,D} + \Sigma k_{nf,D} + k_{ic,D} + k_{DA}$, where $k_{DA}$ is the rate constant for donor to acceptor energy transfer, $k_f$ is the radiative rate constant, $\Sigma k_{nf}$ is the sum of all non-radiative rate constants, and $k_{ic}$ is the rate constant for intersystem crossing to the triplet state (which is pertinent to the discussion below, but otherwise could be lumped together with the other non-radiative processes). If energy transfer does not occur (i.e., $k_{DA}$ is zero), then the last term in the equation for $1/\tau_D$ is dropped. The equation predicts a shortening of $\tau_D$ when energy transfer occurs, which is what is observed for the long $\tau$ of Trp-19, which is 3.94 ns for the W99F mutant (where Trp-19 to Trp-99 energy transfer can not exist) and is 3.36 for the wild type. In principle, the data can be fitted with a kinetic model to determine the magnitude of $k_{DA}$ (which is related to the efficiency of energy transfer, $E_T$, as $E_T = k_{DA}/(k_{f,D} + \Sigma k_{nf,D} + k_{DA})$). However, the fact that Trp-99 also has a decay time in the 3 ns range would make such an analysis difficult.

Whereas one expects the decay time of the donor to decrease when energy transfer occurs, the decay time of the acceptor should remain the same as $1/\tau_A = k_{f,A} + \Sigma k_{nf,A} + k_{ic,A}$. However, the apparent pre-exponential for the acceptor's lifetime should increase. Again, the fact that the intensity decay of Trp-99 is not a mono-exponential in isolation (i.e., in W19F), makes it difficult to tell whether evidence for energy transfer can be extracted from the $\alpha_i$ values of the wild type. Nevertheless, the intensity decay data for wild type *trp* aporepressor is consistent with the existence of Trp-19 to Trp-99 energy transfer, in terms of the values of the donor lifetime component.

*Support from phosphorescence data:* Additional support for the existence of resonance energy transfer between Trp-19 and Trp-99 in *trp* aporepressor comes from low temperature phosphorescence spectra. Essentially, this method uses the triplet state population to monitor the energy transfer that has occurred at the singlet level.

Shown in Figure 9.3 are the low temperature phosphorescence spectra of wild type *trp* aporepressor and the two single tryptophan mutants.[3] Wild type shows resolved 0-0 vibrational peaks at 407 nm and 415 nm, but these two peaks do not have similar intensities. The 407 nm peak is much smaller than the 415 nm peak. From studies with the single tryptophan mutants, it is clear that the 407 nm peak arises from Trp-19 in W99F and the 415 nm peak arises from Trp-99 in W19F.

At low temperature all the temperature dependent non-radiative processes have been frozen out, so that the phosphorescence quantum yield of a tryptophan residue has reached its maximum value. In other words, if two tryptophan residues in a protein do not undergo energy transfer, then one would expect to see equal phosphorescence contributions from the two residues (which would show up as nearly equal 0-0 transition peaks, if such peaks are resolved). However, if there is energy transfer at the singlet level

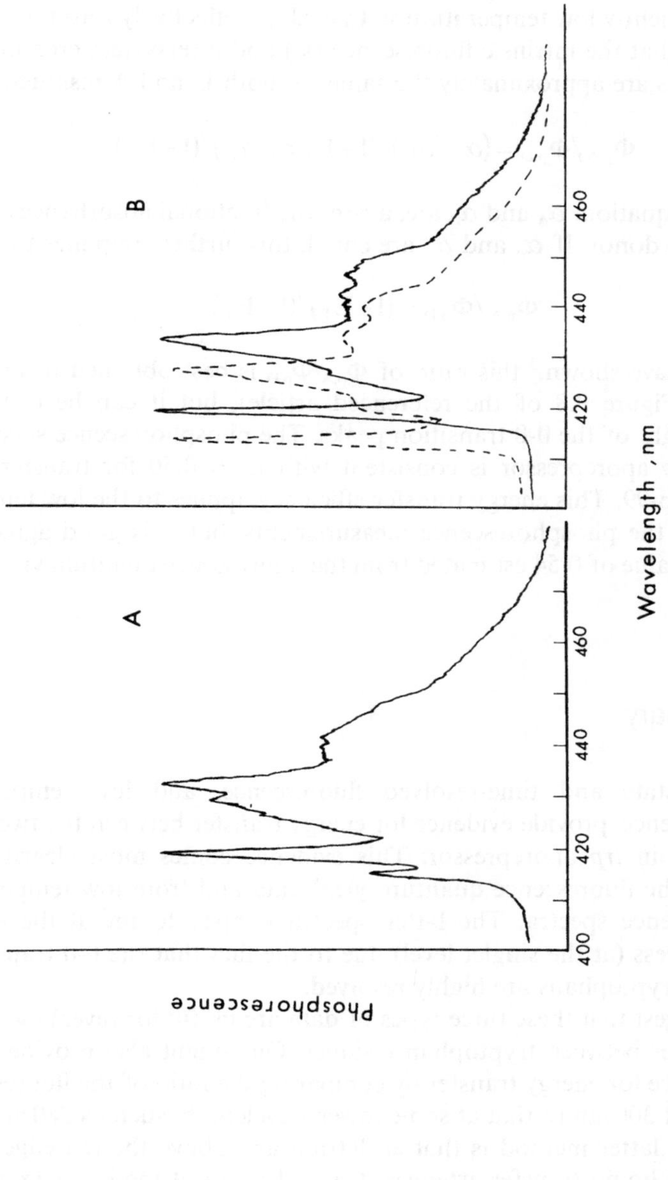


Figure 9.3. Low temperature phosphorescence spectra of wild type (A, left), W99F (B, right, dashed curve), and W19F (B, right, solid curve) *trp* aporepressor at 80 K in a 50% glycerol: phosphate buffer (pH 7.5) glass. Wavelength axis is uncorrected.

(thus increasing the population of excited acceptor), one would expect the acceptor 0-0 peak to be larger than the donor 0-0 peak.

From Scheme I the following relationship can be derived for the ratio of the phosphorescence yields of the donor and acceptor tryptophan residues.[3] The assumptions for deriving this relationship are that the measurements are made a sufficiently low temperature so that $\Sigma k_{nf}$ is effectively zero for both D and A, and that the intrinsic fluorescence ($k_f$) and intersystem crossing ($k_{ic}$) rate constants are approximately the same for both D and A residues.

$$\Phi_{p,A}/\Phi_{p,D} = (\alpha_A/\alpha_D)\cdot(1+E_T\alpha_D/\alpha_A)/(1-E_T) \tag{9.3}$$

In this equation, $\alpha_A$ and $\alpha_D$ are, again, the fractional absorbances of the acceptor and donor. If $\alpha_A$ and $\alpha_D$ are equal, this further simplifies to

$$\Phi_{p,A}/\Phi_{p,D} = (1+E_T)/(1-E_T) \tag{9.4}$$

As we have shown,[3] this ratio of $\Phi_{p,A}/\Phi_{p,D}$ is best obtained from curve fitting (see Figure 9.8 of the referenced article), but it can be estimated from the height of the 0-0 transition peaks. The phosphorescence spectra of wild type *trp* aporepressor is consistent with $E_T = 0.50$ for transfer from Trp-19 to Trp-99. This energy transfer efficiency applies to the low temperature used in the phosphorescence measurements, but it is good agreement with the $E_T$ value of 0.54 estimated from the fluorescence quantum yield data at 20 °C.

## 9.3. Summary

Steady-state and time-resolved fluorescence, and low temperature phosphorescence, provide evidence for energy transfer between the two tryptophan type in *trp* aporepressor. This evidence comes most clearly from analysis of the fluorescence quantum yield data and from low temperature phosphorescence spectra. The latter spectra happen to reveal the energy transfer process (at the singlet level) due to the fact that the 0-0 transitions for the two tryptophans are highly resolved.

We suggest that these three types of data are useful for revealing energy homo-transfer between tryptophan residues. One might also provide qualitative evidence for energy transfer by comparing the ratio of the fluorescence anisotropy at 300 nm to that at some lower wavelength, such as 280 nm. The basis for the latter method is that at 300 nm and above the red-edge effect limits energy homo-transfer, whereas it would occur at the lower excitation wavelength. Provided that electronic oscillators for donor and acceptor fluorophores are not aligned, then energy transfer should lead to depolarization at the lower excitation wavelength.

Recognizing the existence of energy homo-transfer between tryptophan residues is critical to understanding the fluorescence parameters for multi-tryptophan residues. Such energy transfer occurs in *trp* aporepressor to an extent of about 50% from the internal Trp-19 to the more solvent exposed Trp-99. Similar data for other tryptophan proteins show varied results. There appears to be no energy transfer between the two tryptophans of liver alcohol dehydrogenase,[6,7] whereas energy transfer occurs between tryptophan residues in T4 lysozyme.[8,9]

## References

1. Zhang, R.-G., Joachimiak, C. L., Lawson, R. W., Schevitz, W., Otwinowski, Z., and Sigler, P. B. "The crystal structure of *trp* aporepressor at 1.8 Å resolution shows how binding tryptophan enhances DNA affinity" *Nature (Lond.)* **377**, 591–597 (1987).
2. Mann, C. J., Royer, C. A., and Matthews, C. R. "Tryptophan replacements in the *trp* aporepressor from *Escherichia coli*: Probing the equilibrium and kinetic folding models" *Protein Sci.* **2,** 1853–1861 (1993).
3. Eftink, M. R., Ramsay, G. D., Burns, L., Maki, A. H., Mann, C. J., Matthews, C. R., and Ghiron, C. A. "Luminescence studies with *trp* repressor and its single-tryptophan mutants" *Biochemistry* **32**, 9189–9198 (1993).
4. Royer, C. A. "Investigation of the structural determinants of the intrinsic fluorescence of the *trp* repressor using single tryptophan mutants" *Biophys. J.* **63**, 741–750 (1992).
5. Royer, C. A., Mann, C. J., and Matthews, C. R. "Resolution of the fluorescence equilibrium unfolding of *trp* repressor using single tryptophan mutants" *Pro. Science* **2**, 1844–1852 (1993).
6. Ross, J. B. A., Schmidt, C. J., and Brand, L. "Time-resolved fluorescence of the two tryptophans in horse liver alcohol dehydrogenase" *Biochemistry* **20**, 4369–4377 (1981).
7. Eftink, M. R. "Luminescence studies with horse liver alcohol dehydrogenase" in *Adv. Biophys. Chem.* **2**, 81–114 (1992).
8. Ghosh, S., Zang, L.-H., and Maki, A. H. "Relative efficiency of long range nonradiative energy transfer among tryptophan residues in bacteriophage T4 lysozyme" *J. Chem. Phys.* **88**, 2769–2775 (1988).
9. Harris, D. L. and Hudson, B. S. "Photophysics of tryptophan in bacteriophage T4 lysozymes" *Biochemistry* **29**, 5276–5285 (1990).

# 10

# Heme-Protein Fluorescence

Rhoda Elison Hirsch

## 10.1. Introduction

The discovery of amino acid and protein fluorescence revolutionized protein structural analysis (Weber, 1953; Teale and Weber, 1957). *Intact* heme-proteins* had been excluded for more than 20 years from this sensitive and highly informative direct spectroscopic structural probing, because it was generally assumed that the fluorescence emission from the Tyr and Trp residues was effectively quenched by nearby heme moieties (Weber and Teale, 1959; Teale and Weber, 1959). Despite the quenching effects, clever utilization of the phenomena of fluorescence quenching by the hemes is seen in a broad range of early informative ligand binding studies to intact heme-proteins and apoproteins (heme-proteins without the heme) (e.g., Nagel and Gibson, 1967, 1971; Benesch *et al.*, 1976). This is expanded below.

The choice of optics in fluorescence detection also played a significant role in perpetuating the dogma, to the exclusion of fluorescence applications to these complex and vital proteins, some of which include hemoglobins, myoglobins, cytochromes, nitric oxide synthases, peroxidases, catalases, heme binding proteins of the Z-class, hemopexin, and heme oxygenase. Standard fluorescence measurements are typically made using right-angle optics, wherein inner-filter effects become significant with a highly absorbant sample such as a heme-protein. It has been estimated that the hemes give rise to ~99% non-radiative quenching of the aromatic intrinsic fluorophores (Weber and

* *Intact heme-protein* refers to the protein with its heme moiety (ies) and subunits required for functionality.

*Rhoda Elison Hirsch* • Department of Medicine (Hematology) and Department of Anatomy & Structural Biology, Albert Einstein College of Medicine of Yeshiva University, Bronx, New York 10461.

*Topics in Fluorescence Spectroscopy, Volume 6: Protein Fluorescence*, edited by Joseph R. Lakowicz. Kluwer Academic / Plenum Publishers, New York, 2000

Teale, 1959), but this may have to be re-evaluated by including inner-filter effect corrections.

In order to understand heme quenching mechanisms, the availability of a synchrotron light source allowed for the first direct fluorescence lifetime decay measurements in the nanosecond (ns) range, [albeit as noted by the authors, with poor precision (±0.25 ns)], of hemoglobin and its subunits compared to that of the apoprotein (Alpert and Lopez-Delgado, 1979). In 1980, the simultaneous discovery of significant steady-state hemoglobin intrinsic fluorescence emission, by independent laboratories, using more sensitive detectors (Alpert *et al.*, 1980) or non-right angle optics (Hirsch *et al.*, 1980a), facilitated the application of fluorescence principles and methodology to provide a powerful tool to probe this fascinating and multifunctional class of proteins. With this data in mind, the issue of heme-protein fluorescence was revisited by Weber and colleagues (Alpert *et al.*, 1980):

> *... there is no reason a priori that the tryptophan emission in hemoglobin should be totally quenched. In a Forster-type energy transfer process the quenching efficiency is determined, in part, by the angle between the emission dipole of the donor and absorption dipole of the acceptor (Forster, 1948). We now understand that proteins are dynamic structures. Detectable fluctuations in the protein matrix occur within the nanosecond times, i.e., the timescale of the fluorescence process (Lakowicz and Weber, 1973; McCammon et al., 1977).*

Hence, as explained by Fontaine *et al.* (1980), consideration of motions of groups involved in energy transfer mechanisms may dramatically reduce the transfer energy resulting in the observed unquenched steady state emission (Fontaine *et al.*, 1980).

This chapter focuses on: (1) techniques employed to resolve the fluorescence emission from intact heme-proteins; (2) issues related to the origins and assignments of the intrinsic fluorescence signal; (3) the employment of extrinsic fluorescence probes to explore non-aromatic site-specific microdomains; (4) binding assays using fluorescence quenching by heme or heme-proteins; and (5) some examples of fluorescence applications to unravel the interrelationships of structure and function in heme-proteins.

## 10.2. Techniques to Detect Heme-Protein Fluorescence

More sensitive detectors (single photon counting spectroscopy with standard right angle optics (Alpert *et al.*, 1980) or the use of front-face optics (Hirsch *et al.*, 1980a) (Figure 10.1) resulted in the detection of steady-state hemoglobin emission. In order to detect heme-protein fluorescence using right angle optics, low protein concentrations are required. In the case of intact HbA, a tetramer with $2\alpha$ and $2\beta$ chains, significant dissociation to $\alpha\beta$

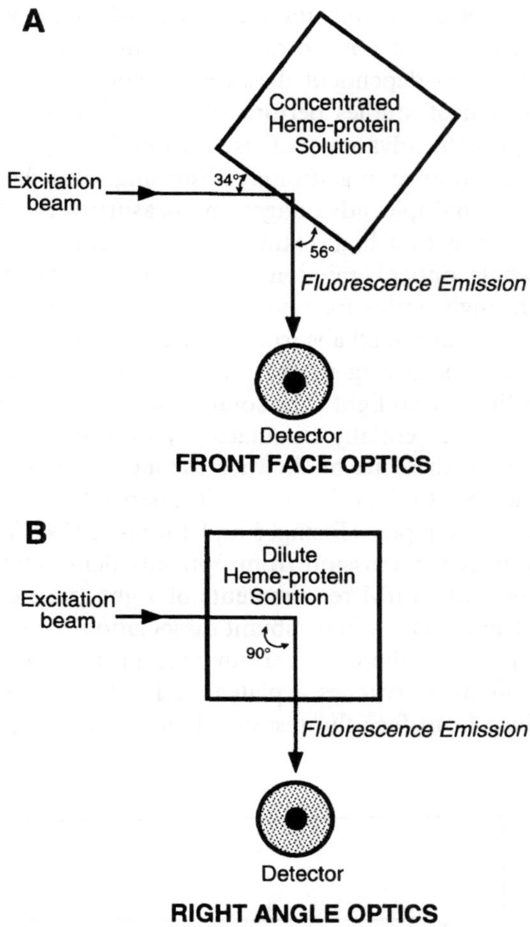


Figure 10.1. A comparison of optical designs for fluorescence measurements: (A) front-face; (B) right-angle optics.

dimers occurs at low concentration in the oxy or R-state forms: under conditions of moderate ionic strength (~0.1 M NaCl), the $K_D$ is $1.0 \times 10^{-6}$ M. In the case of deoxy hemoglobin, dissociation is significantly less ($K_D = 2.0 \times 10^{-11}$ M) (Imai, 1982; Ackers *et al.*, 1976). High salt concentration or mutation will affect the dissociation equilibrium (Antonini and Brunori, 1971; Bunn and Forget, 1986, Herskovits *et al.*, 1977).

Percent of dimer dissociation ($\alpha$) is calculated by:

$$\alpha = \left\{ K_D M_4 \left[1 + (16c/K_D M_4)\right]^{1/2} - K_D M_4 \right\} / 8c \quad \text{(Herskovits } et\ al., 1977) \qquad (10.1)$$

where $K_D$ is the dissociation constant in molarity, $M_4$ is the molecular weight of the tetramer, and c is the concentration in grams/liter.

This concentration dependent dissociation complicates the interpretation and comparison of studies performed under different solution conditions, and highlights the advantage of using front-face optics which reduces inner-filter effects that arise in a strongly absorbing solution. Front-face fluorometry provides multiple advantages in measuring the fluorescence of any protein solution with a high extinction coefficient of absorption. With standard right-angle optics, emission is detected at right angles from the exciting beam through optics focused to the center of the cuvette. With a strongly absorbing solution, all absorption takes place near the front surface, with little excitation occurring in the center of the cuvette, and thus, the detector receives little or no light. The solution itself acts as an "inner filter". Inner filter effects are essentially eliminated by front-face fluorescence measurements. Optimally, these are made when the incident light makes an angle of 34° with the normal to the cell face, or 56°, depending on the orientation of the front-face cell adapter (Eisinger and Flores, 1979). This permits the detection of fluorescence emission from optically dense concentrated solutions (mM as opposed to $\mu$M requirements of right angle optics) of hemeproteins, which is important when subunit dissociation is a significant factor. Unlike right-angle optics, there is a certain concentration of protein wherein the fluorescence intensity reaches a plateau using front-face optics. This is advantageous since front-face fluorescence is not sensitive to concentration

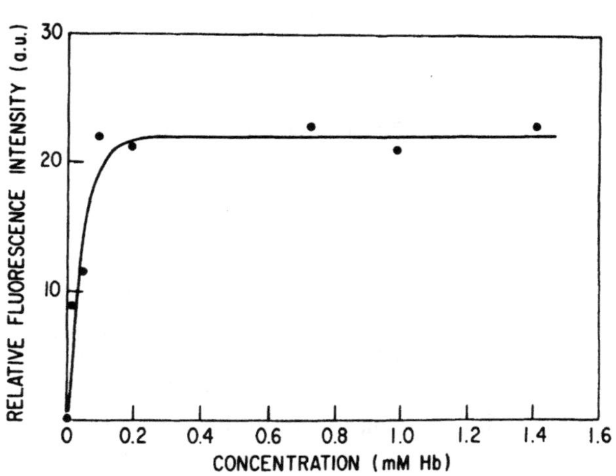


Figure 10.2. The concentration dependence of hemoglobin fluorescence emssion intensity plateaus when using front-face optics. *From Hirsch et al., 1980.* Excitation wavelength, 280 nm. Oxy HbA: 0.07 mM tetramer, 0.05 M potassium phosphate buffer, 25 °C.

increases within this plateau that may occur during titration studies, change of ligand state, or small pipetting errors. For HbA (Figure 10.2), this concentration-independent plateau is reached at concentrations greater than 0.3 g% (~0.19 mM heme or ~0.05 mM tetramer) (Hirsch *et al.*, 1980a).

Front-face measurements may be simply, but suboptimally, achieved in a right angle configuration with the use of small (mm) rounded cuvettes or triangular cuvettes (Hirsch *et al.*, 1980a; Bucci *et al.*, 1988; Hirsch, 1994), with the latter providing more sensitive detection. A front-face cell, is designed for easy insertion into a standard cuvette holder for $1 \times 1$ cm cells, and orients the sample for the optimal angle requirement (Figure 10.3a). The small volume required for this cell (100–200 $\mu$l) becomes advantageous when studying heme-proteins with limited availability (e.g., scarce mutants or recombinant mutants). However, an instrument designed with a horizontal orientation of the light source slit may preclude use of this cell. Most companies now offer the option of temperature controlled front-face adapters designed specifically for the fluorometer.

Novel variations of front-face optical designs provided further advantage in the study of heme-protein fluorescence. The *rhombiform optical cell* (Figure 10.3b) designed by Horiuchi and Asai (1980), simultaneously measures absorption and fluorescence, allowing for the direct and continuous measurements of the binding of a fluorescent allosteric effector to hemoglobin (at a limited range of concentration) while assessing variations in the partial pressure of oxygen during deoxygenation. The solution is gently stirred for gas-exchange. Caution must be exercised when stirring any proteins, and especially hemoglobins, which may be subject to mechanical instability [e.g., HbS (Asakura *et al.*, 1973)]. With the purpose of eliminating reflections and stray emissions (which may become significant for the relatively low fluorescence emission of heme-proteins), Bucci and colleagues developed an optimized shielded cuvette and also designed an optical cell with a front-face configuration that operates on a free liquid surface (Bucci *et al.*, 1992; Gryczynski *et al.*, 1997a) (Figure 10.3c). This also avoids any possible protein conformational changes induced by a protein-solid interface. However, air-water interfaces do have the potential to induce protein unfolding for some proteins and hemoglobin mutants (Elbaum *et al.*, 1976a & b; Hirsch *et al.*, 1980b).

## 10.3. Origin and Assignment of the Steady-State Fluorescence Signal

With the present capability of measuring steady-state and time-resolved heme-protein fluorescence emission, solution-active and dynamical structural

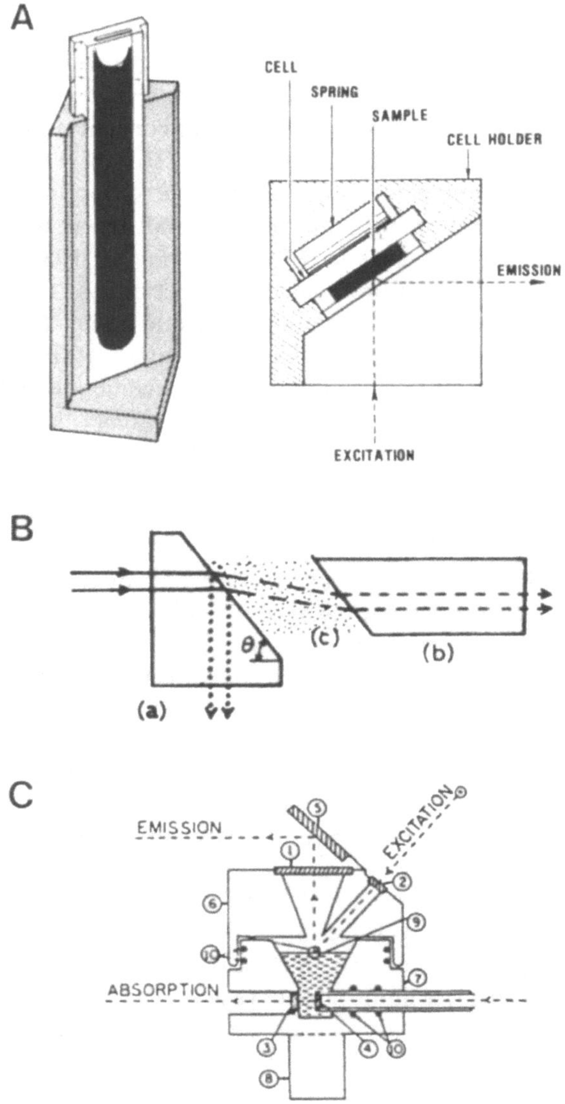


Figure 10.3. Novel fluorescence optical designs useful for the detection of heme-protein fluorescence. (A) *From: Eisinger & Flores, 1979.* Front-face optics is achieved by an insert placed into a standard right-angle cuvette holder. The baseplate shown on the left is removable. The key feature is that the exciting light makes an angle of 34° with the normal to the cell face or, by inverting, one may make the angle of incidence 56°. The central rays of the excitation and emission beams intersect normally at the center of the cuvette holder for either configuration. The front window of the cell is 0.5 mm thick and the sample thickness is 1 mm. This cuvette is advantageous for rare samples, requiring ~100–200 μl. (B) *From: Horichi and Asai, 1980*. Shown is the schematic of the rhombiform optical cell compartment. (a) and (b) are made of quartz; and (c) is the hemoglobin sample. The solid line depicts the incident excitation light beam; the broken and dotted lines show the transmitted and the emitted light, respectively. $\theta$ is 52.4°, avoiding light direct excitation beam reflectance. (C) *From: Bucci et al., 1992*: A side view of the free-surface cuvette. (1–3) fixed quartz windows; (4) sliding quartz window; (5) metallic mirror; (6) body of the cover; (7) body of the cuvette; (8) supporting stem; (9) liquid sample; (10) O-rings.

and functional interrelationships may be revealed by molecular site-specific probing. *Intrinsic fluorescence* makes use of natural aromatic amino acids tryptophan (Trp) and tyrosine (Tyr), when part of the primary protein structure, while *extrinsic fluorescence* employs fluorescent probes that bind or are covalently attached to specific residues or microdomains of the molecule. To date, of the heme-proteins, hemoglobin fluorescence is the most extensively characterized, and thus, serves in this chapter as the exemplary model to discuss the details and complex considerations of heme-protein fluorescence.

### 10.3.1. Intrinsic Fluorescence

Intrinsic protein fluorescence arises from tryptophan and tyrosine residues. Phe also fluoresces, but its low extinction coefficient (especially at 280 nm) and resonance energy transfer make it essentially undetectable. In a protein containing both tryptophan(s) and tyrosine(s) (Class B proteins), the Trp and Tyr signals may be distinguished by the use of specific excitation wavelengths. It is generally assumed that 280 nm excitation results in fluorescence that predominantly arises from Trp, because of resonance energy transfer from the Tyr → Trp. However, with 280 nm excitation, the tyrosine contribution may be dissected out: steady-state front-face fluorometry of myoglobin Tyr mutants with invariant tryptophans exhibited alterations in the emission maximum attributed to the presence or absence of particular Tyr residues (Hirsch and Peisach, 1986). It is well established that the exclusive selection of Trp emission may be made by 296 nm excitation (Eisinger, 1969). Nevertheless, the challenge in fluorescence studies of multi-tryptophan proteins is to assign the source of the emission in both steady-state and time-resolved measurements.

*Inherent limitations*: The nature of heme-proteins precludes the ability to *reliably* calculate corrected spectra for wavelength-dependent effects, absolute intensities or quantum yields. This is *not* problematic when relative comparisons are made on the same fluorometer. Corrected spectra become necessary in the calculation of quantum yields and the overlap integrals needed for Förster energy transfer calculations. [For discussion of corrected spectra, see Parker (1968) and Lakowicz (1999)].

With heme-protein solutions exhibiting high extinction coefficients, a "true background", required for the blank, is difficult to create or attain. The translucent buffer system does not represent the filter effects that occur in an optically dense solution. Likewise, apoprotein (the protein without the heme moiety) cannot be used as the standard of non-heme quenched protein emission, because once the hemes are removed, the protein becomes a new species, structurally distinct from the native intact heme-protein.

### 10.3.2. Apoglobins

Apohemoglobin dimerizes and exhibits altered helical properties with respect to the intact protein (Antonini and Brunori, 1971; Sassaroli *et al.*, 1984). It is well established that the steady-state emission maximum reflects the Trp environment: Trps in a hydrophobic microenvironment emit around ~325–330 nm; surface Trps in a high polar aqueous environment emit around ~350 nm; and those Trp in an intermediate environment, with limited aqueous contact, emit around ~340–342 nm (Burstein *et al.*, 1973). Apohemoglobin, compared to intact human HbA, exhibits an emission maximum shifted about 14 nm to longer wavelengths (at ~344 nm) (Hirsch, unpublished results) consistent with the exposure of the buried $\beta$37 Trp upon dimerization (Chothia *et al.*, 1976). Horse apomyoglobin, compared to intact myoglobin, exhibits a 6 nm emission maximum shift to longer wavelengths (~339 nm) (Hirsch and Peisach, 1986) which cannot be attributed to dissociation since myoglobin is a monomer. These observations indicate a structural change in apomyoglobin where at least one of the two Trps of myoglobin becomes relatively more exposed, but in a limited way, to the aqueous solvent. Hence, the addition of heme induces the natural globin fold to the native hemoglobin conformation, and recent studies take advantage of this to compare subunit folding and association in normal and mutant hemoglobins (Vasudevan and Macdonald, 1977; Chiu *et al.*, 1998). Other heme-proteins, such as horseradish peroxidase and cytochromes, also display significant structural differences as apoproteins (Hamada *et al.*, 1993; Das *et al.*, 1995; Lasagna *et al.*, 1999). This becomes quite significant in the interpretation of fluorescence lifetime data.

The problem of finding the non-quenched structure for purposes of calculations persists. This necessitates an estimate of tryptophan donor emission and lifetimes in the absence of acceptor (in this case, heme). The assumptions employed, while useful as a first approach, become an issue in the interpretation of time-resolved data in light of resonance energy transfer rates and proposed mechanisms to explain the observed intrinsic lifetimes. (An expanded discussion is found below.) Nonetheless, *relative* fluorescence studies are informative.

### 10.3.3. Steady-State Fluorescence of Intact Heme-Proteins

The hemoglobin tetramer ($\alpha_2\beta_2$) contains a total of 6 Trp residues, with each $\alpha\beta$ dimer containing 3 Trp: $\alpha$14 Trp, $\beta$15 Trp, and $\beta$37 Trp. A comparison of the *relative* fluorescence properties of hemoglobin tryptophan variants provides a means to verify the significance of the signal, while suggesting the source of the emission (Figure 10.4) (Hirsch *et al.*, 1980a). It was also demonstrated that the *relative* emission maximum intensity increased in variants containing additional Trp residues: HbH >> HbF > HbA > HbRC (Table 10.1). Secondly, because HbRC ($\beta$37 Trp → Arg) emission, under the conditions applied, approached that of the baseline using 296 nm excitation, it was suggested that $\beta$37 Trp is the primary (but not exclusive) source of fluorescence. This was soon confirmed by another laboratory using Hb Kempsey ($\beta$99 Asp → Asn) and the modified nes-des Arg Hb (Itoh *et al.*, 1981). Noteworthy is the observation that Hb Rothschild, when excited at 280 nm, exhibits an emission maximum 10 nm blue-shifted towards the region of Tyr emission. This is explained by the released resonance energy transfer constraint upon the nearest neighbor, $\beta$35 Tyr (Hirsch *et al.*, 1980a).

As with other Trp proteins, the wavelength of the hemoglobin tryptophan fluorescence emission maximum will shift dependent upon exposure to aqueous solvent (Burstein *et al.*, 1973; Callis and Burgess, 1997). Intact hemoglobin and myoglobin fluoresce maximally at respectively, 325–330 nm (uncorrected, depending upon the instrument) and 331–334 nm depending

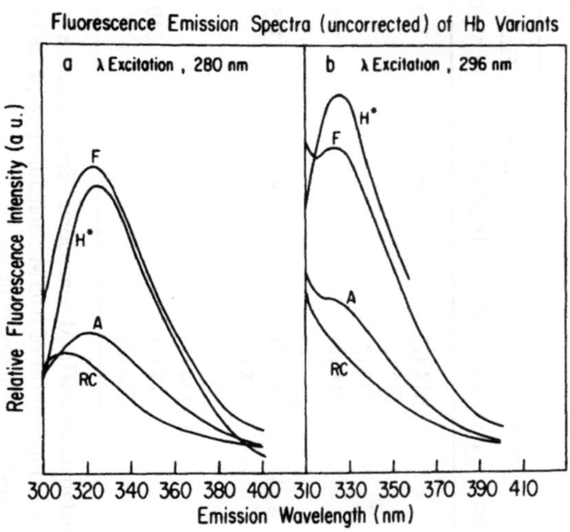


Figure 10.4. The front face steady-state intrinsic fluorescence emission (uncorrected) of oxy hemoglobin tryptophan variants. *From: Hirsch et al., 1980.* H*, HbH where the sensitivity of the recorder is 1/3 less than that recorded for the other hemoglobins (i.e., the relative intensity is three times that shown). F, HbF; A, HbA; RC, Hb Rothschild. More recently, a low intensity, defined emission maximum near 330 nm has been observed for RC (296 nm excitation, different conditions and preparation), while the emission spectrum with 280 nm excitation appears the same (panel a). See Table II for the tryptophan content and chain composition of these hemoglobin variants.

Table 10.1. Relative Intensities of Fluorescence from Intact Hemoglobins

| | | | | Excitation $\lambda$, 280 nm | | Excitation $\lambda$, 296 nm | |
|---|---|---|---|---|---|---|---|
| Hb Variant | Chain Composition | Tryptophans (*) | Total # of Tryptophans | Emission max. (nm) | Ratio of Intensities Hb variants: Hb A | Emission max. (nm) | Ratio of Intensities Hb variant: Hb A |
| Hb A | $\alpha_2\beta_2$ | $\alpha$14 (2)<br>$\beta$15 (2)<br>$\beta$37 (2) | 6 | 325 | 1.0 | 325 | 1.0 |
| Hb H | $\beta_4$ | $\beta$15 (4)<br>$\beta$37 (4) | 8 | 325 | 8.8 | 325 | 9.1 |
| Hb F | $\alpha_2\gamma_2$ | $\alpha$14 (2)<br>$\gamma$15 (2)<br>$\gamma$37 (2)<br>$\gamma$130 (2) | 8 | 325 | 1.5 | 325 | 2.0 |
| Hb RC | $\alpha_2\beta_2^{RC}$ | $\alpha$14 (2)<br>$\beta$15 (2) | 4 | 310 | 0.7 | no max. | — |

Concentration of the hemoglobins are 0.07 mM hemoglobin. Temperature is maintained at 25°C. Slit widths for both excitation and emission light are 6 nm. The hemoglobin solutions consist of 0.05 M potassium phosphate buffer at pH 7.35. All solutions are oxygenated. The values presented here are averages of several independent measurements.

*Figure between parenthesis corresponds to the number of each Trp residue per tetramer.

FROM: Hirsch RE, Zukin RS, and Nagel RL (1980) Biochem Biophys Res Commun 93:432–439. See caption to Fig. 10.4.

upon the myoglobin (Hirsch, 1994a). The 325–330 nm fluorescence emission maximum and the inability to quench hemoglobin fluorescence with 1 M KI support the conclusion that the primary emitting fluorophore lies in a hydrophobic environment in the interior of the protein. $\beta$37 Trp is the only Trp in the protein interior, specifically at the $\alpha_1\beta_2$ interface, the major site of quaternary change during the R $\rightarrow$ T transition. Consistently, independent laboratories further demonstrated that fluorescence emission intensities vary as a function of heme ligand binding (Figure 10.5) and serve as a reporter of the allosteric R $\rightarrow$ T transition and dissociation state (Fontaine *et al.*, 1980; Hirsch and Nagel, 1981; Itoh *et al.*, 1981; Hirsch *et al.*, 1983; 1985; 1994a & b; 1996; 1999; Bucci *et al.*, 1988; Gryczynski *et al.*, 1997a & b; Sokolov and Mukerji, 1998).

The above findings are based on the assumptions that: (1) in a given sample, the hemes are intact in all the subunits; (2) the sample is pure with no denaturation; and (3) artifacts such as Raman scattering or reflectance do not contribute significantly to the spectrum; (4) the light source does not result in photoreactions or denature the protein as has been reported for lasers (Henry *et al.*, 1986); and (5) Trp and Tyr variant hemoglobins containing aromatic substitutions remain in a conformational state that would not alter the fluorescence relative to HbA.

Investigators attempted to eliminate, control or carefully assess the above assumptions. Cuvette designs and cutoff filters eliminated stray light (Gryczynski *et al.*, 1997a). The employment of 1-anilinonapthalene-8-sulfonic acid (ANS), which becomes significantly fluorescent upon binding to the heme pocket, and calculations comparing the absorption at 280 nm and 540 nm, demonstrated that the sample did not contain apoglobin or subunits

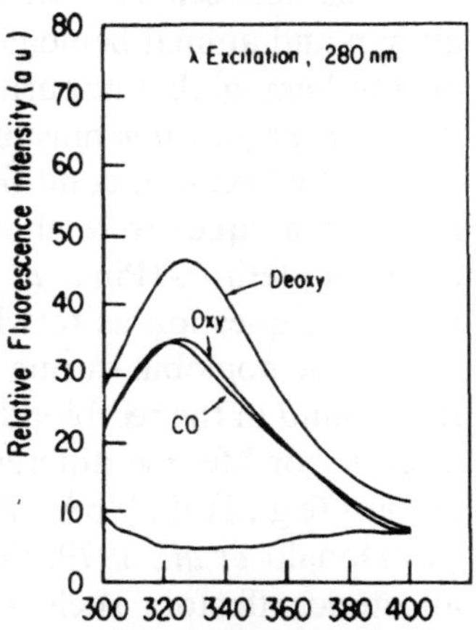


Figure 10.5. The front face intrinsic fluorescence emission of HbA varies as a function of ligand binding. *From: Hirsch and Nagel, 1981.* All solutions are 0.155 mM hemoglobin tetramer, pH 7.35, 0.05 M phosphate, 25 °C. The *lowest curve* is the buffer solution.

without heme, to at least less than 0.5% (Alpert *et al.*, 1980). Moreover, reproducible intensities from different samples disqualified the argument of random impurities. Alterations in intrinsic fluorescence observed as a result of ligand binding to the heme could not occur with met ($Fe^{+3}$), denatured hemoglobin or apoprotein, since the latter do not bind oxygen, CO or other heme ligands. The fluorescence intensity emission of deoxy HbA decreases significantly (~20%) in the oxy liganded state. This alteration in intensity is not observed in the deoxy and oxy forms of hemoglobin mutants (HbH and Hb Kempsey) known not to undergo the R $\rightarrow$ T allosteric transition (Hirsch and Nagel, 1981; Itoh *et al.*, 1981). The consistency of these studies with different allosteric hemoglobin mutants, prepared and studied in different laboratories, refutes the notion that the steady-state fluorescence arises from an impurity or other non-hemoglobin artifact.

It was asserted that the intensity differences observed with HbRC, that were used to assign the major source of the signal to $\beta$37 Trp, arise from the fact that the R-state of HbRC predominates as a dimer (Sharma *et al.*, 1980). However, dissociation of hemoglobin from the native hemoglobin tetramer to dimers results in emission maxima shifts not seen with HbRC (Hirsch *et al.*, 1983). Predictive hemoglobin steady-state fluorescence emission shifts correlate as expected with changes in the tetramer-dimer equilibrium as induced by high salt concentration, and in invertebrate hemoglobins that exist natively as dimers and tetramers or with other known dissociation properties (Hirsch *et al.*, 1983; 1985; 1993; 1994a & b; Harrington and Hirsch, 1991). High pressure techniques coupled with steady-state fluorescence, fluorescence polarization, and fluorescence lifetimes studies of heme-protein fluorescence provided further insight into dissociation properties (Marden *et al.*, 1986; Silva *et al.*, 1989; Pin *et al.*, 1990; Hirsch *et al.*, 1993). There are many correlations between fluorescence parameters and known properties of intact human and animal hemoglobins, and they cannot be dismissed as the result of non-hemoglobin impurities. However, it must be stressed that *precautions in sample preparation* must be taken, and relative or comparative studies must control for solution conditions (Pin *et al.*, 1990; Hirsch, 1994). Chromatography techniques were shown to select for Hb conformational states with different lifetimes (Pin *et al.*, 1990; Bucci *et al.*, 1988; Szabo *et al.*, 1989). This raises the question as to what adducts (i.e., natural or resin derived) might bind to hemoglobin during purification or what hemoglobin ligand(s) (naturally found in the red blood cell) may be removed that could result in altered intensity or lifetime differences. Red cell hemolysates contain minor hemoglobins (e.g., HbF, $HbA_2$, $HbA_{1a}$, $HbA_{1a2}$, $HbA_{1b}$, and glycosylated Hb ($A_{1c}$) (McDonald *et al.*, 1979; Garrick *et al.*, 1980), and other components (i.e., allosteric effectors such as diphosophoglycerate and chloride ions) that alter hemoglobin structure and conformation. Fluorescence, a highly sensitive assay of protein conformational change, may detect these structural/

conformational alterations. In fact, fluorescence methods are a useful reporter of conformational perturbation by allosteric effectors (e.g., Hirsch and Nagel, 1981; Mizukoshi *et al.*, 1982; Sassaroli *et al.*, 1982; Marden *et al.*, 1986; Gottfried *et al.*, 1997; Serbanescu *et al.*, 1998; Hirsch *et al.*, 1996, 1999).

### 10.3.4. Coupling of Diverse Spectroscopic Approaches Confirms Fluorescence Assignments

The coupling of highly sensitive fluorescence techniques with UVRR spectroscopy facilitates the assignment of the source of site-specific fluorescence emission perturbations (Hirsch *et al.*, 1996, 1997, 1999; Wajcman *et al.*, 1996; Sokolov and Mukerji, 1998). UVRR difference spectroscopy of hemoglobins have led to the characterization of band frequencies attributed to specific Trp and Tyr residues (Asher, 1988, 1993; Spiro *et al.*, 1990; Kitagawa, 1992; Cho *et al.*, 1994; Wang and Spiro, 1998; Rodgers and Spiro, 1994; Jayaraman *et al.*, 1995; Hu and Spiro, 1997): The Y8a band at ~1615 $cm^{-1}$ reflects the T to R state loss of the $\alpha$42–$\beta$99 hydrogen bond in the switch region of the interface. A decrease in the intensity of the low frequency shoulder of the W3 band at ~1548 $cm^{-1}$ originates from $\beta$37 Trp in the hinge region of the $\alpha_1\beta_2$ interface. A decrease in intensity without any sizable shifts in peak frequencies in several of the tyrosine and tryptophan resonance Raman bands is attributed to a generalized loosening of the global structure including a weakening of the hydrogen bond between the A-helix tryptophans and their respective bonding partners on the E-helix. An increase in the intensity of these Tyr and Trp resonance Raman bands is ascribed to a strengthening of these H-bonding interactions due to tighter packing between the A and E helices and H and F helices. UVRR differences observed in $\beta$6 mutants compared to HbA in the absence of chloride, show a turn towards greater hydrophobicity in the microenvironment of all three of the Trp residues $\alpha$14, $\beta$15, and $\beta$37 as reflected in the W3 band (Hirsch et al., 1996; Juszczak et al., 1998; Hirsch et al., 1999). Similar findings for T-state fluoromet HbS were reported by another laboratory, also coupling UVRR and front-face fluorometry (Sokolov and Mukerji, 1998). The fluorescence changes observed for the $\beta$6 mutants lead to the conclusion that $\beta$37 is responsible for observed R-T fluorescence differences (Hirsch and Nagel, 1981; Mizukoshi *et al.*, 1982; Sokolov and Mukerji, 1998), while the A-E helix packing changes are responsible for the R-T independent HbA-HbS fluorescence differences, as shown earlier for R-state HbC (Hirsch *et al.*, 1996). Hence, the fluorescence *differences* observed for R-state HbA, C, and S may reflect upon the contribution of the A-helix tryptophans ($\alpha$14, $\beta$15) to the steady-state emission. The coupling of Raman spectroscopy and fluorescence has been successful in other hemoglobin studies (Larsen *et al.*, 1990).

### 10.3.5. Time-Resolved Intrinsic Fluorescence Studies of Heme-Proteins Reveals Complex Data, But Data That Is Consistent with Known Protein Trp Fluorescence

The observation of multiexponential heme-protein fluorescence intensity decays has added to the controversial explanations for the origin of heme-protein fluorescence. The multiexponential decays are consistent with known tendencies for the decay of Trp fluorescence in proteins (Beecham and Brand, 1985). Most importantly, *there is no agreement to date* as to the explanation for multiexponential components observed in *single and multiple* Trp-containing proteins (Brand, 1999; Callis, 1999). Interpretation of the data obtained for heme-proteins often ignores this phenomenon, and is further complicated by the use of different solution conditions by the independent laboratories pursuing this problem (Table 10.2).

A compilation of some representative fluorescence lifetimes observed for heme-proteins (Table 10.3) demonstrates the present need for a systematic comparative study under identical conditions and state-of-the-art instrumentation. Any interpretation of the data requires coupling with the specific conditions and analysis employed. However, for the intact heme-proteins, a commonality of multiexponential decays with lifetimes consisting of a picosecond primary component, a subnanosecond, and a nanosecond component stands out. Changes in the decays as a function of *relative* perturbations become useful, and the reader is advised to refer to the complete articles from which these decays were taken (Table 10.3).

Early hemoglobin and myoglobin fluorescence lifetime decay studies revealed two major lifetime components on the picosecond to nanosecond scale (Itoh *et al.*, 1981; Hochstrasser and Negus, 1984). Recent studies, using more sensitive detectors and improved data analysis, consistently resolve three lifetime components in this timescale (e.g., Szabo *et al.*, 1984, 1989;

Table 10.2. Factors to Control in Heme Protein Fluorescence Measurements

- sample purity and preparation
- buffers that do not alter the fluorescence (e.g., Tris) nor drive the conformational equilibrium (phosphate, chloride)
- native state vs. unfolded or denatured state (pH, temperature, light sources & heating)
- reflectance, Raman, and Rayleigh scattering
- concentration dependent subunit dissociation
- the apoprotein is a new structural species
- mutants may exhibit altered conformation
- altered conformation may arise upon chemical modification with extrinsic fluorescent probes

Bucci *et al.*, 1988; Mizukoshi *et al.*, 1982; Janes *et al.*, 1987; Gryczynski *et al.*, 1997a & b; Gryczynski and Bucci, 1998) (Table 10.3). For hemoglobin, the average lifetimes are reported to vary with the ligation state (Bucci *et al.*, 1988; Gryczynski *et al.*, 1997a). Four lifetime components have been observed for the intrinsic protein fluorescence decay of some heme-proteins, such as in the giant (~4 million Da) acellular dodecameric hemoglobin of the earthworm, *Lumbricus terrestris* (Hirsch *et al.*, 1994a). A best-fit four component exponential decays have also been reported in other proteins and fluorescent oligonucleotides (Dahms and Szabo, 1995; Nordlund *et al.*, 1989; Hochstrasser *et al.*, 1994; Driscoll *et al.*, 1997), including the *single* Trp-containing horse heart apocytochrome-c (Vincent *et al.*, 1988).

#### *10.3.5.1. Interpretation of the Multiexponential Decays Remains Unresolved*

As a first start to evaluating and interpreting these picosecond to nanosecond decays, a number of approximations and average values are required for the calculation of resonance energy transfer. In such evaluations, it is often seen that only 1 or 2 quenching mechanisms are selected for consideration in these calculations. While such restriction may be necessary, given the complexity of contributing factors, the end result is that the assumptions become limited in validity, narrow the interplay of energy transfer mechanisms, and give rise to interpretations that may possibly be misleading.

The rate of energy transfer from a specific donor to a specific acceptor ($k_T$) is given by

$$k_T = (1/\tau_d)(R_0/r)^6 \tag{10.2}$$

where $\tau_d$ is the lifetime of the donor in the absence of the acceptor, r is the distance between the donor and acceptor dipoles, and $R_0$ is the Förster distance at which the efficiency of transfer is 50% (Lakowicz, 1999). At this distance, half of the donor molecules decay by energy transfer and the other half decay by radiative and non-radiative rates.

The transfer rate is calculated by

$$k_T = \left(r^{-6} J \kappa^2 n^{-4} \lambda_d\right) \times 8.71 \times 10^{23}\ \text{sec}^{-1} \quad \text{(Lakowicz, 1983)} \tag{10.3}$$

where $k_T$ is the transfer rate which is equal to the decay rate of the donor in the absence of the acceptor; J is the overlap integral or the degree of spectral overlap between the donor emission and the acceptor absorption; $\kappa^2$ is the factor describing the relative orientation in space of the transition dipoles

Table 10.3. Some Examples of Multiexponential Trp Decays Reported for Heme-Proteins

| Heme Protein | Trp | Conditions | $\tau_1$ (ps) | $f_1$ | $\alpha_1$ | $\tau_2$ (ns) | $f_2$ | $\alpha_2$ | $\tau_3$ (ns) | $f_3$ | $\alpha_3$ | $\tau_4$ (ns) | $f_4$ | $\alpha_4$ |
|---|---|---|---|---|---|---|---|---|---|---|---|---|---|---|
| *Myoglobins*: | | | | | | | | | | | | | | |
| [a]Sperm Whale (SW) Met Mb | 2 (A helix) | 0.025 M Tris | 111 | | 0.958 | 0.872 | | 0.030 | 3.080 | | 0.012 | | | |
| SW met azide Mb | 2 (A helix) | 0.2 M azide | 96 | | 0.853 | 0.222 | | 0.138 | 2.830 | | 0.005 | | | |
| Aplysia Met-azide Mb | 2 (A & H helices) | | 19 | | 0.975 | 0.610 | | 0.015 | 3.190 | | 0.010 | | | |
| [b]SW met Mb | 2 | 0.05 M Na-phos 0.1 M | 80 | | 0.970 | 1.064 | | 0.025 | 8.027 | 0.05 | | | | |
| Tuna Mb | 1 | NaCl, pH 7 | 83 | | 0.966 | 3.332 | | 0.034 | | | | | | |
| [c]SW Mb deoxy | 2 (Trps 7 & 14) | pH 7, 0.1 M phosphate | 18 | | | 0.106 | | | | | | | | |
| SW Mb CO | 2 | pH 7, 0.1 M phos | 23.4 | | | 0.125 | | | | | | | | |
| SW Mb oxy | 2 | pH7, 0.1 M phos | 24.4 | | | 0.122 | | | | | | | | |
| SW Mb met | 2 | pH 7, 0.1 M phos | 21.5 | | | 0.113 | | | | | | | | |
| [d]Horse heart Mb | 2 | pH 7, freq. Domain, global analysis | 40 | 0.149 | 0.509 | 0.116 | 0.394 | 0.463 | 1.363 | 0.21 | 0.021 | 4.822 | 0.247 | 0.007 |
| Horse heart Mb | 2 | pH 7, individual analysis | 35 | 0.142 | 0.550 | 0.130 | 0.407 | 0.425 | 1.491 | 0.198 | 0.018 | 4.894 | 0.253 | 0.007 |
| [e]recomb SW met MbW7F | 1 (Trp 14) | 0.05 M phos, pH 7 | 34 | | 0.968 | 1.368 | | 0.015 | 4.868 | | 0.017 | | | |
| recomb SW CO MbW7F | 1 (Trp 14) | 0.05 M phos, pH 7 | 19 | | 0.983 | 1.723 | | 0.013 | 5.102 | | 0.004 | | | |
| [f]apo horse Mb | 2 | pH 7, bis-Tris | 930 | 30 | | 2.02 | 47 | | 4.94 | 23 | | | | |
| *Hemoglobins*: | | | | | | | | | | | | | | |
| [g]met HbA | 6 | | 90 | 30 | | 1.900 | 37 | | 5.400 | 37 | | | | |
| Oxy HbA | 6 | | 90 | 24 | | 1.900 | 40 | | 5.400 | 37 | | | | |
| Deoxy HbA | 6 | | 70 | 30 | | 1.800 | 41 | | 4.900 | 29 | | | | |
| CO HbA | 6 | | 70 | 30 | | 1.800 | 45 | | 4.900 | 25 | | | | |

| | | | | | | | | | | | | | | |
|---|---|---|---|---|---|---|---|---|---|---|---|---|---|---|
| [h]HbA Subunits | | | | | | | | | | | | | | |
| beta (oxy) | 2 | | 95 | 50 | | 2.650 | 21 | | 6.500 | 29 | | | | |
| beta (CO) | 2 | 0.024 mM; 0.01 M Na-cacodylate, pH 7 | 90 | 41 | | 2.550 | 27 | | 6.450 | 32 | | | | |
| beta (deoxy) | 2 | | 90 | 50 | | 2.330 | 21 | | 6.330 | 29 | | | | |
| alpha (oxy) | 1 | | 80 | 80 | | 2.200 | 20 | | | | | | | |
| alpha (CO) | 1 | | 85 | 65 | | 2.300 | 35 | | | | | | | |
| alpha (deoxy) | 1 | | 65 | 24 | | 1.900 | 40 | 5.4 | 37 | | | | | |
| [i]oxy HbA | 6 | pH 6.6 bis-tris buffer, 0.05 mM conc. Hb | 30 | | 0.300 | 0.580 | | 2.330 | | | | | | |
| oxy HbA | | HPLC purified, pH 8.2, 0.1–0.05 mM heme | 25 | | | | | | | | | | | |
| [j]CO HbA | 6 | 0.2 mg/ml; [buffer not stated] | 12 | 0.217 | 0.617 | 0.031 | 0.272 | 0.326 | 4.405 | 0.119 | 0.001 | | | |
| CO HbA | | 30 mg/ml | 6 | 0.310 | 0.725 | 0.035 | 0.540 | 0.272 | 1.110 | 0.150 | 0.003 | | | |
| oxy HbA | | 30 mg/ml | 3 | 0.230 | 0.863 | 0.035 | 0.540 | 0.183 | 0.670 | 0.230 | 0.004 | | | |
| deoxy HbA | | 30 mg/ml | 2 | 0.250 | 0.852 | 0.028 | 0.660 | 0.147 | 0.820 | 0.090 | 0.001 | | | |
| [k]*L. terrestris* Hb | ~500 Trp per molecule | 0.08 mM Hb, 0.05 M Hepes, pH 7 | 40.6 | | 82.4 | 0.285 | | 12.53 | 0.883 | | 3.7 | 3.782 | | 1.3 |
| *Other heme-proteins*: | | | | | | | | | | | | | | |
| [l]horseradish peroxidase | 1 | pH 6.6 phos buffer | 45 | | 97 | 1.4 | | 2 | 4.6 | | 1 | | | |
| [m]apo-cytochrome C | | 0.139 mM | 210 | 20 | | 1.1 | 27 | | 2.92 | 39 | | 5.08 | 14 | |
| *Tryptophan*: | | | | | | | | | | | | | | |
| [n]single free Trp | 1 | 0.1 M Na-phos, pH7 | 3190 | | 0.70 | 0.850 | | 0.30 | | | | | | |

$\tau$, lifetime; $f$, fractional intensity; $\alpha$, relative amplitude.
[a]Janes *et al.*, 1987; [b]Bismuto *et al.*, 1989; [c]Willis *et al.*, 1990; [d]Gryczynski *et al.*, 1997; [e]Gryczynski & Bucci, 1998; [f]Haouz *et al.*, 1998; [g]Szabo *et al.*, 1984; [h]Albani *et al.*, 1985; [i]Szabo *et al.*, 1989; [j]Gryczynski *et al.*, 1997; [k]Hirsch *et al.*, 1994; [l]Das & Mazumbar, 1995; [m]Vincent *et al.*, 1988; [n]Ross *et al.*, 1981.

of the donor and acceptor; n is the refractive index of the medium; and $\lambda_d = \phi_d/\tau_d$, the quantum yield of the donor in the absence of the acceptor divided by the lifetime of the donor in the absence of the receptor. The Förster distance is calculated by

$$R_0 = 9.79 \times 10^3 (\kappa^2 n^{-4} \phi_d J)^{1/6} \quad (\text{in Å}) \tag{10.4}$$

(For a derivation of these equations, see Lakowicz, 1999).

As pointed out earlier, $\kappa_T$ and $\tau_d$ are dependent on knowing the emission and lifetime for Trp in the protein *without* the acceptor (heme) present. Thus, the nonhomologous structural nature of the apoprotein coupled with the incomplete general understanding of Trp fluorescence emission lifetimes and properties place the assumed values for these components with great uncertainty. This is emphasized by Alpert *et al.* (1980), "Transfer efficiency depends on the degree of overlap between the donor emission and acceptor absorption. In this case, (*re.* 6 Trp in the hemoglobin tetramer), it is difficult to assess the precise degree of overlap since we are not certain that the absorption spectrum of the heme protein is the simple addition of the absorption spectra of the apoprotein and free heme." With these difficulties in mind, several independent efforts have been made to assign a source and calculate expected lifetimes for each of the Trp residues in hemoglobin.

Early lifetime studies and theoretical calculations used to explain the multilifetime components of myoglobin led to the controversial conclusion that the long-lived nanosecond component had to be due to an impurity, resulting in the dismissal of steady-state emission observations (Hochstrasser and Negus, 1984; Janes *et al.*, 1987). However, to date, there is no clear correlation with heme-protein fluorescence lifetimes and steady-state emission. The controversial dismissal of the steady state fluorescence emission was based upon calculations assuming a 20 ns lifetime for free Trp, and computerized simulations, using crystal structures from the Brookhaven Data Base leading to their conclusion that a Trp geometric relation to the heme prohibiting resonance transfer would not occur: calculation of the transfer rate as a function of the angle of rotation about the Trp C($\beta$)—C($\gamma$) bond showed no region where transfer times were not predicted to be subnanosecond (Janes *et al.*, 1987). However, subsequent to this study, three fluorescence lifetime decays were measured for hemoglobin and concluded to arise from different Trp-heme conformations/rotamers (Szabo *et al.*, 1984; Janes *et al.*, 1987).

Given the magnitude of quenching by heme moieties, an explanation for the mechanism of heme-protein fluorescence became paramount. The question of the nanosecond components was revisited by a study of the fluorescence decay of sperm whale myoglobin (2 Trp) and Tuna Mb (1 Trp)

(Bismuto *et al.*, 1989). Frequency-domain fluorometry with data analysis using a continuous Lorenzian distribution of lifetimes yielded two components for the *single* Trp of Tuna myoglobin (83 ps and 3.3 ns) and three components for two Trp containing sperm whale myoglobin (1 in the sub ns time scale and 2 in the ns range). It was concluded that the long lived component may arise from a conformational state (different from the native) in which geometric factors do not allow energy transfer via Forster coupling from Trp to heme. This concept was supported by others: "the probability exists that improbable conformational attitude of the tryptophans substantially reduced the energy transfer to the heme are responsible for ns emission. The impurities may be slowing relaxing conformers of hemoglobin" (Bucci *et al.*, 1988).

Calculating the distance to the hemes in hemoglobin and using average transfer rates, it was estimated that $\beta$15 Trp is the least quenched Trp (~50 fold quenched, followed by $\alpha$14 Trp (~70-fold quenched) and $\beta$37 Trp (~200 times quenched) (Gryczynski *et al.*, 1992). The expected lifetimes were respectively 10 ps, 40 ps, and 30 ps. It was concluded that the proximity of $\beta$37 Trp to 2 hemes, one in the same subunit and the other in the $\alpha$ subunit of the opposite dimer is the cause of this greater estimate of quenching. These energy transfer estimates, requiring numerous assumptions, led to the debated conclusion that emission from $\beta$37 Trp was totally quenched and that $\beta$15 Trp was the primary source of the emission (Gryczynski *et al.*, 1992), contrary to the findings of steady-state emission of hemoglobin Trp and allosteric mutants (Hirsch *et al.*, 1980a; Hirsch and Nagel, 1981; Itoh *et al.*, 1981; Mizukoshi *et al.*, 1982).

Following this report, a detailed quantitative model of heme quenching mechanisms in hemoglobin, myoglobin, and recombinant myoglobins, with consideration of the roles of heme exchange, alterations in heme orientation in the pocket, and heme loss, was presented (Gryczynski *et al.*, 1997a & b; Gryczynski and Bucci, 1998). They showed that measured lifetimes agreed with and could be explained as a function of heme orientation: *Species I:* normal heme as in the crystal structure has the shortest lifetimes (ps); *Species II*: inverted heme rotated 180° around the $\alpha$-$\gamma$—meso-axis of the porphyrin accounts for the few hundred picosecond lifetimes; *Species III*: reversibly dissociated hemes accounts for the nanosecond component. These attempts to unravel the role of heme orientation as a function of the Trp lifetimes provided important insights and provocative theorizing.

While the above model fits their data and theoretical calculations, and provides an excellent start to defining the true nature of heme fluctuations found in proteins, as pointed out earlier, the assumptions employed restrict the viewing of other possible mechanisms. The established concept of heme exchange (Bunn and Jandl, 1966), in addition, implies significant inherent heme mobility that could lead to intermediate orientations unable to act as a donor in Trp energy transfer. Solution studies demonstrate that the heme moiety fluctuates in terms of its structural dynamics and equilibrium (Asher, 1981; Friedman, 1994; Carlson *et al.*, 1994 &1996). An equilibrium of multistate iron-heme conformations are demonstrated for myoglobins by various spectroscopic techniques where environmental changes modulate the equilibrium (Longa, 1998; Chance *et al.*, 1996). In these analyses, the specific kind of heme fluctuations permitted will be a function of the specific protein structure surrounding the pocket.

The calculations presented by Gryczynski *et al.* (1997a) depended on the assumption that the lifetimes of both Trp residues in myoglobin were 4.8 ns in the absence of heme, thereby not accounting for the multi-lifetimes intrinsic to single and multiple Trp heme *and non-heme* proteins. Hence, any model of heme conformation to explain the multiple lifetimes of heme-proteins must be evaluated and discussed in conjunction with the known multiple Trp lifetime decays observed for *non-heme* proteins (discussed below in more detail). While the authors recognized that the simulations did not take into account possible fluctuations of the tryptophan residue, they cited evidence (Hochstrasser and Negus, 1984) supporting the concept that the degrees of freedom are limited. Such concepts and estimates are based upon heme-protein crystallographic structures found in the Brookhaven Data Base. The use of the crystal structure is necessary and legitimate in that the micro-environment of Trp as reported by fluorescence spectroscopy is consistent with that reported for known crystal structures thus supporting the utility of fluorescence spectroscopy (Hasselbacher *et al.*, 1995; Albani, 1998). Nonetheless, if taken absolutely, this approach imposes a restriction upon the molecule and negates the purpose of spectroscopic tools to probe and dissect out *the dynamics of solution-active* protein structure which may allow alternative/additional conformations other than that imposed by crystal packing constraints. Likewise, the concept of Trp side chain conformational heterogeneity (e.g., rotamers) weakens the absolute utility of using a fixed orientation for calculations of energy transfer, but which can be useful as a first approximation (Smith *et al.*, 1986; Ponder and Richards, 1987; Dahms and Szabo, 1995 and 1997). Side chain heterogeneity in crystals and solution and its relationship to function is under extensive investigation, and may have to be considered on an individual protein basis. Such knowledge will help present alternative mechanisms needed to account for empirical steady-state heme-protein fluorescence emission findings.

The problem is further compounded by the intrinsic nature of tryptophan fluorescence emission itself. It was recognized early that free Trp and its derivatives exhibit more than one lifetime (Grinvald and Steinberg, 1976; for reviews, see Beecham and Brand, 1985; Eftink, 1991). The mechanism(s) behind the complex decay of Trp and its derivatives, and the subsequent interpretation of fluorescence data, are subjects of current investigation (Annual Meeting of the Biophysical Society, Fluorescence Subgroup Meeting, February 1999). A creative approach has been in the design of constrained Trp derivatives to relate excited-state properties directly to structure (McLaughlin and Barkley, 1997; McMahon *et al.*, 1997; Chen and Barkley, 1998). As summarized by Callis (1999), evidence has pointed to varying explanations such as: (1) rotamers and other conformational states (Ross *et al.*, 1981; Szabo and Rayner, 1980; Ross *et al.*, 1992; Willis *et al.*, 1994; McLaughlin and Barkley, 1997; Bialik *et al.*, 1998); (2) relaxation models, where the spectra shifts in time because of relaxation about the large $^1L_a$ excited-state dipole; (3) the dark-state model, where the excited electron finds its way back to the ground state; and (4) solvent effects. There is the potential for amino acid quenching by excited state proton transfer (Lys and Tyr) and excited-state electron transfer (Gln, Asn, Glu, Asp, Cys and His) (Chen and Barkley, 1998). Exiplex formation could also contribute to the complex decays (Beecham and Brand, 1985; Eftink, 1991), and it may be possible that all of the above play a role (Brand, 1999). Mechanisms proposed to explain the multiexponential decay of tryptophan are discussed at length in other chapters in this book.

To recapitulate, unraveling the origin and mechanisms of emission from the multiple Trp residues in a protein with a heme moiety becomes extremely complicated. Despite this knowledge, publications still appear with an assignment of each lifetime to a specific Trp residue (Das *et al.*, 1998). Therefore, the following considerations become necessary: *(1) the structural and hence fluorescent inequivalence of the apoprotein and its complementary heme-protein; (2) the intrinsic multiexponential decays of Trp and Trp derivatives seen in single and non-heme multiple Trp proteins, (3) quenching by the hemes; and (4) other quenching mechanisms such as solvent effects*. These factors are discussed in detail by Beecham and Brand, 1985; Eftink, 1991; and in other chapters contained in this book). Hence, caution in the specific assignment of heme-protein time-resolved data is urged.

With respect to solvent effects, it is worth noting here that phosphate has been reported to quench both indole and phenol fluorescence, with the monoanion ($H_2PO_4^-$) more effective in indole quenching than the dianion ($HPO_4^{-2}$) (Williams and Bridges, 1964). Compounded with the role of phosphate as a hemoglobin allosteric effector (Imai, 1982), and the frequent use of phosphate buffer in the purification of hemoglobins and in experimental studies, conflicting data and interpretation may result.

Another source of Trp quenching that may be relevant to heme-protein fluorescence arises from atypical hydrogen bonds that form between an indole amino proton with the proximate phenyl ring, where the two aromatic residues lie within a distance of about 3.5 Å (Nanda and Brand, 1999;

Rouviere *et al.*, 1997; Suywaiyan and Klein, 1989; Levitt and Perutz, 1988). This distance is somewhat larger than that seen for typical hydrogen bonds found in hemoglobin (~2.5 Å and less). The aromatic microenvironments of the 3 Trp composing each $\alpha\beta$ dimer of hemoglobin suggests that this quenching mechanism may be operative (Figures 10.6a–c). While some of the distances are greater than that required for this atypical H-bond formation, energy transfer mechanisms may play a significant role. This hypothesis requires experimental examination.

A hydrogen bond quenching mechanism lends itself to the consideration of recently reported differences in the intrinsic fluorescence of human hemoglobin $\beta 6$ mutants compared to HbA: the relative emission intensity is consistently observed in the order of HbA > HbC ($\beta 6$ Glu → Lys) > HbS ($\beta 6$ Glu → Val) (Hirsch *et al.*, 1999) (Figure 10.7). UV resonance Raman (UVRR) spectroscopic studies by independent laboratories indicate that the H-bond between $\beta 15$ Trp—$\beta 72$ Ser of the A-helix is altered in these $\beta 6$ mutants (Hirsch *et al.*, 1996, 1999; Sokolov and Mukerji, 1998). The possibility exists that the atypical hydrogen bond found in these mutants quench the indole fluorescence in a manner described above, and may serve to explain the differences in fluorescence intensity emitted by these hemoglobin mutants when compared to HbA. While this hypothesis is speculative at this point in time, it highlights the need for multiple factors to be accounted for in the evaluation of heme-protein fluorescence differences.

## 10.4. Extrinsic Fluorescence Probing

*Extrinsic fluorescence* generally refers to the emission of a fluorescent compound bound covalently or non-covalently to a protein, for example, for purposes of probing site-specific residues or microdomains. Usually, extrinsic fluorescence probes offer greater quantum yields and excitation and emission wavelengths that are easier to use and which may serve in resonance energy transfer measurements (Haugland, 1983; Weiss, 1999). As noted earlier, the use of ANS serves to detect the presence of apohemoglobin in hemoglobin preparations (Alpert *et al.*, 1980; Hirsch and Peisach, 1986). Fluorescence studies of ANS coupled to apohemoglobin and apohemoglobin labeled at $\beta 93$ Cys with fluorescein demonstrated that the apohemoglobin dimer (see above) exhibits little change in secondary structure compared to the $\alpha\beta$ dimer of the intact hemoglobin tetramer, except for a slight shrinking of the molecule (Sassaroli *et al.*, 1984). Fluorescence lifetime and high pressure studies of ANS and other similar derivatives serve to characterize conformational substates of different species of apomyoglobins

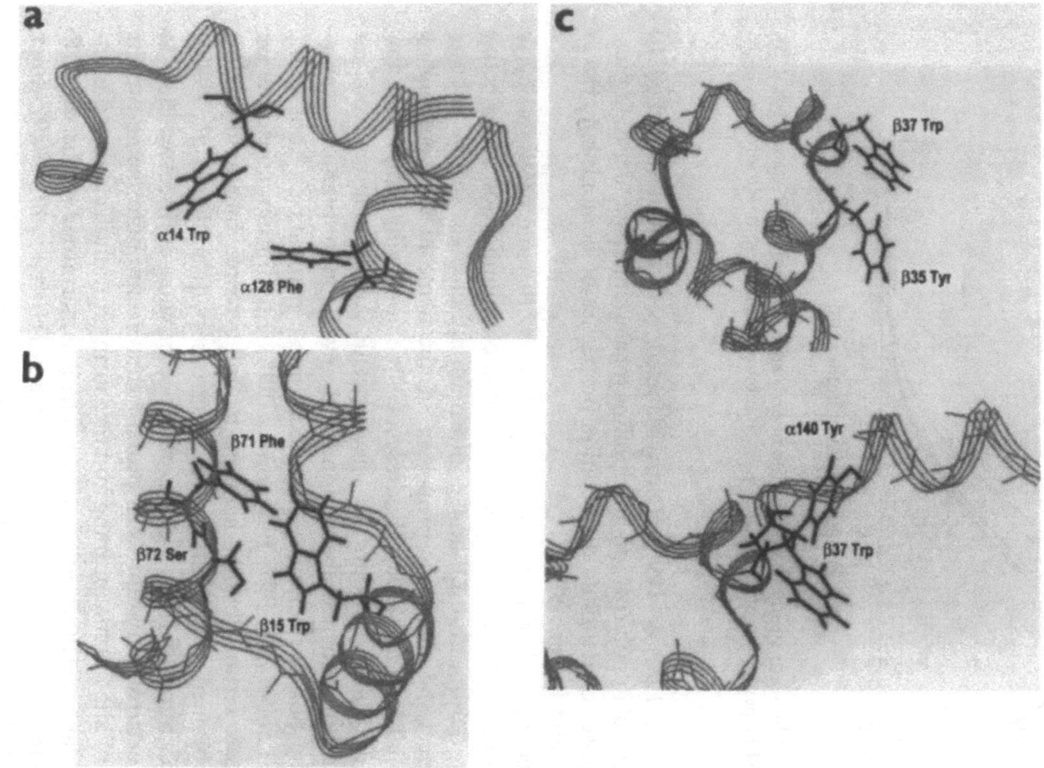


Figure 10.6a–c. The aromatic microenvironment of the tryptophans found in deoxy HbA. (a) $\alpha$14 Trp located ~5Å from Phe 128; (b) $\beta$15 Trp located ~3Å from Phe 71; and $\beta$72 Ser located ~2Å from $\beta$15 Trp; (c) $\beta$37 Trp located ~2Å from $\beta$35 Tyr; and $\alpha$140 Tyr located ~3Å from $\beta$37 Trp. The distances between the side chains may vary up to a few Å depending upon the atom to atom distance measured. Note that the residues depicted are *not proportional* to the protein backbone secondary structures that vary in relative size within the figures presented. These images, courtesy of Dr. Marvin Rich, Department of Biology, New York University, Washington Square, NY, are obtained from the deoxy HbA structure (pdp file 2 hhb of the Brookhaven Data Base, Fermi *et al*., J. Mol. Biol. 175, 159, 1984).

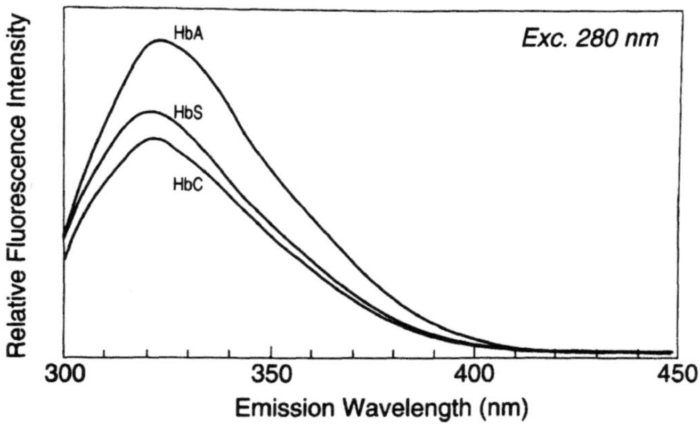


Figure 10.7. The front-face steady -state intrinsic fluorescence emission of R-state HbA, HbS, and HbC (0.05 M Hepes buffer, pH 7.35, 25 °C). Excitation is 280 nm. From: Hirsch *et al.*, 1999.

(Bismuto *et al.*, 1989 & 1996). Dansyl-labeled hemoglobin (attached to amines) are useful in polarization and lifetime measurements under high pressure for purposes of detailing dissociation properties of hemoglobin (Pin *et al.*, 1990; Pin and Royer, 1994).

Fluorescent porphyrins [zinc protoporphyrin (ZPP) and protoporphyrin IX (PPIX)] used to probe heme pocket—globin communication are effective in addressing the question of how conformational changes in one subunit ultimately affect the electronic properties of the heme in the neighboring subunit (Sudhakar *et al.*, 1998). Fluorescence line narrowing, employing low temperature and laser excitation to select specific subpopulations from the inhomogenously broadened absorption band, reveals more than one configuration of the porphyrin moiety in cytochrome-c peroxidase (Anni *et al.*, 1994; Vanderkooi *et al.*, 1997; Fidy *et al.*, 1998). Equilibrium constants for PPIX binding to serum albumin, hemopexin, and cytosolic fatty acid binding protein are obtained using fluorescence spectroscopy (Knobler *et al.*, 1989).

The application of front-face fluorescence provides a direct window to monitor the extrinsic emission of a probe bound to an intact heme-protein for purposes of site-specific probing and measuring intramolecular and intermolecular distances (Hirsch *et al.*, 1986). This has permitted studies of ZPP binding to hemoglobin at non-heme pocket sites (Hirsch *et al.*, 1989), direct monitoring of the $\beta$93 Cys site, and direct monitoring of the central cavity of hemoglobin as a function of allosteric effector binding and perturbation (Hirsch *et al.*, 1986; Gottfried *et al.*, 1997; Hirsch *et al.*, 1999). The fluorescein labeled $\beta$93 site, monitors changes in the R $\rightarrow$ T transition, and provides oxygen dissociation rate constants when used in stopped flow measurements

(Hirsch and Nagel, 1989). Extrinsic probes serve to corroborate implied alterations of the central cavity DPG binding site of $\beta6$ hemoglobin mutants. 8-hydroxy-1,3,6-pyrene trisulfonate (HPT), an established DPG fluorescent analog (MacQuarrie and Gibson, 1971, 1972), provided steady-state fluorescence evidence supporting the hypothesis that the central cavity of $\beta6$ mutants is altered (Hirsch *et al.*, 1999). Furthermore, direct lifetime measurements of HPT binding to hemoglobin defined differences in central cavity crosslinked hemoglobins (designed with the purpose of serving as a therapeutic oxygen carrier) (Gottfried, 1997), and probes the allosteric equilibrium (Marden *et al.*, 1986; Serbanescu *et al.*, 1998). Extrinsic fluorescence probing also yields quantitative and qualitative measurements of polycyclic aromatic hydrocarbon hemoglobin adducts (Day and Singh, 1994).

## 10.5. Quenching of Extrinsic Fluorescence upon Binding by Heme or Heme-proteins

Fluorescence quenching upon binding to heme proteins, when studied in right-angle optical configuration, is a useful tool to calculate binding constants and determine the nature of the interacting species. The quantitative assessment of DPG and IHP binding constants was assessed with the use of the fluorescent HPT upon quenching when bound to hemoglobin (MacQuarrie and Gibson, 1971, 1972). Similarly, quenching of the fluorescent allosteric effector, $\beta$-naphthyl triphosphate upon binding to HbA, revealed evidence in favor of the controversial three-state allosteric model of hemoglobin (Horiuchi, 1982; Horiuchi and Asai, 1983). Hemoglobin binding to the red cell membrane is quantitated by the quenching of fluorescence labeled membranes (Eisinger *et al.*, 1984). Hemoglobin and cytochrome-c interactions with lipids are defined by quenching of a fluorescent probe upon binding to the heme-protein (Gorbenko, 1998).

The finding that haptoglobin binds only to hemoglobin dimers (as opposed to tetrameric hemoglobins) was established by studying the quenching of haptoglobin fluorescence upon binding to hemoglobin. Varying the concentration of hemoglobin suggested that haptoglobin only bound to the hemoglobin dimer. Haptoglobin only interacted with the $\alpha$ chain of the $\alpha\beta$ subunit, and stopped flow fluorescence studies provided accurate binding rates (Nagel and Gibson, 1967, 1971).

The accessibility of various regions of hemoglobin and horseradish peroxidase (HRP) to oxygen diffusion was studied by fluorescence quenching of Trp and a fluorescent porphyrin under elevated pressure of oxygen (Coppey *et al.*, 1981; Jameson *et al.*, 1984; Vargas *et al.*, 1991). It was demonstrated that rapid structural fluctuations occur in the protein matrix of $Hb^{des\ Fe}$ with implications for oxygen escape. Oxygen exhibits a very different entry in HRP compared to hemoglobin (Coppey *et al.*, 1981). These results suggest significant differences in the heme pockets, implying that the heme steric structure differs in these two proteins as confirmed by Vanderkooi and associates (Anni *et al.*, 1994). These findings provide a basis to explain the unique functionalities of these heme-proteins.

## 10.6. Vital Novel Functions of Heme-Proteins Are Now Being Uncovered

The significance and multifunctional role of heme-proteins as regulators in processes other than oxygen transport or storage is first being uncovered and appreciated. Fluorescence studies demonstrated that the Z class of liver cytosolic fatty acid binding proteins preferentially bind heme than other forms of anions, reclassifying them as a heme-protein (Vincent and Eberhard, 1985). Even more provocative is the example of cytochrome c, widely known as an electron carrier in the respiratory pathway and normally present on the outer surface of the inner mitochondrial membrane. Recently, cytochrome-c has been assigned as a key player in apoptosis: upon its release from the mitochondria to the cytoplasm, it serves as a protease activator in the cascade of apototic events involving the cytoplasmic cysteine proteases (Ushamorov *et al.*, 1999). Fluorescence quenching studies are used to provide important structural information regarding cytochrome c folding kinetics: $80\,\mu$sec to 3 ms are detected, using an ultrarapid-mixing continuous flow fluorescence quenching of Trp to heme (Chan *et al.*, 1997).

Cytochrome P-450 helps to convert toxins and foreign lipid-soluble materials into harmless, and easily excreted substances, but converts other substances into carcinogens. Cytochrome P-$450_{BM3}$, from *Bacillus megaterium,* with 5 Trps exemplifies the challenge to define the Trp environments of such a protein: one clever strategy utilizes several fluorescence quenchers with differential environment accessibility as a function of alterations in Trp fluorescence lifetime decays (Khan *et al.*, 1997). It was found that the number of Trp residues accessible to ionic quenchers decreases on interaction of the substrate with the enzyme indicating that some of the Trps move towards the core of the protein upon substrate interaction.

To summarize, the illustrations cited in this chapter (certainly not comprehensive) demonstrate that while mechanisms of heme-protein and non-heme protein Trp fluorescence emission remain a subject of active investigation, fluorescence spectroscopy provides a tool to meet many of the challenging questions concerning solution-active structure-function interrelationships of the diverse, multifunctional, and vital heme-proteins.

## Acknowledgments

The author is grateful to Dr. William R. Laws for his many helpful discussions and critique of the manuscript. A special thanks to Dr. John P. Harrington for reviewing the final versions of the manuscript; and to Dr. Marvin Rich for providing the figures for the aromatic environments of the tryptophans in HbA. This work was supported in part by the National Institutes of Health R01HL58247, R01HL58038 and the AHA-Heritage Affiliate 9950989T.

## References

Ackers G. K., Johnson M. L., Mills F. C. and Ip S. H. Energetics of Oxygenation-Linked Subunit Interactions in Human Hemoglobin. Biochem Biophys Res Commun 1976; 69:135–142.

Albani J., Alpert B., Krajcarski D. T. and Szabo A. G. A Fluorescence Decay Time Study of Tryptophan in Isolated Hemoglobin Subunits. FEBS 1985; 182:302–304.

Albani J. R. Dynamics of Trp Residues in Crystals of Human A1—Acid Glycoprotein (Orosomucoid) Followed by Red-Edge Excitation Spectra. J Flour 1998; 8:213–224.

Alpert B. and Lopez-Delgado R. Fluorescence Lifetimes of Haem Proteins Excited into the Tryptophan Absorption Band with Synchrotron Radiation. Nature 1976; 263:445–446.

Alpert B., Jameson D. and Weber G. Tryptophan Emission From Human Hemoglobin and its Isolated Subunits. Photochemistry and Photobiology 1980; 31:1–4.

Anni H., Vanderkooi J. M., Sharp K. A., Yonetani T., Hopkins S. C., Herenyi L. and Fidy J. Electric Field and Conformational Effects of Cytochrome C and Solvent on Cytochrome C Peroxidase Studied by High-Resolution Fluorescence Spectroscopy. Biochemistry 1994; 33:3475–3486.

Antonini E. and Brunori M. *Hemoglobin and Myoglobin in their Reactions with Ligands*. Amsterdam: North Holland Publ, 1971.

Asakura T., Agarwal P. I., Relman D. A., McCray J. A., Chance B., Schwartz E., Friedman S. and Lubin B. Mechanical Instability of the Oxy-Form of Sickle Haemoglobin. Nature 1973; 244:437–438.

Asher S. Resonance Raman Spectroscopy of Hemoglobin. Methods Enzymol 1981; 76:371–413.

Asher S. A. UV Resonance Raman Studies of Molecular Structure and Dynamics: Applications in Physical and Biophysical Chemistry. Ann Rev Phys Chem 1988; 39:537–538.

Asher S. A. UV Resonance Raman Spectroscopy for Analytical, Physical, and Biophysical Chemistry. Analyt Chem 1993; 65:201A–210A.

Beechem J. M. and Brand L. Time-Resolved Fluorescence of Proteins. Ann Rev Biochem 1985; 54:43–71.

Benesch R. E., Ikeda S. and Benesch R. Reaction of Haptoglobin with Hemoglobin Covalently Cross-Linked Between the Alpha Beta Dimers. J Biol Chem 1976; 251:465–470.

Bialik C. N., Wolf B., Rachofsky E. L., Ross J. B. A. and Laws W. R. Dynamics of Biomolecules: Assignment of Local Motions by Fluorescence Anisotropy Decay. Biophys J 1998; 75:2564–2573.

Bismuto E., Irace G. and Gratton E. Multiple Conformational States in Myoglobin Revealed by Frequency Domain Fluorometry. Biochemistry 1989; 28:1508–1512.

Bismuto E., Sirangelo I. and Irace G. Conformational Substates of Myoglobin Detected by Extrinsic Dynamic Fluorescence Studies. Biochemistry 1989; 28:7542–7545.

Bismuto E., Sirangelo I., Irace G. and Gratton E. Pressure-Induced Perturbation of Apomyoglobin Structure: Fluorescence Studies on Native and Acidic Compact Forms. Biochemistry 1996; 35:1173–1178.

Brand L. Biological Fluorescence Subgroup Meeting, Jablonski Award, Annual Meeting of the Biophysical Society, February 13, 1999. Biophys J 1999, 76 (1, Part 2 of 2).

Bucci E., Malak H., Fronticelli C., Gryczynski I., Laczko G. and Lakowicz J. R. Time-Resolved Emission Spectra of Hemoglobin on the Picosecond Time-Scale. Bio Chem 1988; 32:187–198.

Bucci E., Gryczynski Z., Fronticelli C., Gryczynski I. and Lakowicz J. R. Fluorescence Intensity and Anisotropy Decays of the Intrinsic Tryptophan Emission of Hemoglobin Measured with a 10-Ghz Fluorometer Using Front-Face Geometry on a Free Liquid Surface. J Fluorescence 1992; 2:29–36.

Bunn H. F. and Forget B. G. *Hemoglobin: Molecular, Genetic and Clinical Aspects*. Phila. WB Saunders, 1986.

Bunn H. F. and Jandl J. H. Exchange of heme among hemoglobin molecules. Proc Natl Acad Sci USA 1966, 56:974–978.

Burstein E. A., Vedenkina N. S. and Ivkova M. N. Fluorescence and the Location of Tryptophan Residues in Protein Molecules. Photochem and Photobiol 1973; 18:263–279.

Callis P. R. Origins of Nonexponential Tryptophan Fluorescence Decay in Proteins. Biological Fluorescence Subgroup Meeting, Annual Meeting of the Biophysical Society, February 13, 1999. Biophys J 1999, 76(1, Part 2 of 2):A2.

Callis P. R. and Burgess B. K. Tryptophan Fluorescence Shifts in Proteins from Hybrid Simulations: An Electrostatic Approach. J Phys Chem 1997; 101:9429–9432.

Carlson M. L., Regan R., Elber R., Li H., Phillips G. N. Jr., Olson J. and Gibson Q. H. Nitric Oxide Recombination to Double Mutants of Myoglobin: Role of Ligand Diffusion in a Fluctuating Heme Pocket. Biochemistry 1994; 33:10597–10606.

Carlson M. L., Regan R. M. and Gibson Q. H. Distal Cavity Fluctuations in Myoglobin: Protein Motion and Ligand Diffusion. Biochemistry 1996; 35:1125–1136.

Chan C. K., Hu Y., Takahashi S., Rousseau D. L., Eaton W. A. and Hofrichter J. Submillisecond Protein Folding Kinetics Studied by Ultrarapid Mixing. Proc Natl Acad Sci USA 1997; 94:1779–1784.

Chance M. R., Miller L. M., Fischetti R. F., Scheuring E., Huang W. X., Sclavi B., Hai Y. and Sullivan M. Global Mapping of Structural Solutions Provided by the Extended X-ray Absorption Fine Structure ab initio code FEFF 6.01: Structure of the Cryogenic Photoproduct of the Myoglobin-Carbon Monoxide Complex. Biochemistry 1996; 35:9014–9023.

Chen Y. and Barkley, C. Y. Toward Understanding Tryptophan Fluorescence in Proteins. Biochemistry 1998; 37:9976–9982.

Chiu F., Vasudevan G., Morris A. and McDonald M. J. Fluorescence Studies of Human Semi-Beta-Hemoglobin Assembly. Biochem Biophys Res Commun 1998; 242:365–368.

Cho N., Song S. and Asher S. A. UV Resonance Raman and Excited-State Relaxation Rate Studies of Hemoglobin. Biochemistry 1994; 33:5932–5941.

Chothia C., Wodak S. and Janin J. Role of Subunit Interfaces in the Allosteric Mechanism of Hemoglobin. Proc Natl Acad Sci USA 1976; 73:3793–3797.

Coppey M., Jameson D. M. and Alpert B. Oxygen Diffusion through Hemoglobin and HbdesFe: Quenching of the Tryptophan and Porphyrin Emissions. FEBS Lett 1981; 126:191–194.

Dahms T. E. S. and Szabo A. G. Probing Local Secondary Structure by Fluorescence: Time-Resolved and Circular Dichroism Studies of Highly Purified Neurotoxins. Biophys J 1995; 69:569–576.

Dahms T. E. S. and Szabo A. G. Conformational Heterogeneity in Crystalline Proteins: Time-Resolved Fluorescence Studies. Methods Enzymol 1997; 278:202–221.

Das T. K. and Mazumdar S. pH-Induced Conformational Perturbation in Horseradish Peroxidase Picosecond Tryptophan Fluorescence Studies on Native and Cyanide-Modified Enzymes. J Biochem 1995; 227:823–828.

Das T. K., Mazumdar S. and Mitra S. Characterization of a Partially Unfolded Structure of Cytochrome C Induced by Sodium Dodecyl Sulphate and the Kinetics of its Refolding. Eur J Biochem 1998; 254:662–670.

Day B. W. and Singh K. Analysis of Hemoglobin Adducts by Fluorescence Spectroscopy. Meth Enzymol 1994; 231:674–681.

Driscoll S. L., Hawkins M. E., Balis F. M., Pfleiderer W. and Laws W. R. Fluorescence Properties of a New Guanosine Analog Incorporated into Small Oligonucleotides. Biophys J 1997; 73:3277–3286.

Eftink, Maurice R. "Fluorescence Techniques for Studying Protein Structure" In *Methods of Biochemical Analysis.* New York: John Wiley and Sons, Inc., 1991. Clarence H. Suelter, ed.

Eisinger J. Intramolecular Energy Transfer in Adrenocorticotropin. Biochemistry 1969; 8:3902–3908.

Eisinger J. and Flores J. Front-Face Fluorometry of Liquid Samples. Analytical Biochem 1979; 94:15–21.

Eisinger J., Flores J. and Bookchin R. M. The Cytosol-Membrane Interface of Normal and Sickle Erythrocytes. Effect of Hemoglobin Deoxygenation and Sickling. J Biol Chem 1984; 259:7169–7177.

Elbaum D., Harrington J. and Nagel R. L. Surface Activity of Hemoglobin S. Erythrocyte Structure and Function 1976a; 311–328.

Elbaum D., Harrington J., Roth E. F. Jr and Nagel R. L. Surface Activity of Hemoglobin S and Other Human Hemoglobin Variants. Biochim Biophys Acta 1976b; 427:57–69.

Fermi G., Perutz M. F., Shaanan B. and Fourme R. The Crystal Structure of Human Deoxyhaemoglobin at 1.74 A Resolution. J Mol Biol 1984; 175:159–174.

Fidy J., Laberge M., Kaposi A. D. and Vanderkooi J. M. Fluorescence Line Narrowing Applied to the Study of Proteins. Biochim Biophys Acta 1998; 1386:331–351.

Fontaine M. P., Jameson D. M. and Alpert B. Tryptophan-Heme Energy Transfer in Human Hemoglobin: Dependence Upon the State of the Iron. FEBS Letters 1980; 116:310–314.

Forster T. Intermolecular Energy Migration and Fluorescence. Ann Phys (Leipzig) 1948, 2:55–75.

Friedman J. M. Time-Resolved Resonance Raman Spectroscopy as Probe of Structure, Dynamics, and Reactivity in Hemoglobin. Methods Enzymol 1994; 232:205–231.

Garrick L. M., McDonald M. J., Shapiro R., Bleichman M., McManus M. and Bunn H. F. Structural Analysis of the Minor Human Hemoglobin Components: Hb AIa1, Hb AIa2 and Hb A1b. Eur J Biochem 1980; 106:353–359.

Gorbenko G. P. Resonance Energy Transfer Study of Hemoglobin and Cytochrome C Complexes with Lipids. Biochim Biophys Acta 1998; 1409:12–24.

Gottfried D. S., Juszczak L. J., Fataliev N. A., Acharya A. S., Hirsch R. E. and Friedman J. M. Probing the Hemoglobin Central Cavity by Direct Quantification of Effector Binding using Fluorescence Lifetime Methods. J Biol Chem 1997; 272:1571–1578.

Grinvald A. and Steinberg I. Z. The Fluorescence Decay of Tryptophan Residues in Native and Denatured Proteins. Biochim Biophys Acta 1976; 427:663–678.

Gryczynski Z., Tenenholz T. and Bucci E. Rates of Energy Transfer Between Tryptophans and Hemes in Hemoglobin, Assuming that the Heme is a Planar Oscillator. Biophys J 1992; 63:648–653.

Gryczynski Z., Lubkowski J. and Bucci E. Intrinsic Fluorescence of Hemoglobins and Myoglobins. Methods Enzymol 1997a; 278:538–569.

Gryczynski Z., Beretta S., Lubkowski J., Razynska A., Gryczynski I. and Bucci E. Time-Resolved Fluorescence of Hemoglobin Species. Biophys Chem 1997b; 64:81–91.

Gryczynski Z. and Bucci E. Time Resolved Emissions in the Picosecond Range of Single Tryptophan Recombinant Myoglobins Reveal the Presence of Long Range Heme Protein Interactions. Biophys Chem 1998; 74:187–196.

Hamada D., Hoshino M., Kataoka M., Fink A. L. and Goto Y. Intermediate Conformational States of Apocytochrome C. Biochemistry 1993; 32:10351–10358.

Haouz A., Glandieres J. M., Zentz C., Pin S., Ramstein J., Tauc P., Brochon J. C. and Alpert B. Solvent Effects on Horse Apomyoglobin Dynamics. Biochemistry 1998; 37:3013–3019.

Harrington J. P. and Hirsch R. E. Analysis of the Acid and Alkaline Dissociation of Earthworm Hemoglobin, *Lumbricus terrestris*, by Front-Face Fluorescence Spectroscopy. Biochim Biophys Acta 1991; 1076:351–358.

Hasselbacher C. A., Rusinova E., Waxman E., Rusinova R., Kohanski R. A., Lam W., Guha A., Du J., Lin T. C., Polikarpov I., Boys C. W. G., Nemerson Y., Konigsberg W. H. and Ross J. B. A. Environments of the Four Tryptophans in the Extracellular Domain of Human Tissue Factor: Comparison of Results from Absorption and Fluorescence Difference Spectra of Tryptophan Replacement Mutants with the Crystal Structure of the Wild-Type Protein. Biophys J 1995; 69:20–29.

Haughland R. P. (1983) "Covalent Fluorescent Probes" in *Excited States of Biopolymers* ed. RF Steiner pp. 29–58, Plenum Press, NY.

Henry E. R., Eaton W. A. and Hochstrasser R. M. Molecular Dynamics Simulations of Cooling in Laser-Excited Heme Proteins. Proc Natl Acad Sci USA 1986; 83:8982–8986.

Herskovits T. T., Cavanagh S. M. and San George R. C. Light-Scattering Investigations of the Subunit Dissociation of Human Hemoglobin A. Effects of Various Neutral Salts. Biochemistry 1977; 16:5795–5801.

Hirsch R. E., Zukin R. S. and Nagel R. L. Intrinsic Fluorescence Emission of Intact Oxy Hemoglobins. Biochem Biophys Res Comm 1980a; 93:432–439.

Hirsch R. E., Elbaum D., Brody S. S. and Nagel R. L. Hemoglobin-A and Hemoglobin-S Films at an Air-Water Interface: Absorption Spectra Studies. J Colloid and Interface Sci 1980b; 78: 212–216.

Hirsch R. E. and Nagel R. L. Conformational Studies of Hemoglobins Using Intrinsic Fluorescence Measurements. J Biol Chem 1981; 256:1080–1083.

Hirsch R. E., Squires N. A., Discepola C. and Nagel R. L. The Detection of Hemoglobin Dimers by Fluorescence. Biochem Biophys Res Comm 1983; 116:712–718.

Hirsch R. E., San George R. C. and Nagel R. L. Intrinsic Fluorimetric Determination of the Stable State of Aggregation in Hemoglobins. Anal Biochem 1985, 149:415–420.

Hirsch R. E., Zukin R. S. and Nagel R. L. Steady-State Fluorescence Emission from the Fluorescent Probe, 5-Iodoacetamidofluorescein, Bound to Hemoglobin. Biochem Biophys Res Comm 1986; 138:489–495.

Hirsch R. E. and Peisach J. A Comparison of the Intrinsic Fluorescence of Red Kangaroo, Horse and Sperm Whale Met-Myoglobins. Biochim Biophys Acta 1986; 872:147–153.

Hirsch R. E. and Nagel R. L. Stopped-Flow Front-Face Fluorometer: A Prototype Design to Measure Hemoglobin R → T Transition Kinetics. Analyt Biochem 1989; 176:19–21.

Hirsch R. E., Lin M. J. and Park C. M. The Interaction of Zinc Protoporphyrin with Intact Oxy Hemoglobin. Biochemistry 1989; 28:1851–1855.

Hirsch R. E., Harrington J. P. and Scarlata S. F. The Differential Effects of Carbon Dioxide and Oxygen on the Pressure Dissociation of *Lumbricus terrestris* Hemoglobin. Biochim Biophys Acta 1993; 1161:285–290.

Hirsch R. E. Front-Face Fluorescence Spectroscopy of Hemoglobins. Meth Enzymol 1994a; 232:231–246.

Hirsch R. E., Vidugiris G. J. A., Friedman J. M. and Harrington J. P. Alteration of Tryptophan Fluorescence Properties Upon Dissociation of *Lumbricus terrestris* Hemoglobin. Biochim Biophys Acta 1994a; 1205:248–251.

Hirsch R. E., Friedman J. M., Harrington J. R. and Scarlata S. F. Stability of a Potential Blood Substitute, HbXL99$\alpha$ Under High Pressure. Biochem Biophys Res Commun 1994b; 200:11635–1640.

Hirsch R. E., Lin M. J., Vidugiris G. J., Huang S., Friedman J. M. and Nagel R. L. Conformational Changes in Oxyhemoglobin C (Glu beta 6 → Lys) Detected by Spectroscopic Probing. J Biol Chem 1996; 271:372–375.

Hirsch R. E., Rybicki A. C., Fataliev N. A., Lin M. J., Friedman J. M. and Nagel R. L. A Potential Determinant of Enhanced Crystallization of HbC: Spectroscopic and Functional Evidence of an Alteration in the Central Cavity of Oxy HbC. Br J Haematol 1997; 98:583–588.

Hirsch R. E., Juszczak L. J., Fataliev N. A., Friedman J. M. and Nagel R. L. Solution-Active Structural Alterations in Liganded Hemoglobins C ($\beta$6 Glu → Lys) and S ($\beta$6 Glu → Val). J Biol Chem 1999; 274:13777–13782.

Hochstrasser R. M. and Negus D. K. Picosecond Fluorescence Decay of Tryptophans in Myoglobin. Proc Natl Acad Sci USA 1984; 81:4399–4403.

Hochstrasser R. A., Carver T. E., Sowers L. C. and Millar D. P. Melting of a DNA Helix Terminus within the Active site of a DNA Polymerase. Biochemistry 1994; 33:11971–11979.

Horiuchi K. Some Characteristics of $\beta$-Naphthyl Oligophosphates as Allosteric Effectors of Human Hemoglobin. Int J Biochem 1982; 14:601–608.

Horiuchi K. and Asai H. A Reaction Cell for Simultaneous Measurements of Fluorescence and Absorption in a Hemoglobin Solution at Various Oxygen Pressures. Biochem Biophys Res Comm 1980; 97:811–818.

Horiuchi K. and Asai H. Binding of $\beta$-Naphthyl Triphosphate to Human Adult Hemoglobin accompanying Deoxygenation. Investigated by Simultaneous Measurements of fluorescence, Absorbance and Partial Pressure of Oxygen. Eur J Biochem 1983; 131:613–618.

Hu X. and Spiro T. G. Tyrosine and Tryptophan Structure Markers in Hemoglobin Ultraviolet Resonance Raman Spectra: Mode Assignments via Subunit-Specific Isotope Labeling of Recombinant Protein. Biochemistry 1997; 36:15701–15712.

Imai K. *Allosteric Effects in Haemoglobin*. London: Cambridge University Press, 1982.

Itoh M., Mizukoshi H., Fuke K., Matsukawa S., Mawatari K., Yoneyama Y., Sumitani M. and Yoshihara K. Tryptophan Fluorescence of Human Hemoglobin. I. Significant Change of Fluorescence Intensity and Lifetimes in the T—R Transition. Biochem Biophys Res Comm 1981; 100:1259–1265.

Jameson D. M., Gratton E., Weber G. and Alpert B. Oxygen Distribution and Migration Within Mbdes FE and Hbdes Fe. Multifrequency Phase and Modulation Fluorometry Study. Biophys J 1984; 45:795–803.

Janes S. M., Holtom G., Ascenzi P., Brunor M. and Hochstrasser R. M. Fluorescence and Energy Transfer of Tryptophans in *Aplysia* Myoglobin. Biophys J 1987; 51:653–660.

Jayaraman V., Rodgers K. R., Mukerji I. and Spiro T. G. Hemoglobin Allostery: Resonance Raman Spectroscopy of Kinetic Intermediates. Science 1995; 269:1843–1848.

Juszczak L. J., Hirsch R. E., Nagel R. L., and Friedman J. M. Conformational Differences in CO derivatives of HbA, HbC (E$\beta$6K) and HbS (E$\beta$6V) in the Presence and Absence of Inositol Hexaphosphate (IHP) Detected Using Ultraviolet Resonance Raman Spectroscopy. J Raman Spectroscopy 1998; 29:963–968.

Khan K. K., Mazumdar S., Modi S., Sutcliffe M., Roberts G. C. K. and Mitra S. Steady-State and Picosecond-Time-Resolved Fluorescence Studies on the Recombinant Heme Domain of *Bacillus Megaterium* Cytochrome P-450. Eur J Biochem 1997; 244:361–370.

Kitagawa T. Investigation of Higher Order Structures of Proteins by Ultraviolet Resonance Raman Spectroscopy. Prog Biophys Molec Biol 1992; 58:1–18.

Knobler E., Poh-Fitzpatrick M. B., Kravetz D., Vincent W. R., Muller-Eberhard U. and Vincent S. H. Interaction of Hemopexin, Albumin and Liver Fatty Acid-Binding Protein With Protoporphyrin. Hepatology 1989; 10:995–997.

Lakowicz J. R. and Weber G. Quenching of Protein Fluorescence by Oxygen. Detection of Structural Fluctuations in Proteins on the Nanosecond Time Scale. Biochemistry 1973; 12:4171–4179.

Lakowicz J. R. *Principles of Fluorescence Spectroscopy*. 2nd Edition. New York: Kluwer Academic/Plenum Publishers, 1999.

Larsen R. W., Chavez M. D., Ondrias M. R., Courtney S. H., Friedman J. M., Lin M. J. and Hirsch R. E. Dynamics and Reactivity of HbXL99$\alpha$: A Crosslinked Hemoglobin Derivative. J Biol Chem 1990; 265:4449–4454.

Lasagna M., Gratton E., Jameson D. M. and Brunet J. E. Apohorseradish Peroxidase Unfolding and Refolding: Intrinsic Tryptophan Fluorescence Studies. Biophys J 1999; 76:443–450.

Levitt M. and Perutz M. F. Aromatic Rings Act as Hydrogen Bond Acceptors. J Mol Biol 1988; 201:751–754.

Longa S. D., Pin S., Cortes R., Soldatov A. V. and Alpert B. Fe-Heme Conformations in Ferric Myoglobin. Biophys J 1998; 75:3154–3162.

MacQuarrie R. and Gibson Q. H. Use of a Fluorescent Analogue of 2,3-Diphosphoglycerate as a Probe of Human Hemoglobin Conformation During Carbon Monoxide Binding. J Biol Chem 1971; 246:5832–5835.

MacQuarrie R. and Gibson Q. H. Ligand Binding and Release of an Analogue of 2,3-Diphosphoglycerate from Human Hemoglobin. J Biol Chem 1972; 247:5686–5694.

Marden M. C., Hoa G. H. B. and Stetzkowski-Marden F. Heme Protein Fluorescence Versus Pressure. Biophys J 1986; 49:619–627.

Marden M. C., Hazard E. S. and Gibson Q. H. Testing the Two-State Model: Anomalous Effector Binding to Human Hemoglobin. Biochemistry 1986; 25:7591–7596.

McCammon J. A., Gelin B. R. and Karplus M. Dynamics of Folded Proteins. Nature 1977; 267:585–590.

McDonald M. J., Bleichman M., Bunn H. F. and Noble R. W. Functional Properties of the Glycosylated Minor Components of Human Adult Hemoglobin. J Biol Chem 1979; 254:702–707.

McLaughlin M. L. and Barkley M. D. Time-Resolved Fluorescence of Constrained Tryptophan Derivatives: Implications for Protein Fluorescence. Methods in Enzymology 1997; 278:190–201.

McMahon L. P., Yu H.-T., Vela M. A., Morales G. A., Shui L., Fronczek F. R., McLaughlin M. L. and Barkley M. D. Conformer Interconversion in the Excited State of Constrained Tryptophan Derivatives. J Phys Chem 1997; 101:3269–3280.

Mizukoshi H., Itoh M., Matsukawa S., Mawatari K. and Yoneyama Y. Tryptophan Fluorescence of Human Hemoglobin. II. Effect of Inositol Hexaphosphate on the T—R Transition. Biochim Biophys Acta 1982; 700:143–147.

Nagel R. L. and Gibson Q. H. Kinetics and Mechanism of Complex Formation Between Hemoglobin and Haptoglobin. J Biol Chem 1967; 242:3428–3434.

Nagel R. L. and Gibson Q. H. The Binding of Hemoglobin to Haptoglobin and Its Relation to Subunit Dissociation of Hemoglobin. J Biol Chem 1971; 246:69–73.

Nanda V. and Brand L. Fluorescence Characterization of Homeodomains: Quenching of a Buried Tryptophan by Aromatic Hydrogen Bonding Associations. Biophys J 1999; 76 (1, Part 2 of 2):A448.

Nordlund T. M., Andersson S., Nilsson L., Rigler R., Graslund A. and McLaughlin L. W. Structure and Dynamics of a fluorescent DNA oligomer containing the EcoRI recognition sequence: fluorescence, molecular dynamics, and NMR studies. Biochemistry 1989; 28:9095–9103.

Parker, C. A. Photoluminescence of Solutions. Amsterdam: Elsevier Publishing Company, 1968.

Pin S., Royer C. A., Gratton E., Alpert B. and Weber G. Subunit Interactions in Hemoglobin Probed by Fluorescence and High-Pressure Techniques. Biochemistry 1990; 29:9194–9202.

Pin S. and Royer C. A. High-Pressure Fluorescence Methods for Observing Subunit Dissociation in Hemoglobin. Methods Enzymol 1994; 232:42–55.

Ponder J. W. and Richards F. M. Tertiary Templates for Proteins. Use of Packing Criteria in the Enumeration of Allowed Sequences for Different Structural Classes. J Mol Biol 1987; 193:775–791.

Rodgers K. R. and Spiro T. G. Nanosecond Dynamics of the R $\rightarrow$ T Transition in Hemoglobin: Ultraviolet Raman Studies. Science 1994; 265:1697–1699.

Ross J. B., Rousslang K. W. and Brand L. Time-Resolved Fluorescence and Anisotropy Decay of the Tryptophan in Adrenocorticotropin-(1–24). Biochemistry 1981; 20:4361–4369.

Ross J. B., Wyssbrod H. R., Porter R. A., Schwartz G. P., Michaels C. A. and Laws W. R. Correlation of Tryptophan Fluorescence Intensity Decay Parameters with 1H NMR-determined Rotamer Conformations: [tryptophan2] Oxytocin. Biochemistry 1992; 31:1585–1594.

Rouviere N., Vincent M., Craescu C. T. and Gallay J. Immunosuppressor Binding to the Immunophilin FKBP59 Affects the Local Structural Dynamics of a Surface β-Strand: Time-Resolved Fluorescence Study. Biochemistry 1997; 36:7339–7352.

Sassaroli M., Bucci E. and Steiner R. F. Librational Modes in Liganded and Unliganded Hemoglobin as Seen by Fluorescence Spectroscopy. J Biol Chem 1982; 257:10136–10140.

Sassaroli M., Bucci E., Leisegang J., Fronticelli C. and Steiner R. F. Specialized Functional Domains in Hemoglobin: Dimensions in Solution of the Apohemoglobin Dimer Labeled with Fluorescein Iodoacetamide. Biochem 1984; 23:2487–2491.

Serbanescu R., Kiger L., Poyart C. and Marden M. C. Fluorescent Effector as a Probe of the Allosteric Equilibrium in Methemoglobin. Biochim Biophys Acta 1998; 1363:79–84.

Sharma V. S., Newton G. L., Ranney H. M., Ahmed F., Harris J. W. and Danish E. H. Hemoglobin Rothschild (beta 37 (C3) Trp Replaced by Arg): A High/Low Affinity Hemoglobin Mutant. J Mol Biol 1980; 144:267–280.

Silva J. L., Villas-Boas M., Bonafe C. F. S. and Meirelles N. C. Anomalous Pressure Dissociation of Large Protein Aggregates. J Biol Chem 1989; 264:15863–15868.

Smith J. L., Hendrickson W. A., Honzatko R. B. and Sheriff S. Structural Heterogeneity in Protein Crystals. Biochemistry 1986; 25:5018:5027.

Sokolov L. and Mukerji I. Conformational Changes in FmetHbS Probes with UV Resonance Raman and Fluorescence Spectroscopic Methods. J Phys Chem B 1998; 102:8314–8319.

Spiro T. G., Smulevich G. and Su C. Probing Protein Structure and Dynamics with Resonance Raman Spectroscopy: Cytochrome C Peroxidase and Hemoglobin. Biochemistry 1990; 29:4497–4508.

Sudhakar K., Laberge M., Tsuneshige A. and Vanderkooi J. M. Zinc-Substituted Hemoglobins: Alpha- and Beta-Chain Differences Monitored by High-Resolution Emission Spectroscopy. Biochemistry 1998; 37:7177–7184.

Suwaiyan A. and Klein U. K. A. Picosecond Study of Solute Interaction of the Excited State of Indole. Chem Phys Lett 1989; 159:244–250.

Szabo A. G. and Rayner D. M. The Time Resolved Emission Spectra of Peptide Conformers Measured by Pulsed Laser Excitation. Biochem Biophys Res Commun 1980; 94:909–915.

Szabo A. G., Krajcarski D., Zuker M. and Alpert B. Conformational Heterogeneity in Hemoglobin as Determined by Picosecond Fluorescence Decay Measurements of the Tryptophan Residues. Chemical Physics Letters 1984; 108:145–149.

Szabo A. G., Willis K. J., Krajcarski D. T. and Alpert B. Fluorescence Decay Parameters of Tryptophan in a Homogeneous Preparation of Human Hemoglobin. Chemical Physics Letter 1989; 163:565–570.

Teale F. W. J. and Weber G. Ultraviolet Fluorescence of the Aromatic Amino Acids. Biochem J 1957; 65:476–482.

Teale F. W. J. The Ultraviolet Fluorescence of Proteins in Neutral Solution. Biochem J 1960; 76:381–388.

Ushmorov A., Ratter F., Lehmann V., Droge W., Schirrmacher V. and Umansky V. Nitric Oxide-Induced Apoptosis in Human Leukemic Lines Requires Mitochondrial Lipid Degraation and Cytochrome c Release. Blood 1999; 93:2342–2352.

Vanderkooi J. M., Angiolillo P. J. and Laberge M. Fluorescence Line Narrowing Spectroscopy: A Tool for Studying Proteins. Methods Enzymol 1997; 278:71–94.

Vargas V., Brunet J. E. and Jameson D. M. Oxygen Diffusion Near the Heme Binding Site of Horseradish Peroxidase. Biochem Biophys Res Comm 1991; 178:104–109.

Vasudevan G. and McDonald M. J. Spectral Demonstration of Semihemoglobin Formation During CN-Hemin Incorporation into Human Apohemoglobins. J Biol Chem 1997; 272:517–524.

Vincent M., Brochon J. C., Merola F., Jordi W. and Gallay J. Nanosecond Dynamics of Horse Heart Apocytochrome C in Aqueous Solution as Studies by Time-Resolved Fluorescence of the Single Tryptophan Residue (Trp-59). Biochem 1988; 27:8752–8761.

Vincent S. H. and Muller-Eberhard U. A Protein of the Z Class of Liver Cytosolic Proteins in the Rat that Preferentially Binds Heme. J Biol Chem 1985; 260:14521–14528.

Vincent S. H., Grady R. W., Shaklai N., Snider J. M. and Muller-Eberhard U. The Influence of Heme-Binding Proteins in Heme-Catalyzed Oxidations. Arc Biochem Biophys 1988; 265:539–550.

Wajcman H., Kister J., Galacteros F., Spielvogel A., Lin M. J., Vidugiris G. J., Hirsch R. E., Friedman J. M. and Nagel R. L. Hb Montefiore (126(H9) Asp $\rightarrow$ Tyr). High Oxygen Affinity and Loss of Cooperativity Secondary to C-Terminal Disruption. J Biol Chem 1996; 271:22990–22998.

Wang D. and Spiro T. G. Structure Changes in Hemoglobin upon Deletion of C-Terminal Residues, Monitored by Resonance Raman Spectroscopy. Biochemistry 1998; 37:9940–9951.

Weber G. Rotational Brownian Motion and Polarization of the Fluorescence of Solutions. Adv Protein Chem 1953; 8:415–459.
Weber G. and Teale F. W. J. Electronic Energy Transfer in Haem Proteins. Disc of Faraday Soc 1959; 28:134–141.
Weiss S. Fluorescence Spectroscopy of Single Biomolecules. Science 1999; 283:1676–1683.
Williams R. T. and Bridges J. W. Fluorescence of Solutions: A Review. J Clin Path 1964; 17:371–394.
Willis K. J., Szabo A. G., Zuker J., Ridgeway J. M. and Alpert B. Fluorescence Decay Kinetics of the Tryptophyl Residues of Myoglobin: Effect of Heme Ligation and Evidence for Discrete Lifetime Components. Biochemistry 1990; 29:5270–5275.
Willis K. J., Neugebauer W., Sikorska M. and Szabo A. G. Probing Alpha-Helical Secondary Structure at a Specific Site in Model Peptides via Restriction of Tryptophan Side-Chain Rotamer Conformation. Biophys J 1994; 66:1623–1630.

# 11

# Conformation of Troponin Subunits and Their Complexes from Striated Muscle

Herbert C. Cheung and Wen-Ji Dong

## 11.1. Introduction

Contraction and relaxation in striated muscle (skeletal and cardiac) are regulated by a group of regulatory proteins that are part of the thin filament in the muscle structure. These proteins are tropomyosin (Tm) and the troponin complex (Tn). The thin filament is a pseudodouble helical filament of polymerized actin (F-actin) decorated with the dimeric coiled-coil $\alpha$-helices of Tm and the Tn complex. Each coiled-coil Tm covers the surface of each strand of the actin helix with a stoichiometric ratio of one Tm to seven actin monomers, and each Tm is associated with one Tn. The Tn complex consists of three nonidentical subunits: troponin T (TnT), which binds to Tm; troponin I (TnI), which binds to actin and inhibits actomyosin ATPase; and troponin C (TnC), which binds $Ca^{2+}$ to its N-terminal, regulatory domain to relieve the TnI inhibition. The cycle of contraction-relaxation begins with the binding of activator $Ca^{2+}$ to the TnC regulatory sites within the Tn complex. This binding triggers a series of protein-protein interactions leading to strong interactions between myosin crossbridges of the thick filament and actin that result in force generation. A complete understanding of muscle function requires detailed structural information of the constituent proteins. A great deal is known about the actin-myosin interface because the structures of these two proteins have been solved to high resolution. In contrast, the structure of the regulatory Tm-Tn complex is still unsolved.

---

*Herbert C. Cheung and Wen-Ji Dong* • Department of Biochemistry and Molecular Genetics, University of Alabama at Birmingham, Birmingham, Alabama 35294-2041.

*Topics in Fluorescence Spectroscopy, Volume 6: Protein Fluorescence*, edited by Joseph R. Lakowicz. Kluwer Academic / Plenum Publishers, New York, 2000

Two key questions in muscle regulation are how the initial $Ca^{2+}$ binding signal is relayed to TnI, TnT and actin, and how the signal from TnT is relayed to Tm. A structural consequence of signal transduction among these proteins is a cascade of conformational changes in the proteins resulting in changes of their interactions and forming the structural basis of their functions in the contractile machinery. It is an axiom in structural biology that the structure/function relationship of individual components needs to be understood before the function of a macromolecular assembly can be elucidated from the point of view of structure. With the advent of site-directed mutagenesis and the introduction of polymerase chain reaction (PCR), specific mutants of Tn subunits and Tm have been overexpressed in bacterial systems with good yields. These mutants make it possible for a variety of biochemical and spectroscopic studies that have yielded important insights to the two key questions. This chapter focuses on certain structural aspects of the troponin subunits as related to their functions on the basis of both intrinsic and extrinsic emission properties. The emphasis is on use of single-tryptophan mutants of TnI and TnC for construction of their topography from fluorescence and luminescence resonance energy transfer (FRET and LRET) data.

## 11.2. Topography and Structure of Troponin Subunits

### 11.2.1. Troponin Complex

No three-dimensional structure are available for the heterotrimeric Tn or the Tm-Tn complex from vertebrate muscle that could contribute to the understanding of how these proteins regulate the actin-myosin interaction. Models have been proposed in which the Tm-Tn complex moves laterally on the surface of the actin helix upon $Ca^{2+}$ activation. This movement must involve extensive conformational changes of the component proteins in response to $Ca^{2+}$ binding to the regulatory sites of TnC. An early electron microscopy study of the Tn complex revealed a bipartite structure with a length of 265 Å.[1] The globular domain consists of the TnC and TnI subunits and the long rod-like portion of the structure is part of TnT. A recent single particle analysis of electron micrographs of the Tm-Tn complex obtained from insect flight muscle yielded a 3-dimensional reconstruction of the Tn complex at a 26 Å resolution.[2] The model at this low resolution gives no indication on the topography of individual subunits within the whole complex and provides no clue on potential changes in the overall topography induced by $Ca^{2+}$.

### 11.2.2. Troponin C

Troponin C is the only subunit of the Tn complex whose crystal structure has been solved.[3,4] The crystal structure of TnC from avian vertebrate fast skeletal muscle shows a dumbbell shaped molecule with both the N-terminal and C-terminal segments folded into two globular domains, which are linked by a 22-residue $\alpha$-helix (Figure 11.1). The C-terminal domain has two high-affinity $Ca^{2+}$ sites (sites III and IV) which also bind $Mg^{2+}$. These sites serve to stabilize the protein's structure and have no apparent functional role. The N-terminal domain also has two sites (sites I and II) which bind $Ca^{2+}$ specifically with a low affinity. The crystal structure shows bound $Ca^{2+}$ at sites III and IV, but no bound $Ca^{2+}$ at sites I and II. Since sites III and IV

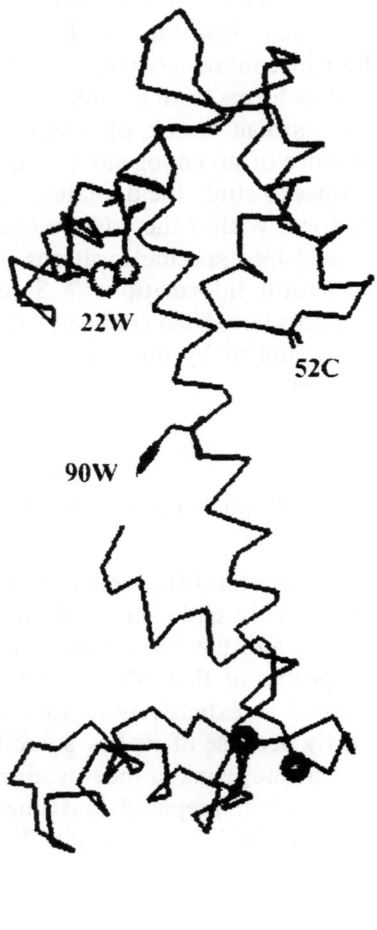


Figure 11.1. A representation of the crystal structure of skeletal TnC containing two bound $Ca^{2+}$ ions (spheres) in the C-terminal domain. The N-terminal domain is devoid of bound $Ca^{2+}$. For the FRET studies described in Sec. 3.2.3, three mutants were used: F22W, N52C, and A90W. The locations of these mutations are indicated.

are expected to be saturated with $Mg^{2+}$ and the regulatory sites (sites I and II) to be unoccupied in relaxed muscle, the crystal structure provides a starting structure to understand potential conformational changes induced in the N-terminal regulatory domain by the binding of activator $Ca^{2+}$ to sites I and II.[5] Sites III and IV each consist of a helix-loop-helix structural motif in which $Ca^{2+}$ is coordinated to the 12-residue binding loop to form the typical EF-hand motif in which the two flanking helices are oriented at an angle close to 90 degrees. In the crystal structure, the flanking helices of the two helix-loop-helix motifs in the N-terminal domain are oriented at angles considerably larger than 90 degrees because of the absence of bound $Ca^{2+}$ at sites I and II. The TnC isoforms from vertebrate slow skeletal muscle and cardiac muscle (cTnC) have identical sequences and only one active regulatory $Ca^{2+}$ binding site (site II) due to a single amino acid insertion and two substitutions in the chelating loop of site I. The crystal structure of cTnC has not been solved, but the solution NMR structures of the two domains of this isoform have been reported.

Most isoforms of TnC, including those from rabbit, chicken, and human contain no tryptophan, although some contain multiple tyrosines. The isoforms from chicken fast skeletal muscle, chicken slow skeletal muscle, and cardiac muscle of several vertebrate species have no tryptophan. The absence of an endogenous tryptophan led to the use of extrinsic fluorescent probes to study the domain conformations of these proteins in early investigations. Within the past several years, single tryptophans have been engineered into specific locations in these isoforms of TnC to obtain specific structural information.[6–11] Several of these engineered single-tryptophan mutants have been studied by time-resolved methods,[12,13] and the others have been studied by the steady-state methods to monitor $Ca^{2+}$ binding to the mutants.

### 11.2.3. Troponin I and Troponin T

Troponin I from most skeletal and cardiac muscles have a single tryptophan. This endogenous fluorophore is highly conserved among several species and has been exploited as a native reporter group on the structural properties of this subunit from both types of muscle.[14–17] TnT has not been studied as extensively as the other two subunits by fluorescence methods, partly because of its low solubility in aqueous solution and the presence of 2–3 tryptophans in most isoforms. Single-tryptophan mutants of TnT have already been prepared, and time-resolved studied have been reported on some of these TnT single mutants.[18]

## 11.3. Conformation of Skeletal Muscle TnC

### 11.3.1 Conformation of the Regulatory Domain of Skeletal TnC

The regulatory N-terminal domain of TnC consists of five helices which are labeled as helix N and helices A-D starting from the N-terminus (Figure 11.2A).[19] The first EF-hand is the $Ca^{2+}$-binding site I and consists of the motif helix A-(loop I)-helix B, and the second EF-hand is the binding site II and consists of the motif helix C-(loop II)-helix D. Helices B and C are linked by a flexible loop (B-C linker), and helix D is linked to the C-terminal domain (not shown in Figure 11.2) via the D/E helical linker (central helix). In the X-ray crystal structure of chicken fast skeletal TnC in which sites I and II are devoid of bound $Ca^{2+}$ (apo N-domain), the A helix is delineated from

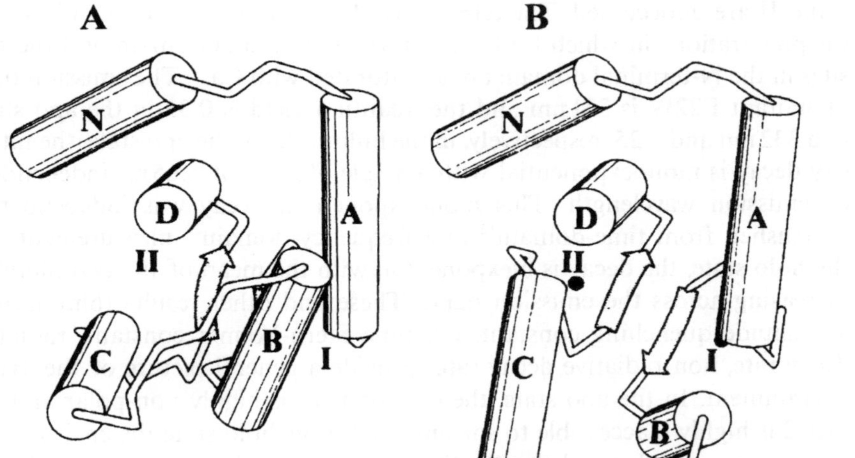


Figure 11.2. A diagram of the proposed $Ca^{2+}$-induced conformational changes in the regulatory N-domain of skeletal troponin C. The five helices are labeled N-helix and helices A-D starting from the N-terminus. Helix D is linked by the central helix to the C-domain which has four helices homologous to helices A-D in the N-domain. The central helix and the C-domain are not shown here. (A) The apo conformation of the N-domain of skeletal TnC, showing the locations of the two unoccupied $Ca^{2+}$ sites (I and II). (B) Proposed conformation of the holo state of the N-domain. In the proposed model, the relative dispositions of helices N, A, and D remain unchanged as in (A), and helices B and C and the linker peptide between B and C move away as a unit from their dispositions in the apo structure. The two closed circles represent the two bound $Ca^{2+}$ ions. The relative dispositions of helices B and C also remain unchanged. (From Ref. 19).

Glu16 to Met28. The corresponding helix in rabbit likely ends at Met25, a residue equivalent to the Met28 of chicken TnC. Chicken mutant F29W[6] and rabbit mutant F26W[9] were generated in bacterial systems and used in several studies of structure/function relationships. More recently, we have reported the steady-state and time-resolved properties of chicken mutants F22W, F52W, and F90W.[11,12]

### 11.3.2. Properties of Single-Tryptophan Skeletal TnC Mutants

#### *11.3.2.1. Structure and Fluorescence of Mutant F22W*

The terms "apo N-domain" and "apo state" are used interchangeably for TnC preparations in which the high-affinity sites III and IV in the C-terminal domain are saturated with $Mg^{2+}$ and the N-terminal regulatory sites I and II are unoccupied. The term "holo TnC" or "holo N-domain" is used for preparations in which both sites in the C-terminal domain and the two sites in the N-terminal domain are all saturated with $Ca^{2+}$. The emission peak of mutant F22W is 331 nm and the quantum yield is 0.33 in the apo state, and 332 nm and 0.25, respectively, in the holo state. In the apo state, the intensity decay is monoexponential with a single lifetime of 5.65 ns, independent of emission wavelength. This monoexponential decay was independently established from time-domain[12] and frequency-domain[20] measurements. In the holo state, the decay is biexponential with the mean of the two lifetimes increasing across the emission band. These and other results (bimolecular acrylamide quenching constant, dynamic Stern-Volmer constant, radiative decay rate, non-radiative decay rate) provide a general picture of the Trp22 environment. In the apo state, the environment is highly non-polar and the Trp22 is highly inaccessible to solvent, and in the holo state the environment becomes more polar and the Trp22 is more accessible to the solvent. F22W in the apo state is among one of very few single-tryptophan proteins that have been shown to decay monoexponentially.

It is of interest to examine conformational differences between the apo and holo states of the N-domain that could account for the observed different intensity decay patterns of F22W. The energy-minimized crystal structure of native TnC reveals that Phe22 is largely buried and not readily accessible to solvent. There is a cavity next to this residue large enough to accommodate a water molecule. Phe22 is in close contacts with five hydrophobic side chains (three methionines and two leucines). A molecular modeling study[11] suggests that the substitution of Phe22 by Trp would retain similar side chain packing as in the native structure, and the Trp22 in the mutant would be similarly inaccessible to solvent. In the modeled holo state of the N-domain,[5] the indole ring appears to be rotated by about 90° about the $C_\beta$-$C_\gamma$ bond and is still within van der Waals contacts with the five hydrophobic side chains. In this holo conformation, the edge of the indole ring is slightly more accessible to solvent than in the apo state. These structural features can explain the high quantum yield in the apo state, and the small red-shift of the emission spectrum and decrease in the quantum yield resulting from $Ca^{2+}$ binding to the N-domain.

A careful analysis of a number of steady-state and time-resolved results has led to the conclusion that solvent relaxation or excited-state reactions are unlikely the dominant origin of the biexponential decay observed in the presence of bound $Ca^{2+}$. An alternative interpretation of the origin of the biexponential decay is ground-state heterogeneity of the Trp22 residue in the holo state. The intensity decay results suggest two Trp22-resolved conformational states. To pursue this possibility, we used the Trp22 decay times determined at several wavelengths across the emission band to construct two decay-associated spectra (DAS) for the Trp22 (Figure 11.3). The dominant spectrum is associated to the longer lifetime with a maximum essentially unaltered as that in the steady-state spectrum (331 nm), and the minor spectrum is associated to the shorter lifetime with a 20-nm red-shift. The dominant emitting species of the $Ca^{2+}$-saturated N-domain is very similar to the homo-

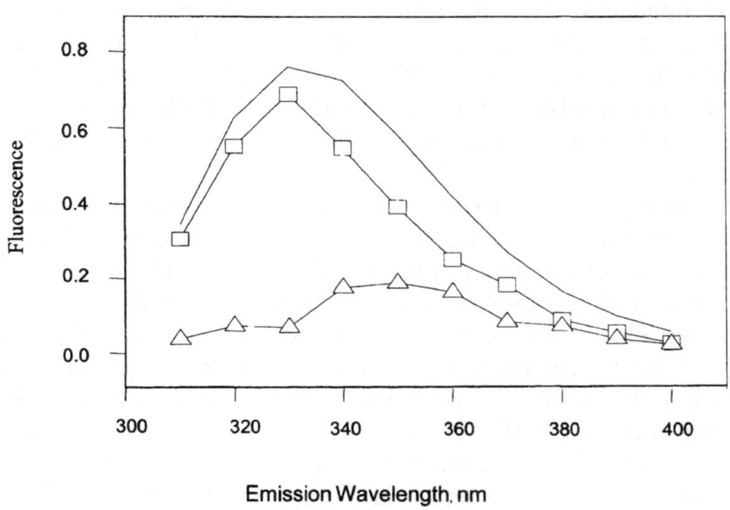


Figure 11.3. Decay-associated (DAS) emission spectra of Trp22 in mutant F22W from skeletal TnC. The top curve is the steady-state spectrum. The other two curves are the DAS spectra associated to the long lifetime (squares) and the short lifetime (triangles). (From Ref. 12).

geneous, one-state apo N-domain in which the fluorophore is largely protected from interaction with the solvent, and the other $Ca^{2+}$-bound emitting species is substantially more exposed. These characteristics of the two DAS spectra are consistent with the very small red-shift of the steady-state spectrum and a substantial decrease in quantum yield which accompany the transition of the N-domain from the apo state to the holo state. Trp22 is located in the middle of the A helix and is in close proximity to the two EF-hands of the $Ca^{2+}$-loaded N-domain. It is possible that one of the two emitting species would reflect the N-domain conformation with only one site occupied, and the other species would correspond to occupation of a second site. A more detailed study involving resolution of the DAS in a $Ca^{2+}$ titration experiment will be needed to address this issue.

#### *11.3.2.2. Fluorescence of Other Single-Tryptophan Mutants*

The steady-state and time-resolved properties of two other chicken skeletal mutants (N52W and A90W) have been reported in some detail.[11,12] The intensity decay of these two tryptophans is more complex than that of Trp22. Even in the apo state of the N-domain, Trp52 has two lifetimes and Trp 90 has three lifetimes. In the holo state, the decay of Trp52 becomes triexponential, whereas the decay of Trp90 is biexponential. Some of these lifetimes, from both the apo and holo states, have a wavelength dependence. These complexities likely are related to the secondary structure in which the residues are located. Trp52 is in the B-C linker, which has no well-defined secondary structure and is flexible. Trp90 is in the central helix, which is known to be flexible with a helix breaker Gly89 adjacent to Trp90. The flexible structural environments likely contribute to the complex decay properties.

The steady-state fluorescence spectra of both chicken skeletal F29W[6] and rabbit skeletal F26W[9] are very similar. This is expected since the two residues are in homologous positions in the A helix. The transition of apo N-domain to holo N-domain is accompanied by a small blue shift of the spectra from 336 nm and an increase in the peak intensity by a factor or 2–3,[6] suggesting that the environment of the two equivalent tryptophans is significantly less polar in the holo state than in the apo state. The intensity decay of chicken mutant F29W was shown to be multiple-exponential in both the apo and holo states.[13] In the X-ray structure, Phe29 is adjacent to the C-terminal end of the A helix and is at the beginning of the loop in the helix A-(loop I)-helix B motif. It is difficult to visualize from the crystal structure how $Ca^{2+}$ binding to the N-domain could induce drastic changes in its environment as the reported fluorescence properties suggest. The answer is found in a recent NMR study of the secondary structure of the N-terminal fragment of chicken skeletal TnC which showed that the A helix ends at Met28 in the apo state, in agreement with the crystal structure. Residue 29 (Phe in the native sequence) is at the beginning of the flexible binding loop and would be highly exposed to the solvent. However, the A helix is found by NMR to be extended by one residue and ends at Phe29 in the holo N-domain.[21] The tryptophan in holo mutant F29W is expected to be incorporated into the C-terminus of the A helix and likely becomes shielded, or at least partially shielded, from solvent. These structural changes explain, at least in part, the large $Ca^{2+}$-induced increase in quantum yield and blue spectral shift of F29W.

### *11.3.2.3. Conformational Change Induced by Activator $Ca^{2+}$*

The N-domain of TnC is the site where the signal of activator $Ca^{2+}$ is transduced to TnI for the enhanced and $Ca^{2+}$-dependent interaction. Since the crystal structure of skeletal TnC containing bound $Ca^{2+}$ in the N-domain is not available, an early modeling study of the holo state of the N-domain structure suggested reorientations of the secondary structural elements in which the B and C helices move as a unit relative to the N, A, and D helices.[5,19] These reorientations would result in an open N-domain conformation and expose a short segment of hydrophobic residues in the B helix. This exposed hydrophobic patch would be the site for the $Ca^{2+}$-dependent interaction with TnI. In this model the $\alpha$-carbon coordinates of the A helix are expected not to change, but the positions of the carbon atoms in helices B and C and the B-C linker (residues 49–54) would move relative to the A and D helices (Figure 11.2B). The holo N-domain is predicted to have an open conformation when compared with the apo structure. We recently tested the possibility of such a $Ca^{2+}$-induced "open" conformation of the N-domain with measurements of FRET between Trp22 (helix A) and Cys52 (B-C linker) and between Trp90 (helix D) and Cys52.[22] (see Figure 11.1 for locations of these residues). Tryptophan was the energy donor and AEDANS linked to Cys52 was the common energy acceptor. Figure 11.4A shows representative intensity decays of Trp22 in the absence and presence of the acceptor. The pronounced curvature displayed in the donor-acceptor sample is due, in part, to an incomplete acceptor labeling (90%). The fast decay component is a clear demonstration of a large energy transfer. Steady-state measurements indicated that the donor quenching was accompanied by an enhancement of acceptor sensitized fluorescence. The decay curve of the donor-acceptor sample obtained in the presence of $Ca^{2+}$ (Figure 11.4B) has a shape indicative of decreased energy transfer.

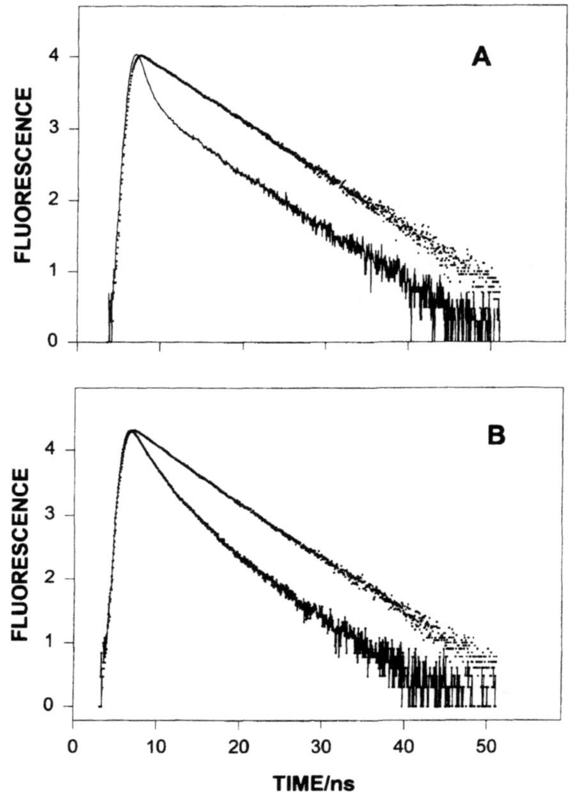


Figure 11.4. Fluorescence intensity decay curves of Trp22 in skeletal TnC mutant F22W containing a single cysteine (Cys52). (A) The decay was determined with the mutant in which the two regulatory sites in the N-domain were not occupied (apo N-domain), and (B) the decay was determined with the mutant saturated with $Ca^{2+}$ in the N-domain (holo N-domain). The top curves in each panel are the decays from the donor-alone samples in which Cys52 was unmodified. The lower curves are the decays from the donor-acceptor samples in which Cys52 was labeled with the energy acceptor IAEDANS and indicate energy transfer between Trp22 and AEDANS-Cys52. (From Ref. 22).

Figure 11.5 shows the peak-normalized distributions of the two distances, residue 22-residue 52 and residue 90-residue 52, and Table 11.1 lists the distance parameters recovered from these distributions. It is clear that the transition of the N-domain from the apo state to the holo state results in an increase in the mean distance between the donor and acceptor sites for both distances. The increases are accompanied by a large narrowing of the distributions. The magnitudes of the $Ca^{2+}$-induced increases of both distances are remarkably similar to the increases predicted by the HMJ model of the

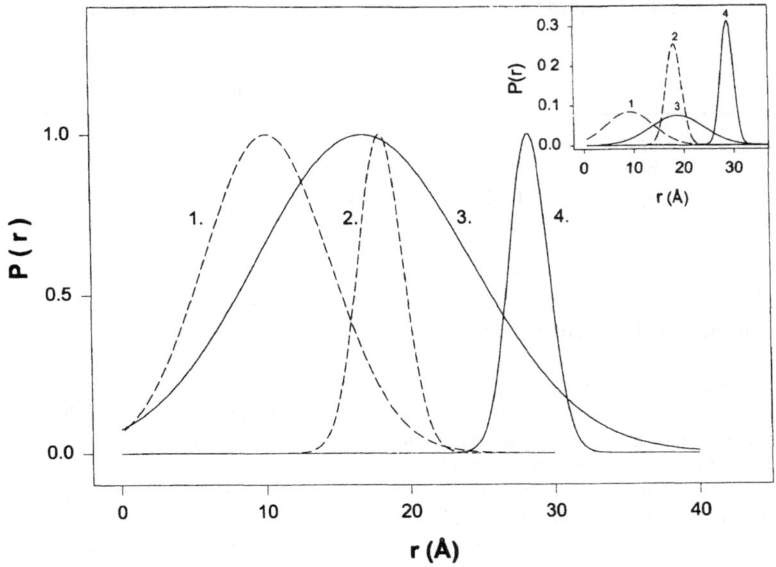


Figure 11.5. The distributions of distances for skeletal TnC mutants in which tryptophan (Trp22 and Trp 90) was the energy donor and AEDANS attached to Cys52 was the energy acceptor. These distributions are peak normalized to facilitate comparison. Broken curves, distance 22–52; solid curves, distance 90–52. Curves 1 and 3 are for apo N-domain and curves 2 and 4 are for holo N-domain. For both distances, the distributions are shifted toward longer distances and become considerably narrower in the holo state (curves 1 vs. 2, and curves 3 vs. 4). The inset shows the same four distribution curves which are area normalized to show the extent of overlaps between the curves from the apo and holo states of each distance. (From Ref. 22).

N-domain.[5,19] As a control, the distance between Trp22 and Cys101 was similarly determined. The effect of $Ca^{2+}$ binding was a small decrease (rather than an increase) in the mean distance and a very small increase of the half-width of the distribution of the distances. The negligible change in the mean distance is consistent with the HMJ model. An interesting feature of the distributions shown in Figure 11.5 is the narrowing of the distributions in the holo state, suggesting a constrained open conformation. An open conformation certainly is needed to expose a critical hydrophobic patch for interaction with TnI as the molecular trigger of the contractile cycle. Whether or not an open conformation is both necessary and sufficient for interaction with TnI is dependent upon the bimolecular rate of interaction between the two proteins and the rate at which the open conformation fluctuates. If the two rates are not compatible, this interaction may be difficult. The constrained

Table 11.1 Distribution of Intersite Distances in Skeletal TnC[a]

| Distance[b] | State[c] | $\bar{r}$ (Å) | hw (Å) | Distance change[d] (Å) | Predicted change[e] (Å) |
|---|---|---|---|---|---|
| 22–52 | apo | 9.2 | 11.1 | 8.9 | 8.8 |
| | holo | 18.1 | 3.7 | | |
| 90–52 | apo | 18.8 | 12.6 | 10.0 | 10.0 |
| | holo | 28.8 | 3.0 | | |
| 22–101 | apo | 25.7 | 13.7 | −1.1 | −0.3 |
| | holo | 24.6 | 14.3 | | |

[a]The parameters of the distribution are the mean distance ($\bar{r}$) and half-width of the distribution (hw).
[b]Distance refers to the donor-acceptor distance between residues indicated.
[c]the apo state refers to the biochemical state in which the two regulatory sites in the N-domain are unoccupied by $Ca^{2+}$, but the two sites in the C-domain are occupied by $Mg^{2+}$. The holo state refers to the biochemical state in which all four sites are saturated by $Ca^{2+}$.
[d]This is the change in the observed mean distance $\bar{r}$ between the holo state and the apo state.
[e]The difference predicted by the HMJ model for the indicated distance between the holo state and the apo state. This prediction refers to changes between the coordinates of the two alpha carbon atoms of the indicated residues.

conformation demonstrated in these studies may provide a mechanism to ensure the bimolecular reaction to take place with rates compatible with physiological demand.

The HMJ model of the holo state of the N-domain conformation is attractive because it provides a simple structural basis for the $Ca^{2+}$-induced trigger of contraction. However, the model provides no insight into the difference in the dynamic nature and potential conformational heterogeneity of the N-domain in the two biochemical states. The area-normalized distributions (inset, Figure 11.5) show overlaps (10%) between the curves for the apo and holo states of both distances. These FRET results suggest that a fraction of the TnC molecules in the apo state may be in the open or partially open conformation, or in transient between the two conformations. There are two potential paths by which activator $Ca^{2+}$ confers a constrained and open conformation. One possibility is that the binding of $Ca^{2+}$ to the closed/partially open conformations forces a domain opening and imposes an open rigid structure of the domain. The half-widths of the distribution of the holo state are less than 4 Å, and this is within the range of the apparent half-widths of severely constrained conformations.[23] The other possibility is that $Ca^{2+}$ prefers binding to those apo molecules with an open or partially open structure and this binding shifts the closed $\rightleftharpoons$ open equilibrium and stabilizes the open conformation. This initial $Ca^{2+}$ complex may undergo further conformational rearrangement to yield the final open conformation. These possibilities have not yet been delineated.

## 11.4. The N-domain Conformation of Cardiac Muscle TnC

To investigate the N-domain conformation of cardiac TnC,[24,25] we used three single-tryptophan cTnC mutants and the same acceptor probe (AEDANS) that was previously used for FRET studies of skeletal TnC. The three inter-site distances studied are (1) Trp20-Cys51, (2) Trp12-Cys51, and (3) Trp20-Cys89 and are indicated in Figure 11.6A. The single tryptophan was the energy donor and AEDANS attached to the single cysteine was the acceptor. Residues 20 and 51 in chicken cardiac TnC are homologous

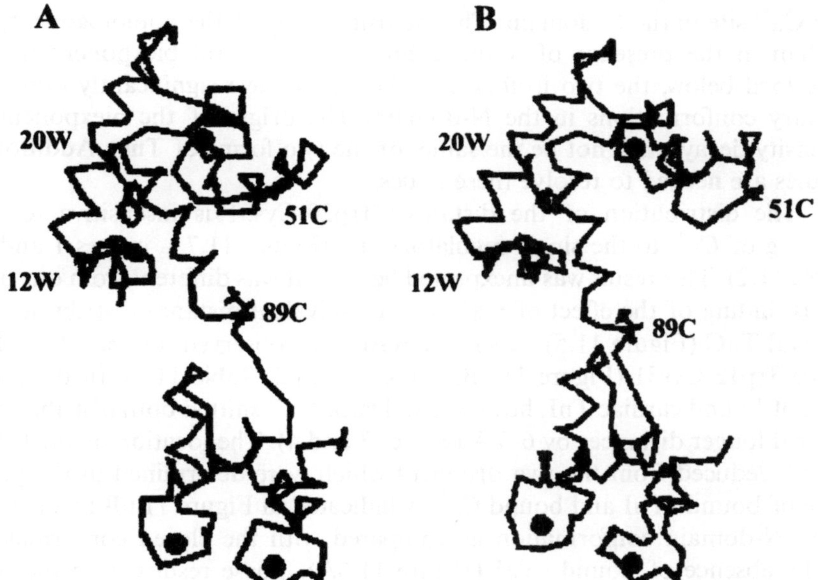


Figure 11.6. A representation of the structure of cardiac TnC. (A) Solution structure of holo cardiac TnC determined by NMR (all three sites are occupied by $Ca^{2+}$, spheres). The four residues which were mutated for FRET studies are indicated in this structure to show their locations (F12W, F20W, N51C, S89C) (B) A representation indicating the position of residue 51 on the basis of FRET distances determined in the holo cTnC-cTnI complex, of the holo structure of cTnC bound to cTnI, showing an opening of the N-domain in the complex compared to the closed holo conformation in the absence of bound cTnI. (Figure 11.6A is from PDB IAJ4).

to residues 22 and 52 in chicken fast skeletal TnC, and the distances Trp20-Cys51 in cardiac TnC corresponds to the distance Trp22-Cys52 in skeletal TnC. The FRET results of these two distances allow a direct comparison of the properties of the N-domain in the two isoforms of TnC.

The intensity decay of Trp20 in cTnC mutant F20W is single-exponential with a lifetime of 4.41 ns in the absence of bound $Ca^{2+}$ at the single regulatory site. The transition of the mutant from the apo state to the holo state results in a change in the intensity decay pattern from monoexponential to biexponential ($\tau_1 = 2.43$ and $\tau_2 = 4.33$ ns), a small red-shift of the emission spectrum, and a decrease of the quantum yield from 0.34 to 0.29. Qualitatively, these time-resolved and steady-state results are very similar to those of Trp22 in the skeletal mutant F22W and suggest that the local environments of the two homologous tryptophans are very similar. In Sec. 3.2.1, we speculate that the two resolved DAS for holo skeletal TnC may reflect two $Ca^{2+}$-loaded TnC conformations, one containing a single bound $Ca^{2+}$ and the other containing both bound $Ca^{2+}$. In the case of cardiac TnC, there is only one $Ca^{2+}$ site in the N-domain. The intensity decay of the homologous tryptophan in the presence of a single bound $Ca^{2+}$ is still biexponential. As described below, the two isoforms of TnC may have significantly different tertiary conformations in the N-domain. The origin of the biexponential intensity decays may not be the same for the two forms of TnC. Additional studies are needed to resolve these issues.

The distribution of the distances Trp20-Cys51 is insensitive to the binding of $Ca^{2+}$ to the single regulatory site (Figure 11.7A, curves 1 and 2; Table 11.2). This result was unexpected because it was different from our previous finding of the effect of $Ca^{2+}$ on an equivalent distance distribution in skeletal TnC (Figure 11.5). A similar result was observed for the cTnC distance Trp12-Cys51 (Figure 11.7B, curves 1 and 2; Table 11.2). In the presence of bound cardiac TnI, however, activator $Ca^{2+}$ shifted both distributions toward longer distances by 6–7 Å (curves 3 and 4). The location of the $C^{\alpha}$ of Cys51 deduced from the two distances which were determined in the presence of bound cTnI and bound $Ca^{2+}$ is indicated in Figure 11.6B to show an open N-domain conformation as compared with the closed conformation in the absence of bound cTnI (Figure 11.6A). These results were the first demonstration that the binding of cardiac TnI is a prerequisite to achieve a $Ca^{2+}$-induced open N-domain in cardiac TnC, and this role of cardiac TnI was not previously recognized.

A distinct feature of the distribution of the distances Trp20-Cys51 is the narrow half-width (2–3 Å) for the apo state of cTnC and its insensitivity to activator $Ca^{2+}$. These hw values (Table 11.2) are a factor of 2–3 smaller than those for the equivalent Trp22-Cys52 distances in apo skeletal TnC (Table 11.1). On the basis of the anisotropy decay data of both donor and

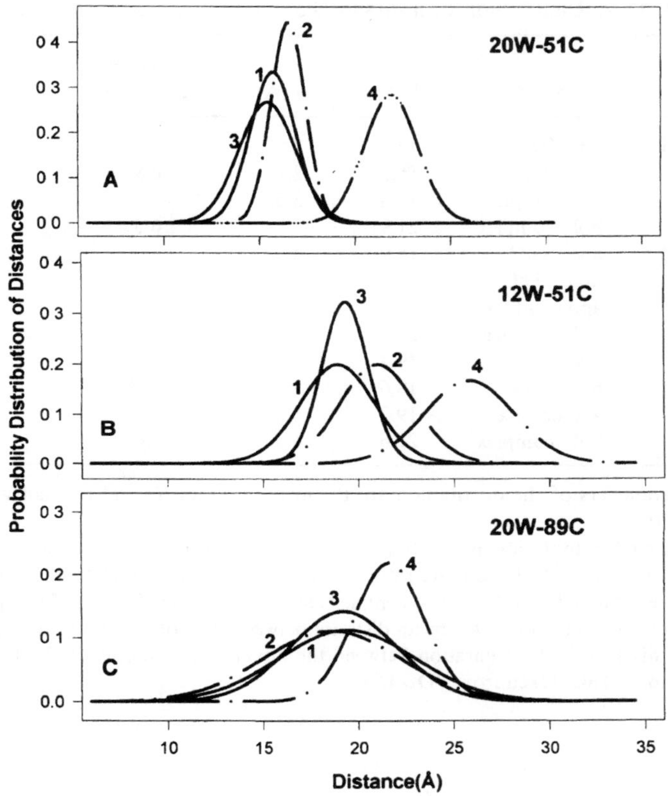


Figure 11.7. The distribution of intersite distances for cardiac TnC mutants. The common donor for all three distances was tryptophan (Trp20 and Trp12), and the common acceptor was AEDANS attached to the single cysteine (Cys51 and Cys89). (A) Distance Trp20-Cys51 (20W-51C), (B) distance Trp12-Cys51 (12W-51C), and (C) Trp20-Cys89 (20W-89C). Four distributions are shown for each donor-acceptor distance. Isolated cTnC: curve 1 (apo N-domain) and curve 2 (holo N-domain). cTnC reconstituted into the cTnC-cTnI complex: curve 3 (apo N-domain of cTnC), and curve 4 (holo N-domain of cTnC). With isolated cTnC, the mean distance and the half-width are not sensitive to activator $Ca^{2+}$ bound to the N-domain (curves 1 vs. 2). In the holo cTnC-cTnI complex, the distributions are shifted toward longer distances for all three distances, although the shift is much smaller for 20W-89C than for the other two distances. (From Ref. 25).

acceptor for the equivalent distances in both isoforms, the narrower distribution of Trp20-Cys51 in cTnC is unlikely related to changes in fluorophore mobilities. The hw of the distribution for Trp12-Cys51 is slightly larger, but still small compared with the hw values of the distributions for the distances in skeletal TnC. Thus, the N-domain of cardiac TnC in the apo state is considerably more constrained than that of apo skeletal TnC. A plausible

Table 11.2. Distribution of Intersite Distances in Cardiac Muscle TnC[a]

| Distance | State[b] | r̄ (Å) | hw (Å) | Distance change[c] (Å) | NMR distance[d] (Å) |
|---|---|---|---|---|---|
| Trp20-Cys51 | apo cTnC | 15.7 | 2.8 | | |
| | holo cTnC | 16.5 | 2.1 | 0.8 | 16.0 |
| | apo complex | 15.4 | 3.5 | | |
| | holo complex | 21.9 | 3.3 | 6.5 | |
| Trp12-Cys51 | apo cTnC | 18.9 | 4.7 | | |
| | holo cTnC | 21.0 | 4.7 | 2.1 | 21.6 |
| | apo complex | 19.3 | 2.9 | | |
| | holo complex | 25.8 | 5.1 | 6.5 | |
| Tr20-Cys89 | apo cTnC | 19.4 | 8.3 | | |
| | holo cTnC | 18.6 | 8.4 | –0.8 | 18.3 |
| | apo complex | 19.2 | 6.6 | | |
| | holo complex | 21.6 | 4.3 | 2.4 | |

[a]The two parameters of the distribution are the mean distance (r̄) and the half-width of the distribution (hw).
[b]The apo state refers to the absence of bound $Ca^{2+}$ at the single regulatory site in the N-domain, but the two sites in the C-domain are saturated with $Mg^{2+}$. The holo state is one in which all three sites are saturated with $Ca^{2+}$. The complex state refers to the cTnC-cTnI complex.
[c]This is the change in the observed mean distance between the holo state and the apo state.
[d]The NMR distance is the separation between the alpha carbon atoms of the two indicated residues in holo cTnC (taken from PDB IAJ4).

explanation for the narrower hw in cTnC likely lies in the difference in the tertiary structure of the N-domain between the skeletal and cardiac isoforms of TnC. The mean distance of 15.7 Å for apo Trp20-Cys51 is significantly longer than the value 9–10 Å for the corresponding distance in skeletal TnC, suggesting a partially open apo conformation. This interpretation is consistent with the mean distance of 18 Å observed with holo skeletal TnC. In holo cTnC, the hw of Trp20-Cys51 decreases by <1 Å and the mean distance increases by <1 Å. The changes of the hw and the mean distance for Trp12-Cys51 between the apo and holo states are also small. These results are strong evidence that the apo N-domain of cardiac TnC is already constrained and partially open and has a different average conformation than the apo skeletal TnC. These surprising results are in agreement with an NMR study which showed that the holo N-domain of cardiac TnC has a closed conformation, very similar to the closed conformation of the apo state.[26] The inability of activator $Ca^{2+}$ to open upon up the N-domain of cardiac TnC has been demonstrated by two very different physical methods.

The FRET results additionally show that the apo N-domain of cTnC is constrained and this constrained conformation is carried over to the holo state.

The distribution of the distance Trp20-Cys89 (Figure 11.7C) requires separate consideration. Residue 89 is at the N-terminal end of the central helix, which is very flexible and its solution NMR structure is undefined.[26] The characteristics of the Trp20-Cys89 distribution may reflect, at least in part, the flexibility of the central helix. This distribution is insensitive to $Ca^{2+}$, just as the distributions of the other two distances are. However, the hw of the apo state is large (>8 Å) and approaching the hw values observed with skeletal TnC. The segmental flexibility of the central helix likely contributes to the large half-width. The half-width decreases by about 2 Å upon formation of the apo complex with cardiac TnI, and by another 2 Å to a final value of 4.3 Å in the holo complex. These changes reflect a more constrained conformation in the region of cardiac TnC involving Trp20 in the A helix and the N-terminal region of the flexible central helix.

## 11.5. Comparison of Cardiac TnC and Skeletal TnC

A modeling study of human cardiac TnC suggested that the holo N-domain of cTnC had an open conformation in which the B and C helices moved away from the D helix,[27] which is similar to the HMJ model of the holo N-domain of skeletal TnC.[5] This suggestion strengthened the general belief that the regulatory domain of TnC from both isoforms would experience similar structural changes in response to the binding of activator $Ca^{2+}$. In contrast to this cTnC model, FRET results clearly indicate substantial conformational differences between the two isoforms in the N-domain in both the apo and holo states. Some of these differences are consistent with the NMR structure of holo cTnC. It is not yet clear what role bound cTnI plays in modulating a $Ca^{2+}$-induced open conformation in cTnC. The previous cTnC model, which is based on the crystal structure of skeletal TnC, has played a central role in the understanding of cTnC function and interpretation of the binding of a group of cardiac therapeutic agents known as $Ca^{2+}$ sensitizers to myofilaments. If the target of these agents is in fact cTnC as has been proposed,[28] it may be necessary to reinterpret the drug binding data on the basis of both the new FRET and NMR results.

## 11.6. Topography of Cardiac Troponin

### 11.6.1. FRET Studies of Cardiac TnI

Troponin I from several species, both skeletal and cardiac, have a single tryptophan which is highly conserved. Cardiac TnI has a unique N-terminal extension (32–34 residues) that is absent in the skeletal isoform and contains two unique sites (Ser23 and Ser24) of phosphorylation by PKA (cAMP-dependent protein kinase). Phosphorylation of these two residues of cTnI within the troponin complex results in a loss of $Ca^{2+}$ sensitivity in cardiac function. The structural basis of this effect is not well understood.

In a time-resolved anisotropy study of the single Trp192 residue in cTnI,[15] we obtained a long rotational correlation time of 24 ns, suggesting an axial ratio of about 4–5 for the protein. Phosphorylation of the two adjacent serines resulted in a decrease of the correlation time to 15 ns, indicative of a more compact or less asymmetric hydrodynamic shape of the phosphorylated cTnI. Four basic residues are located within an 11-residue segment in cTnI that includes Ser23 and Ser24: $^{19}$VRRRSSANYRA.[29] This segment may have an extended conformation due to electrostatic repulsions of the four positively charged side chains. If the two serines are phosphorylated, the two phosphate groups could be involved in electrostatic interactions with the adjacent arginyl side chains leading to a collapse of the extended conformation and a folding of the N-terminal segment towards the C-terminal end. This putative change was recently investigated by determination of the distance between Cys5 and Trp192 using IAANS attached to the cysteine as the energy donor.[16] The mean intersite distance was 45.3 Å in unphosphorylated cTnI, and 35.8 Å in phosphorylated cTnI. This large distance decrease, which is carried over to the cTnC-cTnI complex, supports the idea of a phosphorylation-induced folding of the N-terminal segment. This folding may provide a structural basis to understand how phosphorylation of cTnI by PKA may bring about its physiological effects which include reduction in the affinity of cTnI for cTnC[29] and of cTn for $Ca^{2+}$,[30] and an increase in the rate of $Ca^{2+}$ dissociation from cardiac troponin.[30]

### 11.6.2. The General Shape of cTnI

An early electron microscopy study showed skeletal TnI to be an elongated and extended molecule. This topography has been assumed for the cardiac isoform, but no independent structural information was available for cTnI. We generated nine single-cysteine mutants of cTnI in which the cysteine was located in positions 5, 40, 81, 98, 115, 133, 150, 167, and 192. A combination of the kinetics of sulfhydryl reactivity of these residues and FRET distances from Trp192 to the other eight cysteines was used to gain insights into the topography of individual cTnI and cTnI in complexes with cTnC and cTnC plus cTnT.[31] The results suggest an open and extended conformation of cTnI with a large curvature in which the cysteines are highly exposed to solvent. These structural features are largely retained for phosphorylated cTnI. Upon reconstitution into the trimeric troponin complex, cTnI remains elongated with constrained flexibility from residues 40 to the C-terminus. The highly flexible nature of the N-terminal extension of cTnI, however, is preserved in the complex, suggesting that this segment of cTnI is either not bound or only loosely bound to the C-domain of cTnC.

### 11.6.3. The cTnC-cTnI Complex

The $Ca^{2+}$-binding site III of both skeletal and cardiac TnC is known to have a high affinity for lanthanide ions, and the 12-residue cation binding loop has a single tyrosine located in the 7th position. The residue at this position is involved in cation coordination via its carbonyl oxygen rather than its side chain. We took advantage of this property and replaced the tyrosine in this loop of cTnC with a tryptophan. This mutant, cTnC(Y111W), enabled us to detect the luminescence of $Tb^{3+}$ bound to site III by irradiation of the tryptophan at 295 nm. The 335 nm tryptophan emission band is progressively quenched by increasing $Tb^{3+}$ concentration and the quenching is accompanied by the appearance and enhancement of the three $Tb^{3+}$ bands (Figure 11.8). $Tb^{3+}$ luminescence sensitized by aromatic protein residues has found wide applications in metalloenzymes and EF-hand proteins in which the bound cation is substituted by $Tb^{3+}$. In some systems, the sensitization has been shown to be due to FRET of the Förster type of dipole-dipole interaction, but in other systems the Dexter type of exchange mechanism may be involved.[32] The most intense $Tb^{3+}$band (545 nm) is a $^5D_4 \rightarrow {}^7F_5$ transition and has a very strong spectral overlap with the absorption band of tetramethyrhodamine (TMR). The Förster critical distance $R_o$ of the $Tb^{3+}$-TMR donor-acceptor pair is 60 Å in water.[33] We determined an interdomain distance by LRET from bound $Tb^{3+}$ at site III in the C-domain to Cys35 in the N-domain, using the 545 nm luminescence band as the energy donor and iodoacetamidotetramethylrhodamine (AATMR) attached to the cysteine as an energy acceptor.[34] The intensity decay of $Tb^{3+}$ bound to site III of Y111W

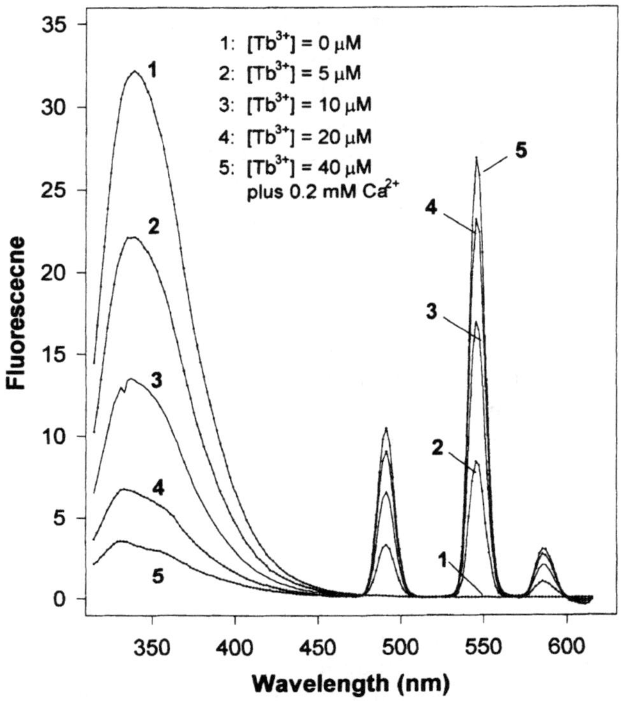


Figure 11.8. Steady-state emission spectra of cTnC mutant Y111W excited at 295 nm, in the presence of $Tb^{3+}$. [cTnC] = 5 μM, [$Tb^{3+}$] as indicated. The 333 nm tryptophan peak is progressively quenched in the presence of increasing [$Tb^{3+}$], and the quenching is accompanied by the appearance and enhancement of the three $Tb^{3+}$ bands (480, 545, and 585 nm). Trp111 is located in the binding loop of site III. These reciprocal changes indicate transfer of donor (Trp) energy to the acceptor $Tb^{3+}$ bound to site III. (From Ref. 34).

is monoexponential, $\tau = 1.47$ ms (top curve, Figure 11.9). In the presence of the acceptor, the decay becomes biexponential with a dominant rapid decay component, followed by a slow component. The fast component arises from donor quenching due to energy transfer from bound $Tb^{3+}$ to the acceptor at Cys35. The distribution of the interdomain distances has a mean distance of 48.0 ± 1.0 Å and a half-width of 9.4 ± 0.6 Å for a sample in which the regulatory site in the N-domain was not occupied.

Interestingly, the mean interdomain distance in apo cTnC is not affected upon formation of the cTnC-cTnI complex (49.2 ± 1.5 Å) or reconstitution with cTnI and cTnT into the three-subunit cardiac troponin complex (47.8 ± 1.2 Å). The corresponding changes in the hw are a decrease of 2–3 Å. The validity of these results is strengthened by the fact that the donor probe

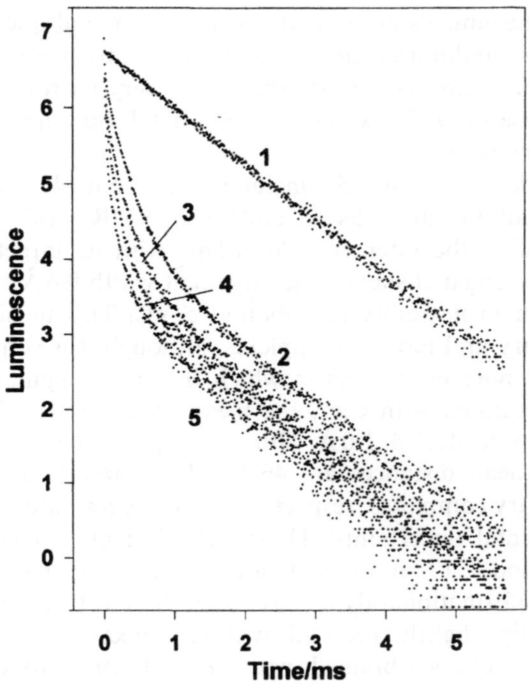


Figure 11.9. Luminescence decay of $Tb^{3+}$ bound to 5 $\mu$M cTnC mutant Y111W, which was labeled with IAATMR at Cys35. Curve 1, $Tb^{3+}$ bound to site III of unlabeled mutant where Trp111 is located, showing a single-exponential decay of the lanthanide in the absence of energy transfer from bound $Tb^{3+}$. The decay time is 1.47 ms. Curves 2–5, the decay of bound $Tb^{3+}$ determined in the presence of the acceptor. $[Tb^{3+}]$/[labeled protein] = 0.4, 1.0, 2.0, and 3.0 (plus 0.2 mM $Ca^{2+}$) for curves 2–5, respectively. The decays in the presence of the acceptor are biexponential. The fast quenched component arises from energy transfer from bound $Tb^{3+}$ to AATMR, and the slow component (~1.43 ms) reflects the decay of unquenched $Tb^{3+}$ due to incomplete acceptor labeling. The experiment was performed with 295 nm excitation generated from a pulsed xenon lamp at 100 Hz, and the $Tb^{3+}$ luminescence was detected at 545 nm. The 295 nm radiation excited the tryptophan residue and, through an energy transfer, enhanced the $Tb^{3+}$ luminescence (see Figure 11.8). The 545 nm band was isolated and its decay was measured. In the presence of an energy acceptor, the sensitized $Tb^{3+}$ became an energy donor to the acceptor. In the experiment described here, the $Tb^{3+}$ signal was collected after a 80-$\mu$s delay. Consequently, the initial amplitude of the decay curves was lost and this loss gave rise to an apparent large amplitude of the long component. This loss did not affect the resolution of the two decay components. The fast (quenched) component of the donor $Tb^{3+}$ decay in curve 4 is 0.37 ms. In a separate experiment with the same sample used for curve 4, the decay of the sensitized acceptor AATMR fluorescence intensity was measured and found to contain a dominant component with a lifetime of 0.37 ms, in agreement with the decay of the quenched donor signal. This agreement confirms that the change in the decay time of the bound $Tb^{3+}$ reflects accurately an energy transfer between the bound lanthanide ion and AATMR attached to Cys35. (From Ref. 34).

is bound $Tb^{3+}$, the luminescence of which arises from multiple electronic transitions acting as randomized donors even in the absence of any rotational motion. The orientation factor ($\kappa^2$) thus has a very narrow range (1/3–4/3), validating the use of $^2/_3$ for $\kappa^2$ in terbium LRET (luminescence resonance energy transfer) studies.

The interdomain distance includes the central helix which has considerable flexibility and has an undefined NMR structure in solution. Unlike calmodulin, the interaction of cTnC with its target proteins does not collapse its elongated shape. The large half-width 9.4Å suggests considerable inter-domain flexibility in unbound cTnC. This flexibility is carried over to the binary and ternary complexes, although it is somewhat reduced (hw = 6–7Å). Under conditions in which the single regulatory site in the N-domain is saturated with $Ca^{2+}$, the mean distance in unbound cTnC is reduced by 5.7Å to 42.3 ± 1.0Å with a negligible change in the hw. The corresponding mean distances are 46.7 ± 1.0Å and 47.0 ± 1.2Å for the binary and ternary complexes, respectively, and the hw values are unchanged within experimental uncertainty. These LRET results suggest that, in the absence of bound $Ca^{2+}$ at the regulatory site, cTnC has a considerable interdomain conformational dynamics. When bound to target proteins, this dynamics is only slightly reduced without effects on an inter-domain separation. The results obtained with reconstituted troponin are more physiologically relevant and show that the binding of activator $Ca^{2+}$ at the regulatory N-domain does not change the interdomain distance or the apparent interdomain flexibility. The longer interdomain distance in the holo troponin complex may reflect a more constrained central helix imposed by bound target proteins. In myofilaments, the $Ca^{2+}$-loaded troponin complex must have optimal structural features for $Ca^{2+}$-dependent interaction between TnC and TnI. Our other studies have suggested that within the holo cardiac three-subunit complex, the N-domain has an open conformation. It is not yet known whether other structural features may additionally facilitate the interaction in cardiac troponin, but some interdomain flexibility as demonstrated in the distribution of the distances may enhance the interaction.

Biochemical studies have established that TnC and TnI are bound to each other in an antiparallel fashion. We investigated the topography of the cTnC-cTnI complex with this antiparallel arrangement. For this purpose, we determined intersite distances across cTnC and cTnI in the complex using both FRET and LRET. The FRET distances were from each of three sites in cTnC (Cys35, Cys84, and Cye89) to eleven cysteines distributed along the sequence of cTnI. The donor 1,5-IAEDANS was attached to the cysteines in the cTnC, and the acceptor DAMBI was linked to the cysteines in the cTnI. These initial results suggest an elongated shape of the cTnC-cTnI complex in which the cTnI may wrap around the cTnC in a spiral fashion.[35] This model is consistent with the interdomain distance of cTnC in the reconstituted troponin complex described in the preceding paragraph.

A second set of distance mapping between cTnC and cTnI was carried out with LRET, using the sensitized emission of bound $Tb^{3+}$ at site III of the cTnC mutant Y111W as the energy donor and AATMR attached to the 11 single cysteines of cTnI as the acceptor. Figure 11.10 shows the distributions of five LRET distances from the cTnC site to the five cysteines located in the C-terminal half of cTnI. The most distal site from the C-domain of cTnC to

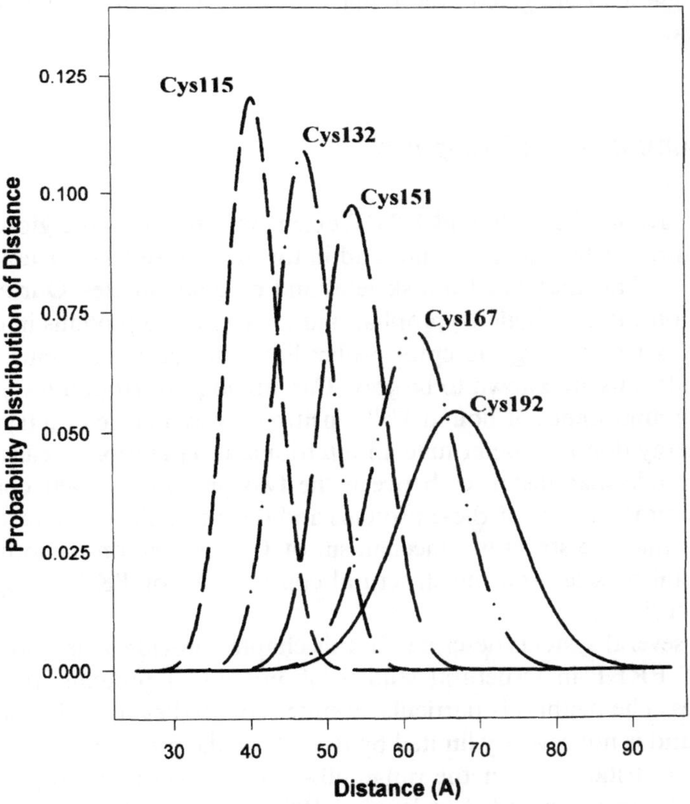


Figure 11.10. Distribution of intermolecular distances between mutants of cTnC and cTnI in the complex cTnC-cTnI, determined by LRET. The luminescence donor is $Tb^{3+}$ bound to site III in cTnC mutant Y111W, and the acceptor is IAATMR covalently linked to single cysteines in cTnI. The experimental protocol was the same as that described in the legend to Figure 11.9. LRET was measured in conditions in which the single regulatory site II in the cTnC was unoccupied.

cTnI is the Cys192, 19 residues from its C-terminus. The pattern of the decreases in the mean distance to the four other cysteines clearly indicates an antiparallel arrangement between the two proteins in which the C-terminal half of cTnI has an extended configuration in the binary complex. The distances to the six residues upstream from Cys115 toward the N-terminus are around 40 Å, within 2–3 Å from one another. A conceptual model based on these results for the cTnC-cTnI complex is one in which the C-terminal half of bound cTnI is extended and adjacent to the N-domain of cTnC and the N-terminal half of cTnI likely folded around the C-domain of cTnC. Our other fluorescence results indicate an elongated conformation of cTnC in its complex with cTnI or in reconstituted cardiac troponin. Taken together, these results suggest an elongated overall shape for the cardiac troponin C-troponin I complex.

## 11.7. Summary and Prospects

We have used FRET and LRET extensively to study the global structural features of troponin subunit, and in this review we have focused on the isoforms of TnC and TnI from skeletal and cardiac muscles. Our emphasis has been on the use single-tryptophan mutants of these proteins in conjunction with extrinsic energy acceptor probes linked to specific cysteine residues. Lanthanide ions are known to be good $Ca^{2+}$ analogs for troponin C. We have used the luminescence of bound $Tb^{3+}$ sensitized by excitation of a tryptophan as the energy donor to determine an interdomain separation in cardiac TnC and intermolecular distances between the two proteins in their complexes. The structural features of these proteins and their complexes are investigated to understand the structural mechanism of $Ca^{2+}$ activation in both cardiac and skeletal muscles and the structural consequence of PKA phosphorylation of cardiac TnI.

The several systems described in this chapter provide a clear idea of the power of FRET in structural studies of individual proteins and protein assemblies. The method is particularly suited for studies of global structural changes and is not severely limited by the size of the proteins. The resolution of intersite distances is in the range 40 Å to 50 Å when tryptophan is the donor, but can be extended to beyond 100 Å with suitable donor-acceptor pairs that are readily available. Unlike NMR or x-ray crystallography, FRET can be successfully applied to an assembly of macromolecules in solution and potentially to reconstituted systems such as contracting muscle fibers in which endogenous troponin subunits are exchanged with the corresponding proteins appropriately modified for structure/function studies. The exchange of endogenous subunits with exogenous proteins is now routinely done with myofibrils and chemically skinned muscle fibers in several laboratories, and such protocols can be used to incorporate appropriately labeled troponin subunits into biologically functioning systems for simultaneous measurements of fluorescence, FRET/LRET, enzymatic activity, and mechanical properties. Such measurements will provide global structural information that can be more directly correlated with functional properties within the same time window and on the same preparation.

## Acknowledgment

The work described herein has been supported, in part, by NIH grant HL52508.

## References

1. S. P. White, C. Cohen, and G. N. Phillips, Structure of co-crystals of tropomyosin and troponin, *Nature 325*, 826–828 (1987).
2. T. Wendt and K. Leonard, Structure of the insect troponin complex, *J. Mol. Biol. 285*, 1845–1856 (1999).
3. O. Herzberg and M. N. G. James, Structure of the calcium regulatory protein troponin-C at 2.8 Å resolution, *Nature 313*, 653–659 (1985).
4. M. Sundaralingam, R. Bergstrom, R. T. Strasburg, P. Rao, P. Roychowdhury, M. Greaser, and B.-C. Wang, Molecular structure of troponin C from chicken skeletal muscle at 3-Å resolution, *Science 227*, 945–948 (1985).
5. O. Herzberg, J. Moult, and M. N. G. James, A model for the $Ca^{2+}$-induced conformational transition of troponin C., *J. Biol. Chem. 262*, 2638–2644 (1986).
6. J. R. Pearlstone, T. B. Borgford, M. Chandra, K. Oikawa, C. M. Kay, O. Herzberg, J. Moult, A. Herklotz, F. Reinach, and L, S. Smillie, Construction and characterization of a spectral probe mutant of troponin C: application to analysis of mutants with increased $Ca^{2+}$ affinity, *Biochemistry 31*, 6545–6553 (1992).
7. G. Trig-Gonzalez, K. Racher, L. Burnick, and T. Borgford, A comparative spectroscopic study of tryptophan probes engineered into high- and low-affinity domains of recombinant chicken troponin C, *Biochemistry 31*, 7009–7015 (1992).
8. M. Chandra, W. D. McCubbin, K. Oikawa, C. M. Kay, and L. B. Smillie, $Ca^{2+}$, $Mg^{2+}$, and troponin I inhibitory peptide binding to a Phe-154 to Trp mutant of chicken skeletal muscle troponin C, *Biochemistry 33*, 2961–2969 (1994).
9. J. Gulati and V. G. Rao, The cardiac $Ca^{2+}$-deficient EF-hand governs the phenotype of the cardiac-skeletal TnC-chimera in solution by $Sr^{2+}$-induced tryptophan fluorescence emission, *Biochemistry 33*, 9052–9056 (1994).
10. C. D. Moyes, T. Borgford, L. LeBlanc, and G. F. Tibbits, Cloning and expression of salmon cardiac troponin C: titration of the low-affinity $Ca^{2+}$-binding site using a tryptophan mutant, *Biochemistry 35*, 11756–11762 (1996).

11. M. She, W.-J. Dong, P. K. Umeda, and H. C. Cheung, Tryptophan mutants of troponin C from skeletal muscle. An optical Probe of the regulatory domain, *Eur. J. Biochem. 252*, 600–607 (1998).
12. M. She, W.-J. Dong, P. K. Umeda, and H. C. Cheung, Time-resolved fluorescence study of the single tryptophans of engineered skeletal muscle troponin C, *Biophys. J. 73*, 1042–1055 (1997).
13. I. D. Clark and A. G. Szabo, A time-resolved fluorescence study of TnC single tryptophan mutant, F29W, *Biophys. J. 64*, A135 (1993).
14. C.-K. Wang, and H. C. Cheung, Proximity relationship in the binary complex formed between troponin I and troponin C, *J. Mol. Biol. 191*, 509–521 (1986).
15. R. Liao, C.-K. Wang, and H. C. Cheung, Time-resolved tryptophan emission study of cardiac troponin I, *Biophys. J. 63*, 986–995 (1992).
16. W.-J. Dong, M. Chandra, J. Xing, M. She, R. J. Solaro, and H. C. Cheung. Phosphorylation-induced distance change in a cardiac muscle troponin I mutant, *Biochemistry 36*, 6754–6761 (1997).
17. M. Chandra, W.-J. Dong, B-S. Pan, H. C. Cheung, and R. J. Solaro, Effects of protein kinase A phosphorylation on signaling between cardiac troponin I and the N-terminal domain of cardiac troponin C, *Biochemistry, 36*, 13305–13311 (1997).
18. X. Zhao, T. Kobayashi, H. Malak, I. Gryczynski, J. R. Lakowicz, R. W. Wade, and J. H. Collins, Calcium-induced troponin flexibility revealed by distance distribution measurements between engineered sites, *J. Biol. Chem. 270*, 15507–15514 (1995).
19. N. C. J. Strynadka and M. N. G. James, Crystal structures of the helix-loop-helix calcium-binding proteins, *Annu. Rev. Biochem. 58*, 951–998 (1989).
20. I. Gryczynski, H. Malak, J. R. Lakowicz, H. C. Cheung, J. Robinson, and P. K. Umeda, Fluorescence spectral properties of troponin C mutant F22W with one-, two-, and three-photon excitation, *Biophys. J. 71*, 3448–3453 (1996).
21. S. M. Gagné, S. Tsuda, M. X. Li, M. Chandra, L. B. Smillie, and B. D. Sykes, Quantification of the calcium-induced secondary structural changes in the regulatory domain of troponin C, *Protein Sci. 3*, 1961–1974 (1994).
22. M. She, J. Xing, W.-J. Dong, P. K. Umeda, and H. C. Cheung, Calcium binding to the regulatory domain of skeletal muscle troponin C induces a highly constrained open conformation, *J. Mol. Biol. 281*, 445–452 (1998).
23. J. R. Lakowicz, I. Gryczynski, W. Wiczk, G. Laczko, F. G. Prendergast, and M. L. Johnson, Conformational distributions of mellitin in water/methanol mixtures from frequency-domain measurements of nonradiative energy transfer, *Biophys. Chem. 36*, 99–115 (1990).
24. Z. Gong, J. Xing, M. Chandra, W.-J. Dong, R. J. Solaro, P. K. Umeda, and H. C. Cheung, Comparison of the regulatory domain conformation of troponin C from cardiac and skeletal muscle, *Biophys. J. 74*, A51 (1998).
25. W.-J. Dong, J. Xing, M. Villain, M. Hellinger, J. M. Robinson, M. Chandra, R. J. Solaro, P. K. Umeda, and H. C. Cheung, Conformation of the regulatory domain of cardiac muscle troponin C in Its complex with cardiac troponin I, *J. Biol. Chem. 274*, 31382–31390 (1999).
26. S. K. Sia, M. X. Li, L. Spyracopoulos, S. M, Gagné, W. Liu, J. A. Putkey, and B. D. Sykes, Structure of cardiac muscle troponin C unexpectedly reveals a closed regulatory domain *J. Biol. Chem. 272*, 18216–18221 (1997).
27. M. Pvaska and J. Taskinen, A model for human cardiac troponin C and for modulation of Its $Ca^{2+}$ affinity by drugs, *Proteins: structure, function, and genetics 11*, 79–94 (1991).

28. M. Pollesello, M. Ovaska, J. Kaiovola, C. Tilgmann, K. Lundstrom, N. Kalkkinen, I. Ulmanen, E. Nissinen, and J. Taskinen, *Binding of a new Ca²+ sensitizer, levosimendan, to recombinant human cardiac troponin C, J. Biol. Chem. 269*, 28584–28590 (1994).
29. R. Liao, C.-K. Wang, and H. C. Cheung, Coupling of calcium to the interaction of troponin I with troponin C from cardiac muscle, *Biochemistry 33*, 12729–12734 (1994).
30. W.-J. Dong, C.-K. Wang, A. M. Gordon, S. S. Rosenfeld, and H. C. Cheung, A kinetic model for the binding of $Ca^{2+}$ to the regulatory site of troponin from cardiac muscle, *J. Biol. Chem. 272*, 19229–19235 (1997).
31. W.-J. Dong, M. Chandra, J. Xing, R. J. Solaro, and H. C. Cheung, Structural mapping of single cysteine mutants of cardiac troponin I, *PROTEINS: structure, function, and genetics* (submitted, 2000),
32. C. W. V. Hogue, J. P. MacManus, D. Banville, and A. G. Szabo, Comparison of terbium (III) luminescence enhancement in mutants of EF hand calcium binding proteins, *J. Biol. Chem. 267*, 13340–13347 (1992).
33. P. R. Selvin and J. E. Hearst, Luminescence energy transfer using a terbium chelate: improvements on fluorescence energy transfer, *Proc. Natl. Acad. Sci. U.S.A. 91*, 10024–10028 (1994).
34. W.-J. Dong, J. M. Robinson, J. Xing, P. K. Umeda, and H. C. Cheung, An Interdomain distance of cardiac troponin C determined by fluorescence spectroscopy, *Protein Sci, 9*, 280–289 (2000).
35. J. Xing, W.-J. Dong, M. Chandra, R. J. Solaro, P. K. Umeda, and H. C. Cheung, Proximity mapping of the cardiac cTnT-cTnI complex, *Biophys. J. 76*, A279 (1999).

# 12

# Fluorescence of Extreme Thermophilic Proteins

Sabato D'Auria, Mosè Rossi, Ignacy Gryczynski, and Joseph R. Lakowicz

## 12.1. Introduction

On the planet earth exist organisms that live and thrive under conditions of extreme temperatures. These are known as thermophiles, and they are important as sources of thermostable enzymes. The macromolecules isolated from thermophiles are not only used in numerous biotechnological applications, but they are also ideal candidates for investigating the correlations between protein structure, function and stability. The understanding of the topological and dynamic aspects of enzyme structure in extremophilic bacteria can clarify the mechanism by which these proteins work in extreme conditions of temperature, hydrostatic pressure and salinity. The same principles that allow such an adaptation represent the basis of the general strategy used for enzyme molecules to pursue folding and function. In fact, the driving forces that are responsible for protein folding reflect the hierarchy of contributions involved in protein stabilization, that is, on the one hand, the nearest neighbor and through-space short-range interactions that optimize packing and minimize cavity volume and, on the other hand, the entropy effects due to water release from hydrophobic surfaces. Both the enthalpic and entropic contributions to the free energy of stabilization are affected by the extreme conditions that we will described in the chapter.

Protein macromolecules, even in their native state, are not in an unique structural state but fluctuate among a large number of conformations

*Sabato D'Auria*[1] and *Mosè Rossi* • Institute of Protein Biochemistry and Enzymology, C.N.R., 80125 Naples, Italy. *Ignacy Gryczynski and Joseph R. Lakowicz* • Center for Fluorescence Spectroscopy, University of Maryland at Baltimore, Baltimore, Maryland 21201.
[1] Also at NCFS. Correspondence to SD. dauria@dafne.ibpe.na.cnr.it

*Topics in Fluorescence Spectroscopy, Volume 6: Protein Fluorescence*, edited by Joseph R. Lakowicz. Kluwer Academic / Plenum Publishers, New York, 2000

differing in small structural details that can influence the emitting properties of tryptophanyl residues. Tryptophans in proteins are often used as spectroscopically active probes to monitor the structural features of macromolecules in solutions. Frequency-domain fluorometry is one of the most common spectroscopic methods to elucidate the structural and dynamics aspects of tryptophan containing proteins. The emission decay may be analyzed in terms of sum of few discrete exponential components or quasi-continuous life time distributions. In a distribution model, the center of the lifetime distribution is indicative of the average microenvironments surrounding the indolic residue, while the distribution width is related to the number of these different microenvironments that the indolic residue experiences during its permanence in the excited state.

In this chapter, we will describe some general aspects of the thermophilic micro-organisms and their enzymes. However, we would like to stress that several researchers are studying the thermophilic micro-organisms, and we cannot hope to cite all the work that has been done. Here, we will focus primarily on the uncommon structural features of the thermophilic enzymes as well as on their unusual functional properties by utilizing fluorescence spectroscopy techniques.

## 12.2. Thermophilic Micro-Organisms

Hyperthermophilic Bacteria and Archaea represent the organisms living at the upper border of life. Hyperthermophiles belong to phylogenetically distant groups and may represent rather ancient adaptations to heat. They have been isolated almost exclusively from environments with in situ temperatures between 80 and 115 °C. Natural biotopes of hyperthemophiles on land are water-containing volcanic areas like solfataric fields and hot springs with low salinity and wide range of pH's values, from 0.5 to 8.5. Marine biotopes are shallow submarine hydrothermal systems, abyssal hot vents (Black Smokers), and active seawounts. These environments contain high concentrations of salt. Although unable to grow, hyperthemophiles may survive for long time at ambient temperature. This ability may be essential for dissemination through the cold atmosphere and hydrosphere.[1]

In Figure 12.1 is depicted the 16S rRNA-based universal phylogenetic tree. As we can see, it shows a tripartite division of the living world, consisting of the domains of Bacteria, Archaea and Eukarya. The root is derived from phylogenetic trees of duplicated genes of ATPase subunits and elongation factors Tu and G. Short phylogenetic branches indicate a rather slow rate of evolution. Deep branching points are evidence for early separation of two groups. For example, the separation of the Bacteria from the

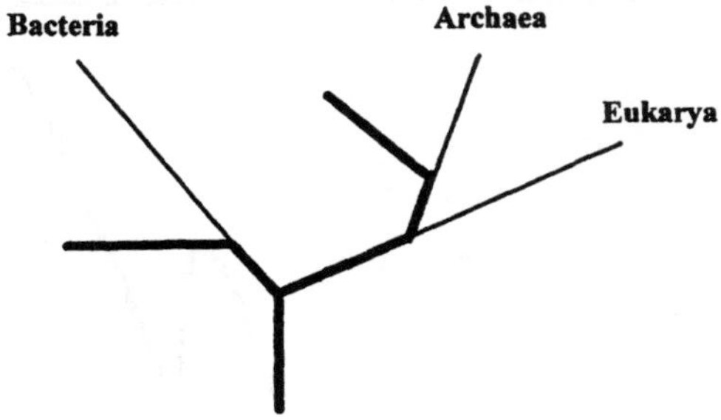


Figure 12.1. 16S rRNA-based universal phylogenetic tree.

Eukarya-Archaea lineage (Figure 12.1) is the deepest and earliest branching point known so far. Surprisingly, all the deepest and shortest lineages within the universal phylogenetic tree are represented by hyperthermophiles, indicating that these microorganisms appear to be the most primitive organisms still existing and the last common ancestror may have been a hyperthermophile. However, recently Galtier *et al.*[2] have shown that the estimates of ancestral G + C contents appear incompatible with a thermophilic life-style of the most recent common ancestror, suggesting that hyperthermophilic species evolved from mesophilic organisms via adaptation to high temperature.

## 12.3. Thermophilic Enzymes

The structural and functional features of proteins and enzymes isolated from thermophilic micro-organisms have attracted the interest of many research groups in the last years.[3–7] These enzymes are active and stable at high temperatures, and possess an unusual stability towards the denaturing action of the common protein denaturants.[8,9]

In Figure 12.2 is depicted the effect of temperature on the catalytic activity of the enolase from mesophilic (*Rabbit*), moderate thermophilic (*Thermus X1*) and hyperthermophilic (*Thermus aquaticus*) sources.[10,11] The hyperthermophilic enzyme is barely active at 30 °C, showing the maximal activity over 80 °C. These particular features prompted the use of thermophilic enzymes as suitable protein models for addressing a number of fundamental problems in current protein research[12,13] as well as their utilization as biocatalysts under rather harsh environmental conditions.[14–16] The investigations on thermophilic proteins and the recent protein engineering studies are leading to a quantitative understanding of the structure/stability/function

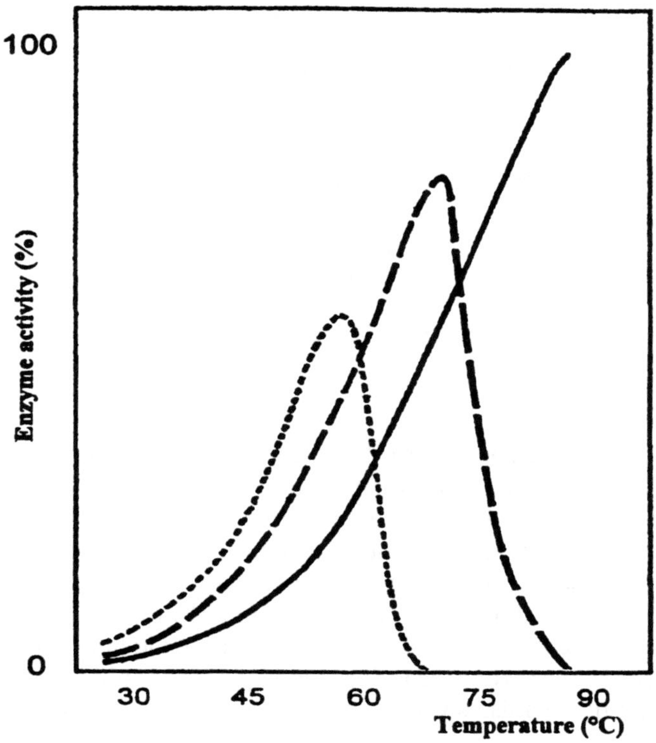


Figure 12.2. Effect of temperature on the catalytic activity of enolase from *Rabbit* (dotted line), *Thermus X1* (dashed line) and *Thermus aquaticus* (continuous line).

relationships in proteins and enzymes and to a larger use of these in pharmaceutical and food industries.

In fact, recently some useful guidelines for improving the protein stability have been deduced from studies on thermophilic enzymes and successfully applied in protein research.[17–19]

Several investigations on the molecular properties of thermostable enzymes pointed out that the uncommon stability of thermophilic proteins could be due to quite subtle differences between mesophilic and thermophilic proteins. The analysis of homologous proteins from mesophilic and thermophilic sources in terms of amino acid sequences and three-dimensional structures showed that the enhanced thermostability is the result of a variety of stabilizing effects, such as hydrophobic interactions, ionic and hydrogen bonding, disulfide bonds, metal binding, etc. Comparison of amino acid sequences of thermophilic and mesophilic molecules provided indications of certain preferences in terms of the amino acid composition of proteins from hyperthermophiles, which could determine a stabilizing effect.

In particular, lower levels of asparagine, glutamine, cysteine, methionine and tryptophan were suggested, which are more susceptible to deamidations or oxidation at high temperatures, and higher levels of isoleucine, alanine and prolines, which should provide tighter packing in hydrophobic cores and extra stability to loops. However, these results were descriptive and did not explain in detail the molecular mechanisms of stabilization.[20]

The increasing number of 3D-structures of enzymes and proteins from hyperthermophiles is making possible to shed some light on the determinants of protein stability.[21–25] One of the most relevant differences found when comparing the 3D-structures of proteins from hyperthermophiles with their mesophilic counterparts is an increased number of ion-pairs organized in large networks that serve to cross-link non-contiguous points of the protein structure. However, it is worth noting that even if the stabilizing effect of salt bridges in proteins was proposed more than 20 years ago,[26] the extent and manner of how electrostatic interactions could contribute to protein stability is still an object of the debate.[27] A variety of physical and chemical reasons that have been recently advanced to explain the enhanced enzyme thermostability have been recently reviewed.[28–30]

## 12.4. Conformational Stability of Extreme Thermophilic Enzymes

It is usually accepted that the tertiary structure of a protein is only marginally stable. The conformational stability of a protein is the sum of a large number of weak interactions, including hydrogen bonds, van der Waal interactions, salt bridges and hydrophobic effects, and the destabilizing forces arising largely from conformational entropy. All of these forces are affected in a complex way by environmental conditions, including, such as, solvent and temperature. In an average protein the sum of stabilizing interactions as well as the destabilizing forces are large and $\Delta G$ is only of the order of 40 $kJmol^{-1}$.[31] A single weak interaction, for example, may contribute up to 25 $kJmol^{-1}$.[31] From these general considerations, the very high stability of proteins from thermophilic sources does not seem so remarkable. Most estimates of $\Delta G$ of various proteins came from spectroscopic methods, and in particular fluorescence measurements have been often used to follow GdnHCl and urea transitions.[32] However, both GdnHCl and urea exert a significant influence on the fluorescence of tyrosine and tryptophan: the dependence of emissions of free tryptophan and tyrosine on the concentrations of GdnHCl or urea were found to be non-linear.[33] A question, therefore, arises: Are we justified in doing a linear extrapolation of pre- and post-transition base lines? Other disadvantages in using this technique to follow denaturation can be found in.[34]

The GdnHCl-induced denaturation of the $\beta$-glycosidase from the hyperthermophilic Archaeon *Sulfolobus solfataricus* was followed by steady-state fluorescence spectroscopy, at 25 °C. The steady-state fluorescence emission spectrum (Figure 12.3) with a maximum at 338 nm (the excitation was set at 295 nm) is dominated by the contribution of the 68 tryptophanyl residues. Figure 12.4 shows the dependence of the steady-state fluorescence intensity at 338 nm on GdnHCl concentration, at 25 °C.

Upon complete denaturation in 6 M GdnHCl, the exposure to water of the tryptophanyl residues leads to a red shift of the maximum to 354 nmn as well as a drastic decrease in the fluorescence intensity (Figure 12.4). The value of GdnHCl concentration corresponding at half-completion of the transition, indicated as $C_{1/2}$, and determined from both fluorescence observables was 2.9 M, with the concentration of the protein fixed at 0.01 mg/ml. The denaturation was completely reversible. Renaturation of the protein by suitable dilution of fully unfolded samples, keeping the concentration of the $\beta$-glycosidase fixed at 0.01 mg/ml, showed a complete recovery of all the native spectroscopic features. The extent of the renaturation did not depend on the

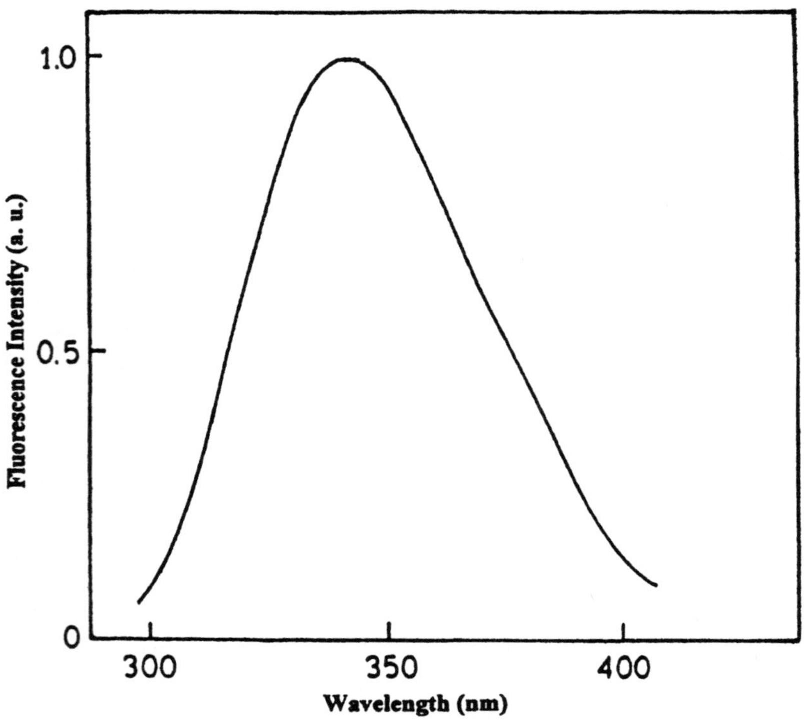


Figure 12.3. Staedy-state fluorescence spectrum of *Sulfolobus solfataricus* $\beta$-glycosidase at 25 °C. Protein concentration 0.05 mg/ml.

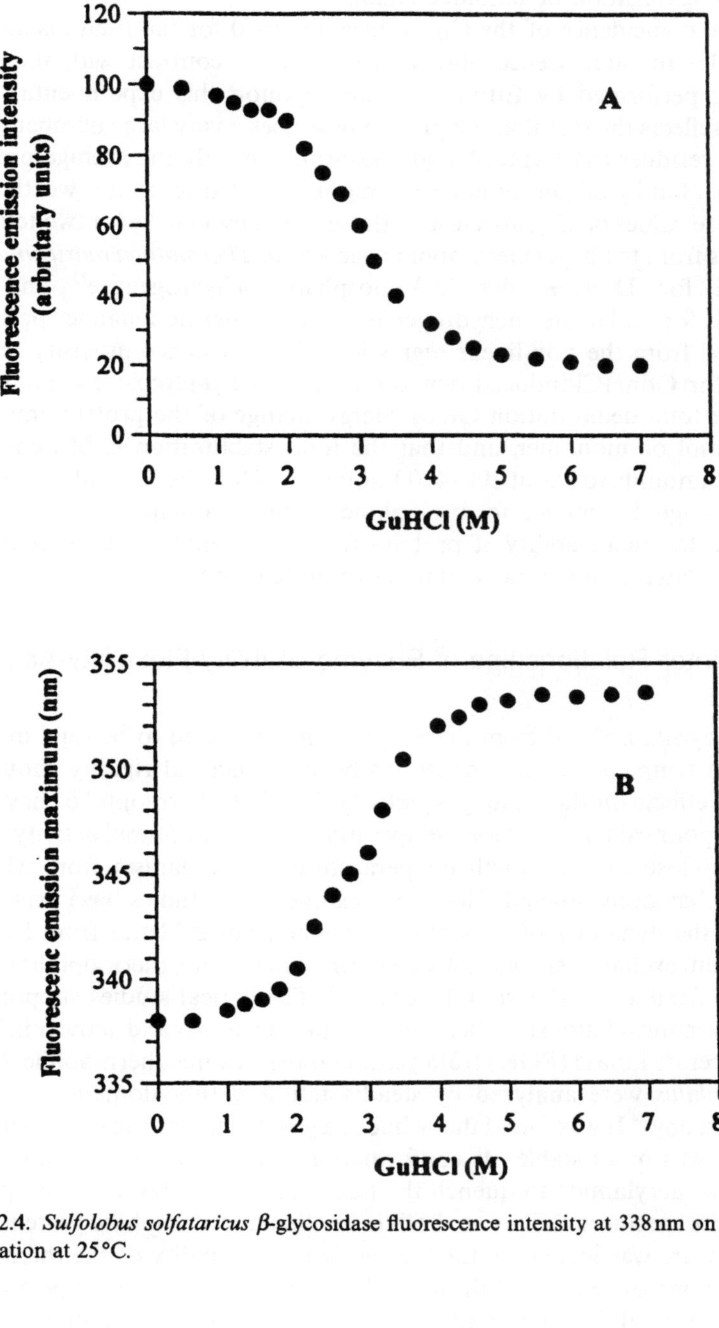


Figure 12.4. *Sulfolobus solfataricus* $\beta$-glycosidase fluorescence intensity at 338 nm on GdnHCl concentration at 25 °C.

incubation time, and in all probability, the low protein concentration used avoided aggregation of unfolded chains.

The coincidence of the $C_{1/2}$, values obtained for the β-glycosidase with two different fluorescence observables seems to contrast with the general analysis performed by Eftink.[34] In our opinion this experimental finding simply reflects the fact that the protein possesses a very large number of tryptophan residues (64 tryptophanyl residues), and only the average properties of such a family of fluorophores are monitored. However, it is worth to note that these values of $C_{1/2}$, are close to those found by Janicke for two tetrameric proteins from the hyperthermophilic bacterium *Thermotoga marittima*: 2.1 M GuHCl for D-glyceraldehyde-3-phosphato dehydrogenase[35] and 2.6 M GuHCl for L-lactate dehydrogenase.[36] The thermodynamic parameters obtained from the non-linear regression of fluorescence intensity measurements for GdnHCl-induced denaturation of the β-glycosidase pointed out that the total denaturation Gibbs energy change of the protein amounts to 196 kJ/mol of monomer, and that the total stabilization Gibbs energy per residue amounts to about 300–400 $J mol res^{-1}$.[37] These figures fall in the middle of the range determined for mesophilic globular protein,[38–40] indicating that the high thermal stability of proteins from thermophilic micro-organisms is not correlated to an extra stability at room temperature.

## 12.5. Inter-Relationships of Enzyme Stability-Flexibility-Activity

Enzymes isolated from thermophiles are expected to be rigid molecules at room temperature and consequently this structural rigidity should have adverse effects on their catalytic activity.[17] In fact, thermophilic enzymes are usually poor catalysts at room temperature and their optimal activity temperature is close to the growth temperature of the organism from which the enzyme has been isolated. However, relatively few studies have been carried out on the dynamics of very stable enzymes, but evidence from hydrogen-deuterium exchange shows that, at a given temperature, thermophilic enzymes are less flexible than mesophilic ones.[41,42] Theoretical studies support this.[43] Moreover, the relationships between stability, dynamics and activity in 3-phosphoglycerate kinase (PGK) from *yeast* and the extreme thermophile *Thermus thermophilus* were analyzed by steady-state and time-domain fluorescence spectroscopy.[44] It was found that while at a given temperature the thermophilic protein was more stable, its conformational dynamics, as measured by the ability of acrylamide to quench the fluorescence of a buried tryptophan, as well as its specific activity were both lower than for mesophilic protein. As the temperature was increased, the thermodynamic stability of the thermophilic protein approached that of the mesophilic one at its working temperature. The conformational dynamics and the specific activity of a thermophilic enzyme increased up to the physiological operational temperature, and they became similar to those of a mesophilic enzyme at its operational temperature. These results suggested a direct relationship and balance holds between thermodynamic stability, dynamics and specific activity in globular proteins.

3-Phosphoglycerate kinase is a mono-tryptophanyl protein. Even if the interpretation of the fluorescence data is more easy in mono tryptophnyl proteins, it should be stated that the fluorescence decay from single tryptophan proteins gives information that is restricted to the local fluorophore surrounding. Therefore, there is only partial confidence that the achieved conclusions are representative for the whole protein structure. Conversely, the emission decay from multi-tryptophan proteins offers a general picture of the dynamic behavior of the overall protein structure, provided that the numerous tryptophanyl residues are quite homogeneously distributed in the primary structure and that each of them contributes to the fluorescence.

## 12.6. Hyperthermophilic β-glycosidase from the Archaeon S. solfataricus

The model enzyme we have chosen to study the relationships between activity, flexibility and stability is the β-glycosidase isolated from the hyperthermophilic Archaeon *Sulfolobus solfataricus* (Sβgly). This enzyme, a tetramer of 240 kDa and composed by four identical subunits, shows a wide substrate specificity, is active at high temperature, is thermostable, and is also stable and active in the presence of detergent and organic solvents. Moreover, its structure has recently been solved at 2.6 Å revealing the positions of the 17 tryptophan residues per subunit.[45] The conformational dynamics of the Sβgly were investigated in a wide range of temperature as well as upon addition of organic solvents and detergents by frequency-domain fluorometry. The data point out a relationship between the enzyme activity and the protein conformational dynamics.

SβGly was crystallized in its native tetrameric form and its structure was solved at 2.6 Å using multiple isomorphous replacement (Figure 12.5).[45] The protein shows the classic (βα)8 fold that was observed in the two mesophilic members of the glycosyl hydrolase family-1 crystallized so far: the cyanogenic β-glucosidase from *Trifolium repens*.[46–47] Several theories, mainly based on the comparison of the amino acid sequences of thermophilic and mesophilic molecules, were proposed in order to explain the molecular origins of the enhanced stability of enzymes from hyperthermophiles. When this analysis is applied to Sβgly it is clear that most of these rules do not hold, since the enzyme shows a lower alanine and isoleucine content and a higher tryptophan and asparagine content than the mean of the mesophiles.[45] In contrast, two structural features might contribute to thermal stabilization in

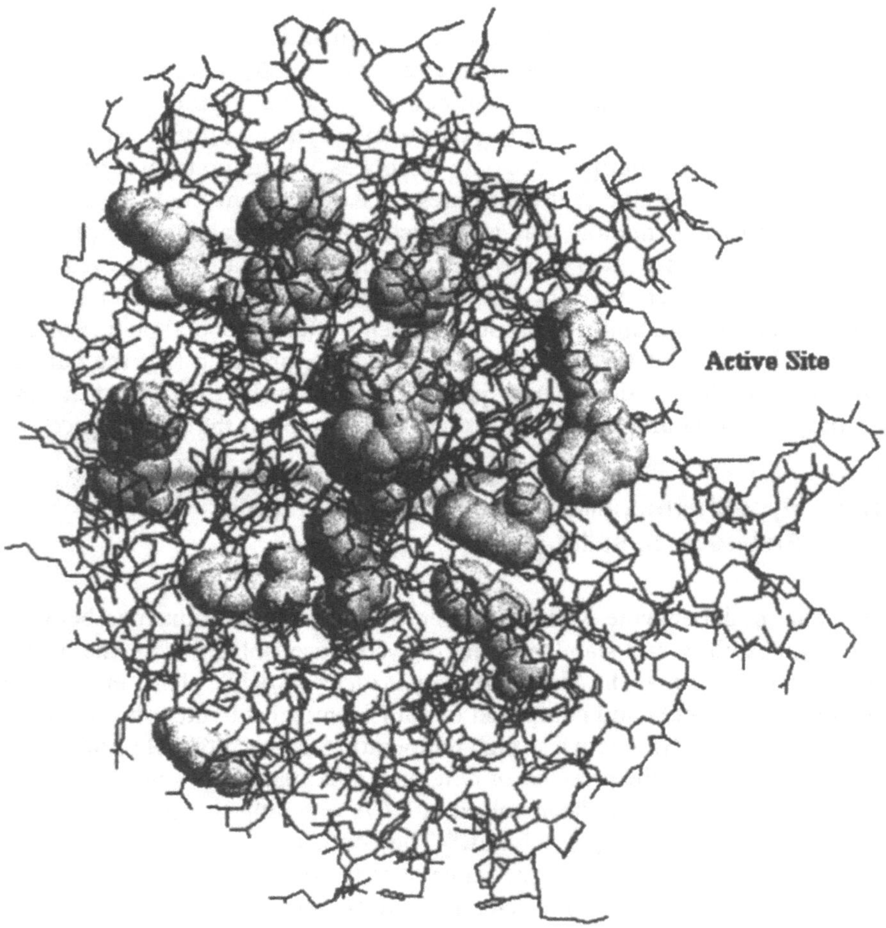


Figure 12.5. The *Sulfolobus solfataricus* β-glycosidase subunit resolved at 2.6 Å. The tryptophanyl residues are shown as spacfill.

Sβgly. This protein maintains a surprisingly higher number of buried hydrophilic cavities than is generally observed,[46,47] with one water molecule every 11.4 amino acid residues versus a 1 : 27 ratio in proteins in general. This feature has only been observed in Sβgly.

The second structural feature is the high number of ionic groups involved in ion-pairs. The Sβgly tetramer contains 524 charged groups including the α-amino and the carboxyl groups; 58% of these are involved in ion-pairs interactions and about 60% of them occur as part of multiple ion-pairs networks with at least three charged centers. As a comparison, the cyanogenic $\beta$-glucosidase from clover has only 41% of its charged residues involved in ion-pairs and over 65% are isolated pairs. The networks tend to occur in non-contiguous positions covering the surface and spanning different domains and subunits. In this way, the networks could act as an electrostatic cross-link between folded structural elements. The abundance of ion-pairs arranged in large networks was found in other enzymes from hyperthermophiles.[46–47] Interestingly, arginines frequently occur in ion-pair networks because of their ability to form multidentate interactions; this might explain the higher number of arginine found in enzymes from hyperthermophiles. These finding suggest that large networks ion-pairs are important for the thermal stability.

Here we report the effect of some perturbant agents on the structure of S$\beta$gly monitored by fluorescence spectroscopy. In particular, we will try to correlate the enzyme activity to the conformational dynamics of the enzyme at different temperature as well as in the presence of detergents, organic solvents, etc.

## 12.7. Effect of Temperature on Tryptophanyl Emission Decay

Figure 12.6 shows the dependence of the S$\beta$gly enzymatic activity on the temperature. In order to study the enzyme activity over 100 °C we used a special stainless cell which allowed to avoid the boiling of the sample.[48] As shown in Figure 12.6, the enzyme shows the maximal activity at 125 °C. Moreover, it is worth to note that the enzyme was still very active at 150 °C.

Figure 12.7 shows the steady-state fluorescence emission spectra of S$\beta$gly at different temperatures. The fluorescence emission of S$\beta$gly with excitation at 295 nm is dominated by the contribution of tryptophanyl residues. The staedy-state emission spectrum of S$\beta$gly at 25 °C shows an emission maximum at 340 nm, that it is blue-shifted compared with the emission maximum of N-acetyltryptophanylamide, centred at 348 nm.[49] The temperature increase to 90 °C causes a marked reduction of the fluorescence emission intensity without appreciable spectral shift (2 nm); further temperature increase to 125 °C results in a spectral shift to 348 nm.

These results suggest that the protein structure does not undergo to dramatic conformational changes and that the shift of the emission maximum could be due to more exposure of the tryptophan residues to the solvent or to a modest increase in the extent of spectral relaxation at high temperatures.

The tryptophanyl-emission decay properties of S$\beta$gly were investigated by frequency-domain fluorometry. Frequency-domain data were obtained with a frequency domain fluorometer operating between 2 and 2000 MHz.[50]

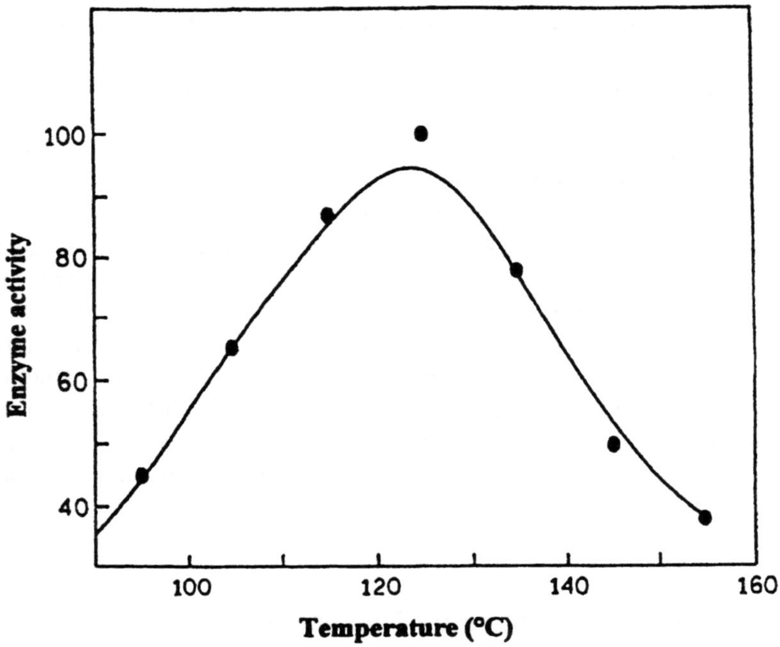


Figure 12.6. *Sulfolobus solfataricus* β-glycosidase activity at different temperatures.

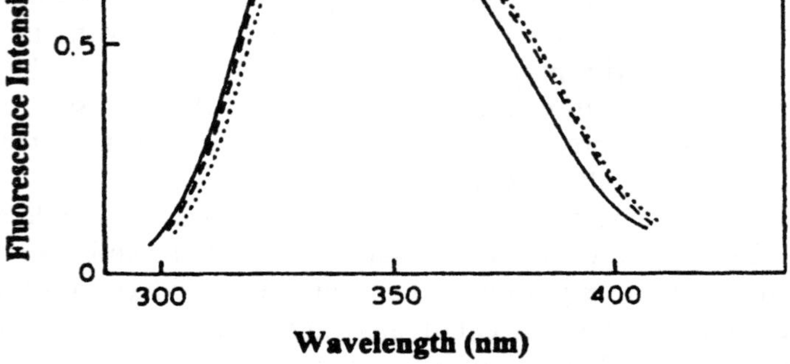


Figure 12.7. Staedy-state emission spectra of *Sulfolobus solfataricus* β-glycosidase at different temperatures.

The modulated excitation was provided by the harmonic content of a laser pulse train with a repetition rate of 3.75 MHz and a pulse width of 5 ps from synchronously pumped and cavity dumped rhodamine 6G dye laser. The dye laser was pumped with a mode-locked argon ion laser (Coherent, Innova 100, USA). The dye laser output was frenquency doubled to 295 nm for tryptophan excitation and the intensity decay measurements were performed by using the magic angle polarizer orientations.[51–53] The fluorescence emission decay was observed through an interference filter at 340 nm. The observed frequency response is complex as consequence of the large fluorescence heterogeneity related to the high tryptophan content of the Sβgly, as well as to the intrinsic protein dynamics. The emission decay was studied as function of temperature and the observed phase shifts and modulation factors are shown in Figure 12.8. The data were analyzed in terms of multi-exponential model. The best fits were obtained using the three exponential models. Figure 12.8 also shows the effect of iodide on the Sβgly emission decay at 125 °C (Figure 12.8D). As we can see, the Sβgly emission decay in the absence and in the presence of 0.2 M iodide at 125 °C are almost the same, indicating that some tryptophanyl residues are not accessible to the quencher molecules even at 125 °C, probably because they are localized in buried regions of the protein macromolecule.

Table 12.1 shows the multi-exponential analysis of the intensity decays of the protein at different temperatures. However, in an attempt to visualize the conformational dynamics of the protein at different temperatures we analyzed the data by the lifetime distribution model.[54] In our opinion, the interpretation of the emission decay in terms of continuous distribution is more satisfying than that obtained by means of discrete components, not only on a statistical basis, but because of the large number of tryptophanyl residues that the protein possesses.

The upper part of the Figure 12.9 shows the bimodal tryptophanyl-lifetime distribution of Sβgly at 20 °C. Two well separated components appear in the lifetime distribution: one corresponding to the short component with a center at 2.2 nsec, and the other corresponding to the long component centered at 7.0 nsec. The short-component is broad, with a width

Table 12.1. Mean Lifetime and Intensity Decay Parameters of S_gly

| | $\tau_1$ (ns) | $\tau_2$ (ns) | $\tau_3$ (ns) | $\alpha_1$ | $\alpha_2$ | $\alpha_3$ | $\chi^2$ |
|---|---|---|---|---|---|---|---|
| Sβgly 25 °C | 0.72 | 2.6 | 7.4 | 0.15 | 0.55 | 0.29 | 1.0 |
| S&gly 90 °C | 0.83 | 2.4 | 6.2 | 0.56 | 0.39 | 0.04 | 1.2 |
| Sβgly 125 °C | 0.17 | 1.0 | 4.3 | 0.61 | 0.36 | 0.018 | 1.3 |

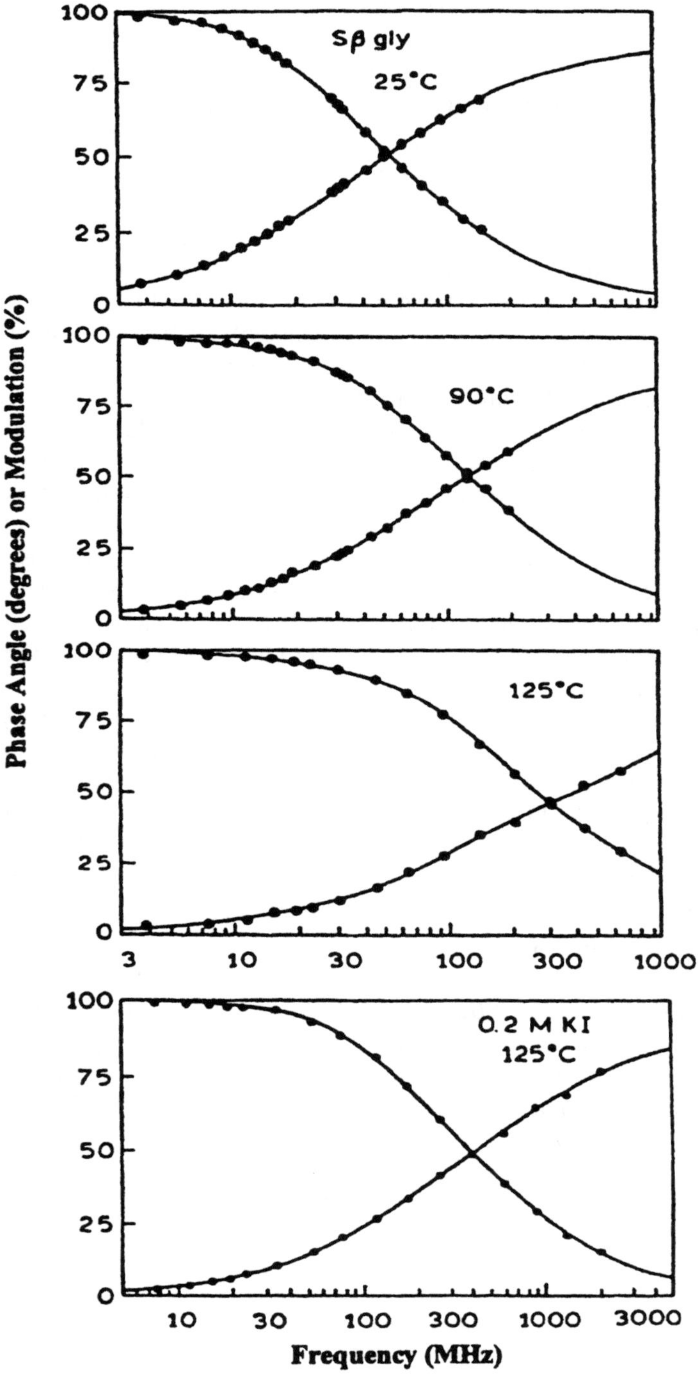


Figure 12.8. Frequency dependence of the phase shift and demodulation factor of *Sulfolobus solfataricus* β-glycosidase fluorescence emission at different temperatures and in the presence of iodide.

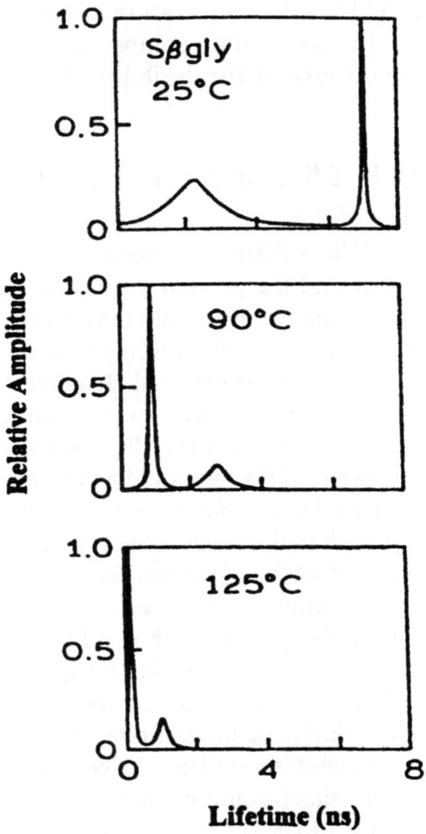


Figure 12.9. Tryptophanyl lifetime distribution pattern of *Sufolobus solfataricus* β-glycosidase at different temperatures.

of 1.5 nsec. In contrast, the long-component is very sharp, the width being 0.1 nsec. The middle part of the Figure 12.9 shows the Sβgly tryptophanyl—lifetime distribution at 90 °C. As we can see, the centers of both components are shortened, being the short- and long-component centered at 0.9 and 2.7 nsec, respectively. Finally, at 125° the short component, centered at 0.14 nsec, become very sharp (0.073 nsec) and the long-component, with a width of 0.28 nsec, is centered at 0.98 nsec (bottom, Figure 12.9).

The quenching and emission decays data point out that Sβgly retains the structural organization in a wide range of temperature and that the flexibility increase of the protein structure may be directly related to the enzymatic activity. The large number of sub-states, characterized by the same energy content but differing in some structural details, are responsible for the broadness of the fluorescence lifetime distributions of the protein at 25 °C, what is a temperature at which the enzyme does not show any activity.[55] Increasing the temperature results in a sharpening of the distribution components, and at 125 °C (the temperature at which the enzyme displays the maximal activity) the distribution components become very short and narrow, indicating a high degree of the flexibility of the protein structure.

## 12.8. Effect of pH on Tryptophanyl Emission Decay of S*β*gly

When S*β*gly is exposed at pH 10.0 its structure is affected to various extent and the protein displays a reduced enzymatic activity. The perturbation is detectable by different spectroscopic techniques.[56] Here, we report the effects of pH 10.0 on S*β*gly tryptophanyl emission decay, at 25 °C.

The conformational dynamics of S*β*gly at pH 7.0 and pH 10.0 were investigated by frequency-domain fluorometry. The best fits were obtained from bimodal-lifetime distributions with Lorentian shapes. Figure 12.10a shows the S*β*gly bimodal tryptophanyl distribution at pH 7.0, 25 °C. Two well separated components appear in the lifetime distribution, suggesting that the tryptophanyl emission decays can be represented from a short-component, centered at 2.2 nsec, and from a long-component, centered at 7.0 nsec. The short component is broad, with a width of 1.5 nsec, while the long component is sharp (the width is 0.1 nsec).

Figure 12.10b shows the S*β*gly bimodal tryptophanyl distribution at pH 10.0, 25 °C. As we can see, the two distribution components become broader, particularly the longest one, whose width changes from 0.1 to 1.2 nsec, with a concomitant shift of the center from 7.0 to 6.2 nsec. Moreover, the center of the short component is essentially unchanged, and the width increases from 1.5 to 2.5 nsec. These observations indicate that the protein at pH 10.0 assumes a more structurally and/or solvent exposed structure. In fact, the width of both components increases, indicating that the number of different microenvironments of the tryptophanyl residues is enhanced. It is likely that the deprotonation of some residues at pH 10, where the protein possesses a net negative charge (isoeletric point is 4.5),[55] introduces electrostatic repulsions that weaken the intramolecular interactions and favor at the same time solvent permeation inside the protein matrix. As consequence, many others sub-states of the conformational space become accessible to the protein.

## 12.9. Effect of Organic Solvents on S*β*gly Tryptophanyl Emission Decay

In a previous investigation[57] we showed the effect of a some aliphatic alcohols on the activity and structure of S*β*gly. The enzyme activity was

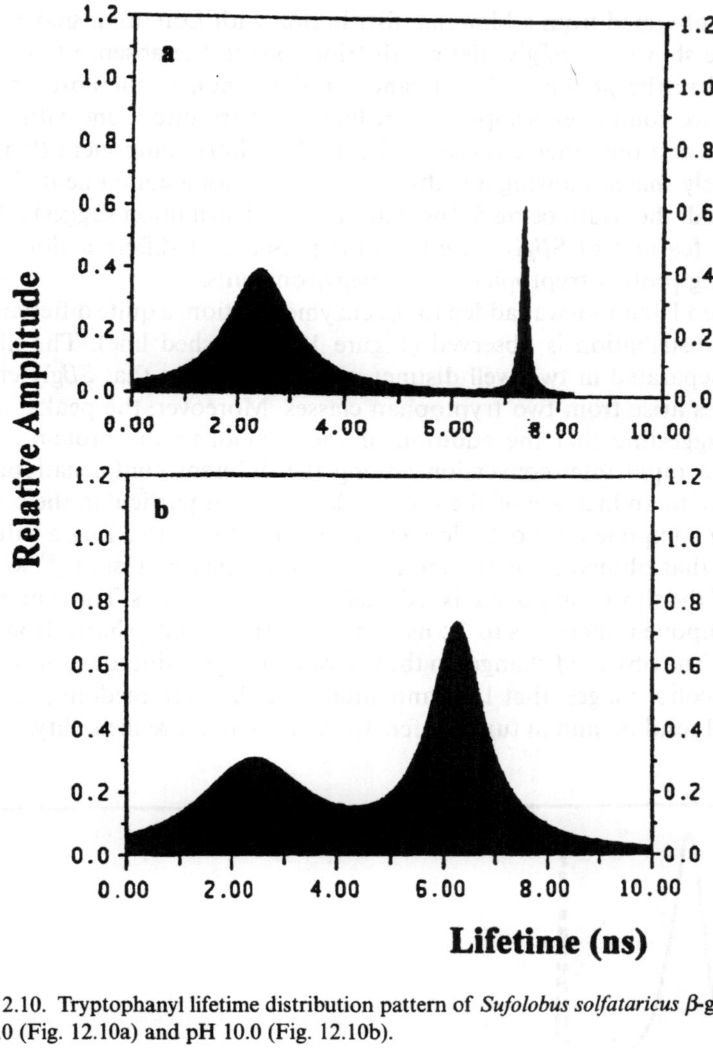


Figure 12.10. Tryptophanyl lifetime distribution pattern of *Sufolobus solfataricus* β-glycosidase at pH 7.0 (Fig. 12.10a) and pH 10.0 (Fig. 12.10b).

stimulated by the addition of alcohols, and in particular the addition of 0.4 M n-butanol to the enzyme solution resulted in the maximal activation. Moreover, we showed that circular dichroism spectra and Fourier Transform Infrared spectroscopy failed to structural variations of the protein in the presence of alcohols. Steady-state fluorescence spectra of Sβgly were also similar both in the presence and in the absence of the alcohol. In an attempt to visualize the conformational dynamics of Sβgly alone and in the presence of 1-butanol we analyzed the data by the lifetime distribution model.[54] The best fits were obtained from a bimodal distribution with Lorentian shape. Figure 12.13 are shows the Sβgly lifetime distributions in the absence (continuous line) and in the presence of 1-butanol (dashed line). In the absence of the alcohol two components appear in the lifetime distribution: one with a center at 0.54 ns and the other centered at 2.5 ns. The short component (0.54 ns) is moderately sharp, showing a width of 0.6 ns. The long component (2.5 ns) is very broad, the width being 5.9 ns. This lifetime distribution suggests that the emission features of Sβgly arise from the presence of different slowly interconverting protein tryptophanyl microenvironments.

When 1-butanol was added to the enzyme solution, a quite different Sβgly lifetime distribution is observed (Figure 12.11, dashed line). The lifetimes appear separated in two well distinct peaks, suggesting that Sβgly emissive properties arise from two tryptophan classes. Moreover, the peaks are very sharp suggesting that the addition of the alcohol to the protein solution induces a rapid inter-conversion among the different conformational sub-states due to an increase of the protein flexibility. In particular, the center of the short component becomes longer, passing from 0.54 to 2.2 ns, a value very close to that observed for the monomeric tryptophanyl residue,[49] while the width of the short component is reduced from 0.6 to 0.01 ns. The center of the long component increases to 7.5 ns and its width becomes sharp (from 5.9 to 0.33 ns). The observed changes in the emission decays induced by the addition of the alcohol suggest that 1-butanol induces additional freedom to the tryptophanyl residues and in turn confers to the protein more flexibility.

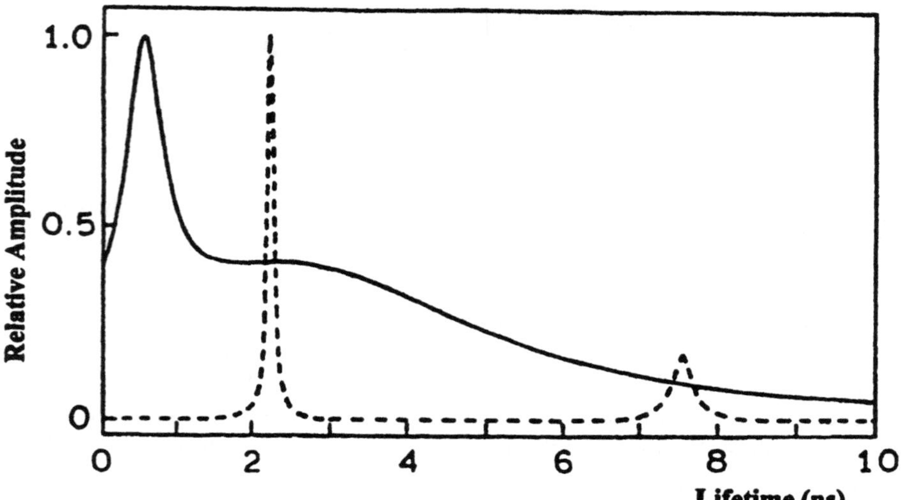


Figure 12.11. Tryptophanyl lifetime distribution pattern of *Sufolobus solfataricus* β-glycosidase in the absence (continuos line) and in the presence of 80 mM n-butanol (dotted line).

In conclusion, understanding protein behavior in biological reactions is fundamental to shedding light on the mechanism governing biochemical processes and to determining the influence that the polypeptide chain exerts on the active site. Biological activity and native structure of a protein are strictly linked: small structural alterations in the macromolecule may produce profound effects on the protein behavior. It is well known that chemical (e.g. organic solvents) and physical (e.g. temperature) perturbants affect both the structural and functional properties of biological macromolecules; multi-component solvents such as aquo-alcohol mixtures were shown to influence significantly the thermodynamic stability of a number of proteins.

Our results show that the presence of different alcohols causes a marked enzyme activation at low temperature. The circular dichroism and infrared spectroscopy analyses point out that the secondary structure of the protein is not affected by the presence of 1-butanol (data not shown).

On the other hand the fluorescence decay data indicate that the addition of 80mM 1-butanol to the protein solution affects the protein microenvironment, inducing a more flexible protein structure which is probably the origin of the increased enzyme activity. Moreover, these results also show the power of the time-resolved fluorescence to detect small environmental changes in the protein structure not observed by the other techniques.

In conclusion, we suggest that the fluorescence measurements on extreme thermophilic enzymes can give original insights in the study of their structure-function relationships with particular relevance to their conformational dynamics. Additional data from other thermophilic proteins are needed, and such work is in progress.

## Acknowledgments

This chapter is dedicated to the Memory of Dr. Mario Milan.

We thank Mr. Carlo Vaccaro for his technical assistance and Dr. Ferdinando Febbraio for the assistance in the preparation of the figures. This work was supported by an EU contract "Extremophiles as Cell Factories" and by the National Center for Research Resources, NIH RR-08119.

## References

1. Woese, C. R., Kandler, O., and Wheelis, M. L. (1990) Towards a natural system of organisms: Proposal for the domains Archaea, Bacteria and Eucarya. Proc. Natl. Acad. Scie. USA, 87, 4576–4579.

2. Galtier, N., Tourasse, N., and Gouy, M. (1999) A Nonhyperthermophilic common ancestror to extant life forms, Science, 283, 220–221.
3. Van den Burg, B., Vriend, G., Veltman, O. R., Venema, G., and Eijsink, V. G. H. (1998) Engineering an enzyme to resist boiling. Proc. Natl. Acad. Sci. USA, 95, 2056–2060.
4. Vielle, C., and Zeikus, J. K. (1996) Thermoenzymes. Biotechnol. Annu. Rev., 2, 1–83.
5. Daniel, R. M., Dines, M., and Petach, H. (1996) The denaturation and degradation of stable enzymes at high temperatures. Biochem. J., 317, 1–11.
6. D'Auria, S., Barone, R., Rossi, M., Nucci, R., Barone, G., Fessas, D., Bertoli, E., and Tanfani, F. (1997) Effects of temperature and SDS on the structure of $\beta$-glycosidase from the thermophilic archaeon *Sulfolobus solfataricus*. Biochem. J., 323, 833–840.
7. Perutz, L., and Raidt, H. (1975) Stereochemical basis of heat stability in bacterial ferredoxins and in haemoglobin A2. Nature, 255, 256–259.
8. Guagliardi, A., Manco, G., Rossi, M., and Bartolucci, S. (1989) Stability and activity of a thermostable malic enzyme in denaturants and water-miscible organic solvents. Eur. J. Biochem., 183, 25–30.
9. D'Auria, S., Rossi, M., Barone, G., Catanzano, F., Del Vecchio, P., Graziano, G., and Nucci, R. (1996) Temperature-induced denaturation of $\beta$-glycosidase from the archaeon *Sulfolobus solfataricus* J. Biochem., 120, 292–300.
10. Barnes, L. D., and Stellwagen, E. (1973) Enolase from the thermophile *Thermus X-1*. Biochemistry, 12, 1559–1565.
11. Stellwagen, E., Crounlund, M. M., and Barnes, L. D. (1973) A thermostable enolase from the extreme thermophile *Thermus aquaticus* YT-1. Biochemistry, 12, 1552–1559.
12. Jaenicke, R. (1991) Protein stability and molecular adaptation to extreme conditions. Annu. Rev. Biophys. Bioeng., 10, 1–67.
13. Adams, M. W. W., Perler, F. B., and Kelly, R. M. (1995) Extremoezymes: expanding the limits of biocatalysis. Ann. Rev. Microbiol., 47, 627–658.
14. Somero, G. N. (1995) Proteins and temperature Ann. Rev. Physiol., 57, 43–68.
15. Zamost, B. L., Nielsen, H. K., and Starves, R. L. (1991) Thermostable enzymes for industrial applications. J. Ind. Microbiol., 8,71–82.
16. Danson, M. J., Hough, D. W., Russel, R. J. M., Taylor, G., and Peral, L. (1996) Enzyme thermostability and thermoactivity. Protein Eng., 9, 629–630.
17. Fersht, A. R., and Serrano, L. (1993) Pronciples of protein stability derived from protein engineering experiments. Curr. Opin. Struct. Biol., 3, 75–83.
18. Fagain, C. O. (1995) Understanding and increasing protein stability Biohim Biophys Acta, 1252, 1–14.
19. Daniel, R. M., Cowan, D. A., Morgan, H. M., and Curran, P. (1982) A correlation between protein thermostability and thermoactivity. Protein Eng., 9, 629–630.
20. Menendez-Arias, L., and Argos, P. (1989) Engineering protein thermostability sequence statistics point to residue substitutions in $\alpha$-helices. J. Mol. Biol., 206, 397–406.
21. Baumann, H., Knapp, S., Lundback, T., Ladenstain, R., and Hrd, T. (1994) Solution structure and DNA-binding properties of a thermostable protein from the archaeon *Sulfolobus solfataricus*. Nature Struct. Biol., 1, 808–819.
22. Russel, R. J. M., Hough, D. W., Danson, M. J., and Tajlor, G. L. (1994) The crystal structure of citrate synthase from thermophilic archaeon *Thermoplasma acidiphilum* Structure, 2, 1157–1167.
23. Henning, M., Dairmont, B., Sterner, R., Kirschner, K., and Jansonius, J. N. (1995) 2.0 Å structure of the indole-3-glycerophosphate synthase from the hyperthermophile *Sulfolobus solfataricus*: possible determinants of protein stability Structure, 1295–1306.

24. Korndfer, I., Steipe, B., Huber, R., Tomschy, A., and Jaenicke, R. (1995) The crystal structure of holo-glyceraldehyde-3-phosphate dehydrogenase from hyperthermophilic bacterium *Thermotoga marittima* at 2.5 Å resolution, J. Mol. Biol., 246, 511–521.
25. Yip, K. S. K., Stillman, T. J., Britton, K. L., Artymiuk, P. J., Baker, P. J., Sedelnikova, S. E., Engel, P. C., Pasquo, A., Chiaraluce, R., Consalvi, V., and Rice, D. W. (1995) The structure of Pyrococcus furiosus glutamate dehydrogenase reveals a key role for ion-pair networks in mantaining enzyme stability at extreme temperatures, Structure, 3, 1147–1158.
26. Perutz, M. F., and Raidt, H. (1975) Stereochemical basis of heat stability in bacterial ferredoxins and haemoglobin A2. Nature, 255, 256–259.
27. Goldman, A. (1995) How to make my blood boil, Structure, 3, 1277–1279.
28. Querol, E., Perez-Pons, J. A., and Mozo-Villaris, A. (1996) Analysis of protein conformational characteristics related to thermostability Protein Eng., 9, 265–271.
29. Vogt, G., and Argos, S. (1997) Protein thermal stability: Hydrogen bonds or internal packing? Folding & Design 2, 540–546.
30. Vogt, G., Woel, S., and Argos, P. (1997) Protein thermal stability, hydrogen bonds and ion pairs. J. Mol. Biol., 269, 631–643.
31. Ahmed, F. (1993) Measuring the conformational stability of enzymes. In Thermostability of enzymes Edior MN Gupta, pp. 96–122, Springer-Verlag, NY, USA.
32. Pace, C. N. (1986) Methods Enzymol, 131, 266–280.
33. Schmid, F. X. (1989) In Protein Structure: a practical approach, Creighton, T. E. (Ed) IRL Press, 251–285.
34. Eftink, M. R. (1994) Biophys J., 66, 482–501.
35. Wrba, A., Schweiger, A., Schultes, V., Jaenicke, R., and Zavodszky, P. (1990) Biochemistry, 29, 7584–7592.
36. Ostendorp, R., Auerbach, G., and Jaenicke, R. (1996) Protein Sci., 5, 862–873.
37. Catanzano, F., Graziano, G., De Paola, B., Barone, G., D'Auria, S., Rossi, M., and Nucci, R. (1998) Guanidine-induced denaturation of $\beta$-glycosidase from *Sulfolobus solfataricus* expressewd in *Escherichia coli* Biochemistry, 37, 41, 14484–14490.
38. Johnson, C. R., Morin, P. E., Arrowsmith, C. H., and Freire, E. (1995) Biochemistry, 34, 5309–5316.
39. Boudker, O., Todd, M., and Freire, E. (1997) J. Mol. Biol., 272, 770–779.
40. Neet, K., and Timm, D. E. (1994) Protein Sci., 3, 2167–2174.
41. Hiller, R., Zhoug, Z. H., Adams, M. W. W., and Englander, S. W. (1997) Stability and dynamics in a hyperthermophilic protein with melting temperature close to 200 °C. Proc. Natl. Acad. Sci. USA, 94, 11329–11332.
42. Daniel, R. M., Dines, M., and Petach, H. (1996) The denaturation and degradation of stable enzymes at high temperatures. Biochem. J., 317, 1–11.
43. Vihinen, M. (1987) Protein Eng., 1, 477–480.
44. Varley, P. G., and Pain, R. H. (1991) Relation between stability, dynamics and enzyme activity in 3-phosphoglycerate kinases from yeast and *Thermus thermophilus*. J. Mol. Biol., 220, 531–538.
45. Aguilar, C., Sanderson, I., Moracci, M., Ciaramella, M., Nucci, R., Rossi, M., and Pearl, L. (1997) Crystal structure of the $\beta$-glycosidase from the hyperthermophilic arcaeon *Sulfolobus solfataricus*: resilience as a key factor in thermostability. J. Mol. Biol., 271, 789–802.
46. Barret, T., Suaresch, C. G., Tolley, S. P., Dodson, E. J., and Hughes, M. A. (1995) The crystal structure of a cyanogenic $\beta$-glucosidase from white clover, a member of the family 1 glycosil hydrolase. Structure, 3, 951–961) and the phospho-$\beta$-glucosidase from *Lactococcus lactis*.

47. Weismann, C., Beste, G., Hengstenberg, W., and Schultz, G. E. (1995) The three dimensional structure of 6-phospho-β-galactosidase from *Lactococcus lactis*. Structure, 3, 961–968.
48. D'Auria, S., Nucci, R., Rossi, M., Grycniski, I., Grycniski, Z., and Lakowicz, J. R. (1999) β-Glycosidase from the hyperthermophilic Archaeon *Sulfolobus solfataricus*: Enzyme activity and conformational dynamics above 100 °C. Biophys. Chem. 81, 23–31.
49. Lakowicz, J. R. (1986) Principles of fluorescence spectroscopy Plenum Press, NY, USA.
50. Lakowicz, J. R., and Gryczynski, I. (1991) Frequency-domain fluorescence spectrscopy in Lakowicz, J. R. (Ed) Topics in fluorescence spectrscopy, Plenum Press, NY, USA, pp. 293–331.
51. Lakowicz, J. R., Gryczynski, I., and Limkeman, M. (1986) J. Biol. Chem., 261, 2240–2248.
52. Lakowicz, J. R., Laczko, G., and Gryczyniski, I. (1986) Rev. Sci. Instrum., 57, 2499–2504.
53. Laczko, G., Gryczyniski, I., Wiczk, W., Malak, H., and Lakowicz, J. R. (1990) Rev. Sci. Instrum., 61, 9233–9237.
54. Lakowicz, J. R., Gryczyniski, I., Laczko, G., Wiczk, W., and Johnson, M. L. (1994) Distribution of distances between the tryptophan and the N-terminal residue of mellitin in its complex with calmodulin, troponin C and phospholopids Protein Sci., 3, 628–637.
55. Pisani, F. M., Rella, R., Raia, C., Rozzo, C., Nucci, R., Gambacorta, A., De Rosa, M., and Rossi, M. (1990) Thermostable β-glycosidase from the archaebacterium *Sulfolobus solfataricus*. Purification and properties. Eur. J. Biochem., 187, 321–328.
56. D'Auria, S., Moracci, M., Febbraio, F., Tanfani, F., Nucci, R., and Rossi, M. (1998) Structure-function studies on β-glycosidase from *Sulfolobus solfataricus*. Molecular bases of thermostability, Biochimie, 80, 949–957.
57. D'Auria, S., Nucci, R., Rossi, M., Tanfani, F., Bertoli, E., Malak, H., Gryczynski, I., and Lakowicz, J. R. (1999) The β-glycosidase from the Archaeon *Sulfolobus solfataricus*: structure and activity in the presence of alcohols. J. Biochem. 126, 545–552.

# Index

MIX
Papier aus verantwortungsvollen Quellen
Paper from responsible sources
FSC® C105338

FSC
www.fsc.org

If you have any concerns about our products,
you can contact us on
**ProductSafety@springernature.com**

In case Publisher is established outside the EU,
the EU authorized representative is:
**Springer Nature Customer Service Center GmbH**
**Europaplatz 3, 69115 Heidelberg, Germany**

Printed by Libri Plureos GmbH
in Hamburg, Germany